Lecture Notes in Control and Information Sciences

Edited by M. Thoma and A. Wyner

79

Signal Processing for Control

Edited by
K. Godfrey, P. Jones

Springer-Verlag
Berlin Heidelberg New York Tokyo

Editors
Keith Godfrey
Peter Jones
Department of Engineering
University of Warwick
Coventry, CV4 7AL

ISBN 3-540-16511-8 Springer-Verlag Berlin Heidelberg New York Tokyo
ISBN 0-387-16511-8 Springer-Verlag New York Heidelberg Berlin Tokyo

Printed in Germany

Offsetprinting: Mercedes-Druck, Berlin
Binding: B. Helm, Berlin
2161/3020-543210

FOREWORD

The last decade has seen major advances in the theory and practice of control engineering. New algorithms such as self-tuning regulators have been accompanied by detailed convergence analysis; graphical work-stations allow a designer to explore a wide range of synthesis methods; microprocessors have enabled the practical realization of advanced control concepts. This growth of techniques has meant that only a few universities with large departments could train research students over the whole spectrum of control. Students in smaller departments could specialize in their research topics yet fail to appreciate developments in related areas.

The U.K. Science and Engineering Research Council (SERC) has for many years sponsored a set of six Vacation Schools designed to bring together research students working in control and instrumentation and to broaden their perspective of the field. The schools are all one week long and held at six-monthly intervals over a three-year cycle. Recently the scheme has been modified slightly to provide three 'basic' courses and three 'advanced' courses, the idea being that a student whose research topic is within a certain area would attend the advanced course relating to his topic and the basic courses outside his topic. Attendance at the schools is restricted to some 50 to 60 and industrial participants are allowed to take up spare places to encourage interaction between the students and practising engineers.

The introductory schools in the cycle are Deterministic Control I (state-space methods, classical control, elements of multivariable frequency-response design methods), Computer Control (sampled data theory, computer control technology and software, elements of instrumentation) and Signal Processing for Control. The advanced schools are Deterministic Control II (optimization, numerical methods, robustness and multivariable design procedures), Instrumentation (basic technology, sensor development and application studies) and Stochastic Control (stochastic systems, adaptive control, identification and pattern recognition). Each school has lectures, examples classes and experimental sessions. Case studies showing the application of the ideas in practice are presented, often by industrial engineers.

This volume consists of the lecture notes for the school on Signal Processing for Control. This school, held every three years at the University of Warwick, has proved to be popular with the students as it successfully combines the educational role of introducing many important ideas with the motivation provided by the wide range of interesting application examples. Whilst no multi-author book can ever be completely comprehensive and consistent, the editors are to be congratulated in providing an excellent introduction and overview of an increasingly important and practical discipline.

D.W. Clarke

Oxford University

(Chairman, Control and Instrumentation Subcommittee, SERC)

PREFACE

These lecture notes are from a Vacation School held at the University of Warwick (Coventry, England) from Sunday 15th to Friday 20th September 1985. The School, sponsored by the U.K. Science and Engineering Research Council (SERC), aimed to provide an introduction to the theory and application of signal processing in the context of control systems design.

There were 42 participants, 32 of whom were research students in the area of control engineering (the majority on SERC-funded studentships), the remaining 10 being industry-based engineers involved in control engineering and related topics. Some prior knowledge of classical control theory was assumed, involving familiarity with calculus, differential equations, Fourier series, Fourier and Laplace transforms, z-transforms, frequency domain methods of linear systems analysis, and basic matrix techniques.

The School was based on a complementary set of lectures, case studies and practical sessions covering the following topics:

(i) analytical and computational techniques for characterising random signals and their effect on dynamic systems;
(ii) system identification and parameter estimation;
(iii) digital filtering and state estimation;
(iv) state/parameter estimation in feedback control.

CURRICULUM OF THE SCHOOL

The School consisted of three Revision Lectures (R1 to R3), eleven further Lectures (L1 to L11) and four Case Studies (C1 to C4). The Revision Lectures were presented on the Sunday afternoon at the start of the School and contained material which most participants would have encountered at undergraduate level; attendance at these was optional. The "main-stream" Lectures (L1 to L11) were presented from Monday through to Friday. These covered the topics listed in (i) to (iv) above, building from the material in R1 to R3 through to more advanced techniques. The four Case Study lectures were designed to illustrate the practical application of the more theoretical material in L1 to L11. Outlines of R1 to R3, L1 to L11 and C1 to C4 are given later in this Preface.

Facilities for interactive dynamic data analysis were provided via the PRIME 550 computer system installed at the University of Warwick as a part of the SERC

Interactive Computing Facility. In addition, the MATRIX-X analysis and design package was available on a SYSTIME 8780 computer at the University. Students were able to perform a series of experiments involving the analysis of random data and the modelling of dynamic systems based on accessible data files chosen to illustrate representative applications and problems.

A hardware demonstration of data analysis techniques in both the time domain and frequency domain was given on a Hewlett Packard 5420B Digital Signal Analyzer. The demonstration was devised and run by Professor W.A. Brown of the Department of Electrical Engineering, Monash University, Australia, who was on sabbatical leave in the Department of Engineering at the University of Warwick at the time of the School.

On the Wednesday afternoon of the School, participants went on an industrial visit to the Lucas Research Centre at Shirley (near Birmingham) to hear presentations of relevant research and development projects in the area of automotive systems control and to tour the engine test and other experimental facilities.

The Vacation School Dinner was preceded by a keynote address given by Professor Thomas Kailath of the Electrical Engineering Department of Stanford University, California. Professor Kailath entitled his address "Signal Processing and Control" and dealt with numerical computation aspects of signal processing (in particular, square root algorithms) together with implementation considerations involving parallel processing and VLSI. Traditionally, keynote addresses at Vacation Schools of this type are intended as an up-to-date overview of some aspects of the topic of the School. As such, lecture notes are not sought and none are available for Professor Kailath's talk.

MATERIAL COVERED IN THE NOTES

A. Revision Lectures R1 to R3

In R1, *Signal Analysis I*, basic analytical and computational techniques that are available for the characterisation of dynamic signals and data are reviewed. These are Fourier series and the Fourier transform, the Discrete Fourier transform (including the Fast Fourier Transform algorithm), the Laplace transform, sampled data and the z-transform and a brief overview of random signal analysis and estimation errors.

Methods for characterising dynamic systems are discussed in R2, *Systems Analysis I*. These include differential equation representation, impulse response and convolution in the time domain, frequency response and methods of determining frequency responses, and the (Laplace) transfer function. Sampled data systems are also covered, with material on difference equations, pulse transfer functions, zero order hold elements, convolution sum and the estimation of unit pulse response using cross-correlation.

One of the primary aims of R3, *Matrix Techniques*, is to standardise notation and terminology of basic matrix concepts for subsequent lectures at the School. The use of vector-matrix concepts in studying dynamic systems is discussed, in particular the transfer function matrix and the state transition matrix. Vector-matrix difference equations for sampled data systems are described and the notes conclude with discussions of quadratic forms and diagonalisation, Taylor series, maxima and minima and multiple linear regression.

B. <u>Lectures L1 to L11</u>

In L1, *Relevant Probability Theory*, the main concepts of probability theory applied to the characterisation of scalar and vector random variables and random signals are outlined. Both discrete and continuous random variables are considered and, as well as single variable probability distributions and density functions, joint and conditional distributions are defined and illustrated with examples. Uses of the characteristic function are described and aspects of vector random variables are discussed, including marginal densities, vector moments and normal random vectors. The notes conclude with a brief discussion of stochastic processes, including aspects of stationarity.

Basic concepts in mathematical statistics and some of their applications in the analysis of signals and dynamic systems are described and illustrated in L2, *Relevant Statistical Theory*. Bias, variance, consistency and efficiency of an estimate are defined and methods of hypothesis testing and establishing confidence intervals are described, with illustrative examples. The Cramer-Rao bound and maximum likelihood estimation are discussed and the notes conclude with a discussion of optimal estimation techniques.

The emphasis in L3, *Systems Analysis II*, is on the use of autocorrelation and crosscorrelation in the time domain and the corresponding Fourier-transformed quantities, the power spectral density and cross-spectral density function in the frequency domain. The response of linear systems to stationary random excitation is considered, in particular methods for determining output power spectrum for a system with a specified (Laplace) transfer function excited by an input signal with a specified power spectrum. Corresponding quantities for discrete-time systems are also described.

An important problem in experiment planning is that of deciding in advance how much data must be collected to achieve a given accuracy. The considerations that affect the question are discussed in L4, *Signal Analysis II*, for a number of data analysis procedures and it is shown how a quantitative analysis leads to useful guidelines for the design of experimental procedures involving random data. The relationships between record characteristics and probable errors are described both for time domain and frequency domain analyses.

In L5, *Design and Implementation of Digital Filters*, both finite-impulse-response (FIR) filters (also known as moving average (MA) filters) and infinite-impulse-response (IIR) filters (also known as autoregressive-moving average (ARMA) filters) are considered. Impulse-invariant design of IIR filters is described. It is shown how aliasing can affect the frequency response of such designs and a method of avoiding this inaccuracy by use of bilinear transformation is discussed. The design of FIR filters by Fourier series and windowing is described and computer-optimised FIR filters are discussed. Problems of quantisation and rounding, which are of such practical importance in digital filtering, are also considered.

Statistical techniques for the estimation of parameters of dynamic systems from input-output data are described in L6 *Parameter Estimation*. In the section on non-recursive estimation, emphasis is placed on maximum-likelihood estimation and a problem in linear regression, that of estimating the pulse response sequence of a system, is considered in some detail. Recursive least squares is discussed, in particular, how to avoid direct matrix inversion. The notes conclude with a brief discussion of nonlinear regression.

The theme of recursive methods is continued in L7 *Recursive Methods in Identification*. Recursive forms of standard off-line techniques are described, in particular least squares and instrumental variables. Stochastic approximation and the stochastic Newton algorithm are discussed and this is followed by sections on the model reference approach and Bayesian methods and the Kalman filter. The problems with the various approaches when the system is time-varying are described and the convergence and stability of the different algorithms are considered.

Frequency domain analysis of dynamic systems is considered in L8, *Spectral Analysis and Applications*. In the first part of the notes, several examples of autocorrelation functions and corresponding (continuous) power spectra of waveforms are given and spectral relationships in closed loop systems are considered. The problems of digital spectral analysis are then reviewed. Some of the statistical properties of spectral estimates are discussed and the notes conclude with a brief description of cepstral analysis.

In the first part of L9, *Observers, State Estimation and Prediction*, the Luenberger observer is described in some detail, with asymptotic and reduced order observers being discussed. The closed loop properties of a system in which a stable asymptotic observer is applied to an otherwise stable control system design are considered. The Luenberger observer arose with regard to state estimation for deterministic, continuous-time systems; the emphasis of the notes now switches to discrete time systems, in which any noise that affects the system is directly taken into account. Successive sections of the notes deal with the Kalman filter, prediction and smoothing.

The problems introduced by nonlinearities are considered in L10, *Introduction to Nonlinear Systems Analysis and Identification*. Static nonlinearities are discussed in the first part of the notes. Nonlinear systems with dynamics are then considered, in particular the Volterra series representation. The inherent complexity of the analysis has led to the development of approximation methods based on linearisation techniques and these are described. Identification algorithms for nonlinear systems, considered next, can be categorised as functional series methods, algorithms for block oriented systems and parameter estimation techniques. Some of the ideas presented are illustrated by a practical application in which the relationship between input volume flow rate and level of liquid in a system of interconnected tanks is identified. The notes conclude by considering control of nonlinear sampled data systems.

The final lecture, L11, *An Introduction to Discrete-time Self-tuning Control*, provides a tutorial introduction to self-tuning control in its traditional discrete-time setting. The notes start by considering a slightly modified version of the self-tuning regulator of Åström and Wittenmark, the modifications including control weighting and set-point following. A weighted model reference controller is then considered and finally a pole placement self-tuning controller is discussed. All three approaches are viewed within a common framework, namely that of emulating unrealisable compensators using a self-tuning emulator.

C. Case Studies C1 to C4

In C1, *Exploring Biological Signals*, some applications of systems techniques to biomedicine are described; in the examples described, signal processing and modelling are confined to one-dimensional time series. In the first part of the notes, the modelling of signals is considered. This is illustrated by the application of Fast Fourier Transforms, Fast Walsh Transforms, autoregressive modelling, phase lock loops and raster scanning to electrical signals from the gastrointestinal tract and by the analysis and subsequent modelling of the blood pressure reflex control system (part of the cardiovascular system). In the second part, the modelling of systems (as distinct from signals) is illustrated by two examples, the first the determination of lung mechanics and the second the identification of muscle relaxant drug dynamics. The latter is part of studies aimed at achieving on-line identification and control in the operating theatre.

Engineering surfaces have in their manufacture a large proportion of random events, and the study of surfaces, either for understanding of tribology or as a means of manufacturing control, provides a very interesting application of random process theory and spectral estimation. A range of such applications is illustrated in C2, *Stochastic Methods and Engineering Surfaces*. After a review of methods of modelling surfaces, subsequent sections deal with profile statistics, roughness parameters and profile filtering. Surface classification techniques are then described and these include the shape of autocorrelation functions, the first two even moments of the power spectral density and the skew and kurtosis of the amplitude probability density function. The notes conclude with a more detailed discussion of spectral analysis of surfaces.

Experiences gained in six applications of identification are described in C3, *Practical Problems in Identification*. The processes ranged from a steelworks blast furnace to a gas turbine engine, from an oil refinery distillation column to a human being. It is shown that while useful estimates of the dynamics of systems in industry can sometimes be obtained from simple step responses, noise is often at such a level that signals with impulse-like autocorrelation functions are needed, but that direction-dependent dynamic responses can then be a problem. If normal operating records are used, problems can arise if feedback is present and this may not be very obvious in some instances. For sampled records, the spacing of samples may mean that some parameters of a model are estimated with low accuracy. Finally, when trying to estimate the parameters of an assumed nonlinearity, it is essential that the data available adequately span the nonlinear characteristic.

The final Case Study, C4, *LQG Design of Ship Steering Control Systems*, is concerned with the control of course of a ship in the face of disturbances from ocean currents and sea waves. Modelling of the ship, wind, wave and steering gear and then the combined model of ship and disturbance are described. The cost function is formulated and the existence of the solution to the LQG (Linear, Quadratic, Gaussian) problem is investigated. The Kalman filter and controller design are then described and then simulation results are presented. It was found that one of the main problems was to design a Kalman filter which would estimate the ship motions; with the disturbance model changing significantly in different sea conditions, a fixed gain Kalman filter may not give an adequate estimation accuracy.

ACKNOWLEDGEMENTS

We would like to take this opportunity to thank the contributors to these lecture notes for their cooperation which greatly eased our editing task. In particular, we express our thanks to Professor John Douce and Dr. Mike Hughes, our colleagues at Warwick for their help and encouragement throughout the planning, preparation and editing of these notes.

We also thank Ms Terri Moss for her excellent typing of the manuscript and Mrs. Alison Negus for her invaluable secretarial support.

University of Warwick

December 1985

Keith Godfrey

Peter Jones

LIST OF CONTRIBUTORS

Dr. S.A. Billings	Department of Control Engineering, University of Sheffield, Mappin Street, Sheffield S1 3JD.
Dr. D.G. Chetwynd	Department of Engineering, University of Warwick, Coventry CV4 7AL.
Professor J.L. Douce	Department of Engineering, University of Warwick, Coventry CV4 7AL.
Dr. P.J. Gawthrop	Department of Control, Electrical and Systems Engineering, University of Sussex, Falmer, Brighton BN1 9RH.
Dr. K.R. Godfrey	Department of Engineering, University of Warwick, Coventry CV4 7AL.
Professor M.J. Grimble	Industrial Control Unit, Department of Electronic and Electrical Engineering, University of Strathclyde, 204 George Street, Glasgow G1 1XW.
Dr. M.T.G. Hughes	Department of Engineering, University of Warwick, Coventry CV4 7AL.
Dr. R.P. Jones	Department of Engineering, University of Warwick, Coventry CV4 7AL.
Dr. M.R. Katebi	Industrial Control Unit, Department of Electronic and Electrical Engineering, University of Strathclyde, 204 George Street, Glasgow G1 1XW.
Professor D.A. Linkens	Department of Control Engineering, University of Sheffield, Mappin Street, Sheffield S1 3JD.

Dr. J.P. Norton	Department of Electronic and Electrical Engineering, University of Birmingham, PO Box 363, Birmingham B15 2TT.
Dr. K. Warwick	Department of Engineering Science, University of Oxford, Parks Road, Oxford OX1 3PJ.
Dr. P.E. Wellstead	Control Systems Centre, UMIST, PO Box 88, Manchester M60 1QD.

NOMENCLATURE

One of the main drawbacks to the usefulness of many multiple-author texts is that different authors may use different symbols for the same quantity, which confuses most readers who are relatively new to the field and can even cause confusion among those familiar with the field. To remove this drawback from this text, the editors asked all authors to adhere to a specified nomenclature which is listed below.

1. GENERAL

$j = \sqrt{-1}$; F^* = complex conjugate of F

s = Laplace transform operator

F(s) = Laplace transform of f(t)

(alternatively (script) $\mathcal{L}[f(t)]$).

z = z-transform and shift operator

$F(z)$ = z-transform of $f(nT) = \sum_{n=0}^{\infty} f(nT).z^{-n}$ (or (script) $\mathcal{Z}(\ \ )$).

T = sampling interval

$t = 0^+$: time immediately after (but not including) t = 0.

$t = 0^-$: time immediately before t = 0.

$\delta(\ )$ = Dirac delta function.

u = system input (see also x)

y = system output

n = order of system

ω_0, ζ = 2nd. order system parameters

ϕ = phase angle

h(t) = unit impulse response of a system

x = state variable (also used for system input where no confusion with state variable is possible).

2. MATRICES

Matrices and vectors not usually underlined, except where confusion between these and scalar quantities might occur (e.g. in describing the Kalman filter).

Vectors are column vectors, except where specifically designated as row vectors.

Superscript T for transpose.

det A or $|A|$ = determinant of A

μ_{ij} = minor of element a_{ij}

γ_{ij} = cofactor of element a_{ij}

AdjA = Adjugate (adjoint) of A.

Tr(A) = Trace of A.

J = Jacobian matrix, with element $\{i,j\} = \dfrac{\partial y_i}{\partial x_j}$.

$\Phi(t)$ (or $\Phi(t,t_0)$ when appropriate)

= state transition matrix

λ = eigenvalue

Λ = diagonal matrix with eigenvalues along principal diagonal

v = column eigenvector

w = row eigenvector

Q = Quadratic form = $x^T Ax$

C = Curvature matrix, with element $\{i,j\} = \dfrac{\partial^2 f}{\partial x_i \partial x_j}$

Matrices for Kalman filter: See Section 6.

3. PROBABILITY AND ESTIMATION

P(A) = Probability of A.

P(A|B) = Conditional probability of A, given B

$P(X_i,Y_j)$ = joint probability distribution = $P(x = X_i$ and $y = Y_j)$.

Binomial distribution: p = probability of success in one trial

q = probability of failure in one trial

P(r) = Probability of r successes in n trials = $\dfrac{n!}{r!(n-r)!}\, p^r q^{n-r}$

Poisson distribution: ν = average number of events in unit time

P(r) = Probability of r events in time interval T = $\dfrac{(\nu T)^r}{r!}\, e^{-\nu T}$

$\bar{x}$ or μ_x = Mean value of x = E[x]

E[ ] = Expected value of quantity in square brackets.

Var[x] or σ_x^2 = Variance of x = $E[(x-\bar{x})^2]$

$f(X)$ = probability density function of x

$F(X)$ = cumulative distribution function = $\int_{X_{min}}^{X} f(X)dX$

$f(X|A)$ = Conditional probability density function

where $f(X|A)dX = P(X \leq x < X+dX$, given the event A)

C_k = k'th central moment of p.d.f. about the mean

$$= \int_{X_{min}}^{X_{max}} (X-\mu_x)^k f(X)dX = E[(x-\mu_x)^k]$$

m_{kr} = joint moment = $E[x^k y^r]$

$Cov[x,y]$ or σ_{xy}^2 = covariance between x and y

$$= E[(x-\mu_x)(y - \mu_y)].$$

ρ_{xy} = correlation coefficient = $\frac{\sigma_{xy}^2}{\sigma_x . \sigma_y}$

$\Phi(\omega)$ = characteristic function of a continuous variable x

$$= E[\exp(j\omega x)] = \int_{X_{min}}^{X_{max}} f(X)\exp(j\omega X)dX.$$

ε_t = Noise sequence

θ = Unknown parameter vector

$\hat{\theta}$ = estimate of θ

$L(z,\theta)$ = Likelihood function of observations z

(script) $\mathcal{L}(z,\theta)$ = Log. likelihood function of observations z.

4. TIME DOMAIN

τ = Time shift

$R_{xx}(\tau)$ or $R_x(\tau)$ = autocorrelation function of x(t)

$$= E[x(t).x(t+\tau)].$$

$R_{xy}(\tau)$ = crosscorrelation function between x(t) and y(t)

$$= E[x(t).y(t+\tau)]$$

$C_{xx}(\tau)$ or $C_x(\tau)$ = autocovariance function of x(t)

$= E[(x(t)-\mu_x)(x(t+\tau)-\mu_x)]$ for stationary x(t)

$\rho_{xx}(\tau)$ or $\rho_x(\tau)$ = normalised autocovariance function = $C_{xx}(\tau)/C_{xx}(0)$

$C_{xy}(\tau)$ = crosscovariance between x(t) and y(t)

$= E[(x(t)-\mu_x)(y(t+\tau)-\mu_y)]$

λ = Basic interval (bit interval, clock pulse interval) of discrete interval random binary signal or pseudo-random binary signal.

V = amplitude (i.e. $\pm$ V) of binary signal

$\oplus$ = modulo 2 addition

$Q_n(a)$ = n'th order Hermite polynomial

5. FREQUENCY DOMAIN

Fourier transform pair:

$$F(j\omega) = \int_{-\infty}^{\infty} f(t)e^{-j\omega t}dt \text{ or } F(jf) = \int_{-\infty}^{\infty} f(t)e^{-j2\pi ft}\,dt$$

$$f(t) = \frac{1}{2\pi}\int_{-\infty}^{\infty} F(j\omega)e^{j\omega t}d\omega \text{ or } f(t) = \int_{-\infty}^{\infty} F(jf)e^{j2\pi ft}\,df.$$

Fourier transform can also be written as (script) $\mathcal{F}[f(t)]$.

N.B! Fourier transform taken as two-sided except where specifically stated otherwise.

Discrete Fourier transform (DFT):

$$F_D\,(k\Omega) = \sum_{n=0}^{N-1} f(nT)\exp(-j\Omega Tnk).$$

$S_{xx}(\omega)$ or $S_x(\omega)$ = Power spectral density function, i.e. Fourier transform of $R_{xx}(\tau)$ (or $R_x(\tau)$).

(or $S_{xx}(f)$ or $S_x(f)$)

$S_{xy}(j\omega)$ (or $S_{xy}(jf)$) = Cross-spectral density function

= Fourier transform of $R_{xy}(\tau)$.

$H(j\omega)$ = System frequency response = Fourier transform of h(t).

$\gamma_{xy}^2(\omega)$ = coherence function = $|S_{xy}(j\omega)|^2/[S_{xx}(\omega).S_{yy}(\omega)]$.

B = cyclic bandwidth (Hz).

6. KALMAN FILTERS AND SELF-TUNING REGULATORS

6.1. Continuous time Kalman filter

Plant model: $\dot{x}(t) = Ax(t) + Dw(t)$

$$z(t) = Cx(t) + v(t)$$

$E[x(0)] = m_0$; $Cov[x(0), x(0)] = \Sigma_0$

$Cov[w(t),w(t)] = Q$; $Cov[v(t),v(t)] = R$

Filter: $\dot{\hat{x}}(t) = A\hat{x}(t) + K(t)[z(t) - C\hat{x}(t)]$

$$K(t) = P(t)C^T R^{-1}$$

6.2 Discrete time Kalman filter

Plant model: $x(k+1) = F\,x(k) + hw(k)$

$$z(k) = Hx(k) + v(k)$$

$E[x(0)] = m_0$; $Cov[x(0), x(0)] = \Sigma_0$

$Cov[w(k),w(k)] = Q$; $Cov[v(k),v(k)] = R$

Filter: $\hat{x}(k) = F\hat{x}(k-1) + K(k)[z(k) - H\hat{x}(k-1)]$

$$K(k) = P(k)H^T[R + HP(k)H^T]^{-1}$$

6.3 Self tuning regulator

Plant model: $y = \dfrac{z^{-k}\,B(z^{-1})}{1+A(z^{-1})}\,u + \dfrac{C(z^{-1})}{1+A(z^{-1})}\,e$

Regulator: $u = \dfrac{G(z^{-1})}{1+F(z^{-1})}\,y.$

CONTENTS

REVISION LECTURES

UNIVERSITY OF WARWICK

Department of Engineering

With the compliments of

Keith Godfrey

Peter Jones

COVENTRY CV4 7AL TELEPHONE COVENTRY (0203) 523523 TELEX 311904

SIGNAL ANALYSIS I

Prof. J.L. Douce

1. INTRODUCTION

In this lecture, we review the basic analytical and computational techniques that are available for the characterisation of dynamic signals and data.

Some familiarity with the Fourier, Laplace, and z transformations is an essential prerequisite, and it is hoped that those who are new to the subject will find these introductory notes helpful as a guide to further study of the subject. Students who are already familiar with these subjects might well regard these notes simply as an introduction to the notation which will be used throughout the vacation school.

2. FOURIER SERIES AND FOURIER TRANSFORM

A repetitive signal x(t) with repetition period 2T can be expressed in terms of the Fourier series (see Ref. (1) in the Suggestions for Further Reading, at the end of these notes)

$$x(t) = \frac{1}{2} a_o + \sum_{r=1}^{\infty} a_r \cos r\omega_o t + \sum_{r=1}^{\infty} b_r \sin r\omega_o t \qquad (1)$$

where ω_o is the fundamental frequency $\frac{\pi}{T}$ (rad/sec.) and the Fourier coefficients a_r, b_r are given by

$$a_r = \frac{1}{T} \int_{-T}^{T} x(t) \cos r\omega_o t \, dt \qquad (2)$$

$$b_r = \frac{1}{T} \int_{-T}^{T} x(t) \sin r\omega_o t \, dt \qquad (3)$$

The signal is thus expressed as the sum of sinusoidal components of frequency ω_o, $2\omega_o$ ... $k\omega_o$ ...

EXAMPLE 1

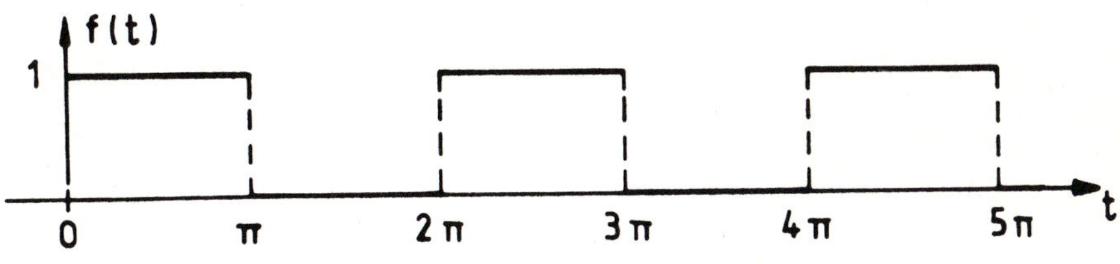

Figure 1

It is easily verified that for this signal, the Fourier series is given by:

$$f(t) = \frac{1}{2} + \frac{2}{\pi}\left(\sin t + \frac{1}{3}\sin 3t + \frac{1}{5}\sin 5t + \frac{1}{7}\sin 7t + \ldots\right)$$

EXAMPLE 2

(Try this for yourselves). The Figure below shows a portion of a function f(t):

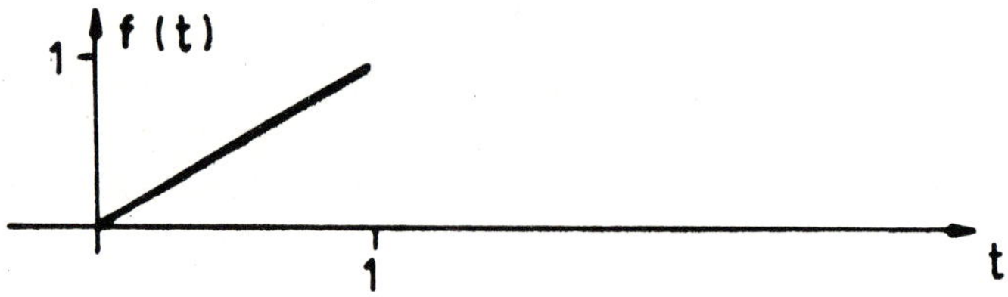

Figure 2

Sketch the rest of the function if its Fourier series is given by:

(a) $\frac{1}{2} a_0 + a_1 \cos \pi t + a_2 \cos 2\pi t + \ldots$

(b) $b_1 \sin \pi t + b_2 \sin 2\pi t + \ldots$

(c) $\frac{1}{2} a_0 + b_1 \sin 2\pi t + b_2 \sin 4\pi t + \ldots$

A non-repetitive signal can similarly be expressed by the Fourier transform:

$$x(t) = \frac{1}{2\pi}\int_{-\infty}^{\infty} X(j\omega)e^{j\omega t}\, d\omega \tag{4}$$

$$X(j\omega) = \int_{-\infty}^{\infty} x(t)e^{-j\omega t}\, dt. \tag{5}$$

$X(j\omega)$ is called the *Fourier spectrum* of $x(t)$.

EXAMPLE 3

Find the Fourier spectrum of a rectangular pulse of height A and width T centred about the origin.

$$F(j\omega) = \int_{-\infty}^{\infty} f(t)e^{-j\omega t}\, dt$$

$$= A\int_{-T/2}^{T/2} e^{-j\omega t}\, dt = -\frac{A}{j\omega}\left[e^{-j\omega t}\right]_{-T/2}^{T/2}$$

$$= - \frac{A}{j\omega} \{e^{-j\omega T/2} - e^{j\omega T/2}\}$$

$$= \frac{2A}{\omega} \sin(\omega T/2) = AT \frac{\sin(\omega T/2)}{\omega T/2}$$

$$F(0) = \int_{-\infty}^{\infty} f(t)dt = AT$$

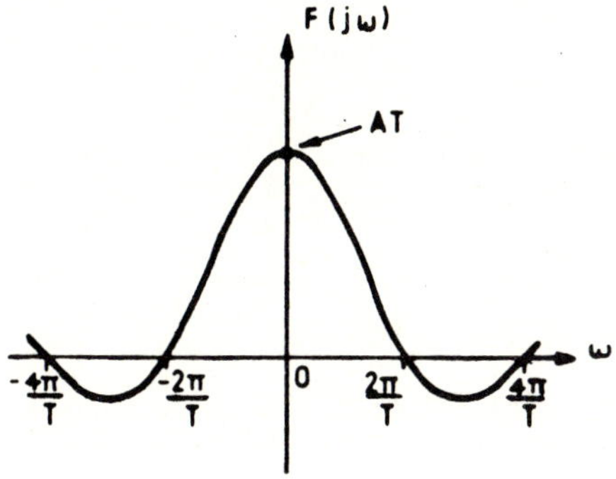

Figure 3

(Note that $F(j\omega)$ is purely real, because $f(t)$ is an *even* function of t).

EXAMPLE 4

Find the Fourier spectrum of a rectangular pulse of height A starting at $t = 0$ and finishing at $t = T$.

$$F(j\omega) = A \int_0^T e^{-j\omega t}\, dt = - \frac{A}{j\omega} [e^{-j\omega t}]_0^T$$

$$= \frac{A}{j\omega} (1 - e^{-j\omega T})$$

$$= e^{-j\omega T/2} \quad \frac{2A}{\omega} \sin(\omega T/2).$$

Note that this is the same as in Example 3, except for the multiplying factor $e^{-j\omega T/2}$. The amplitude $|F(j\omega)|$ is thus the same as before, but there is now a phase lag proportional to the frequency.

EXAMPLE 5

Find the Fourier spectrum of the function shown

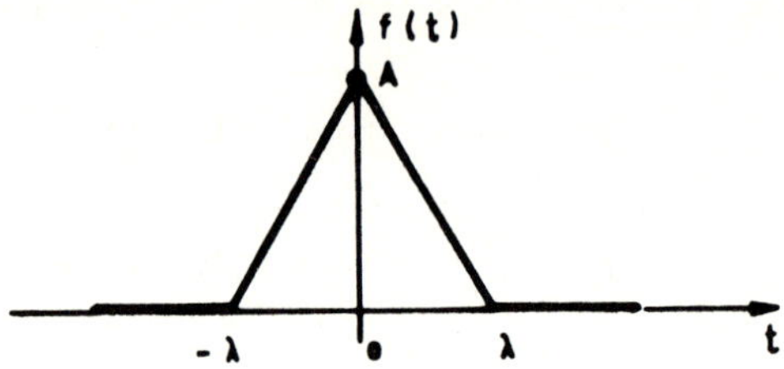

Figure 4

Since f(t) is an even function, and since $e^{-j\omega t} = \cos \omega t - j \sin \omega t$,

$$F(j\omega) = 2 \int_0^{\infty} f(t) \cos \omega t \, dt$$

$$= 2A \int_0^{\lambda} (1 - t/\lambda) \cos \omega t \, dt$$

$$= 2A\{ \frac{\sin \omega\lambda}{\omega} - \frac{\sin \omega\lambda}{\omega} - \left[\frac{1}{\lambda} \; \frac{\cos \omega t}{\omega^2}\right]_0^{\lambda} \}$$

$$= \frac{2A}{\lambda\omega^2} (1 - \cos \omega\lambda) = A \lambda . \; \frac{\sin^2 \frac{\omega\lambda}{2}}{\left(\frac{\omega\lambda}{2} \right)^2}$$

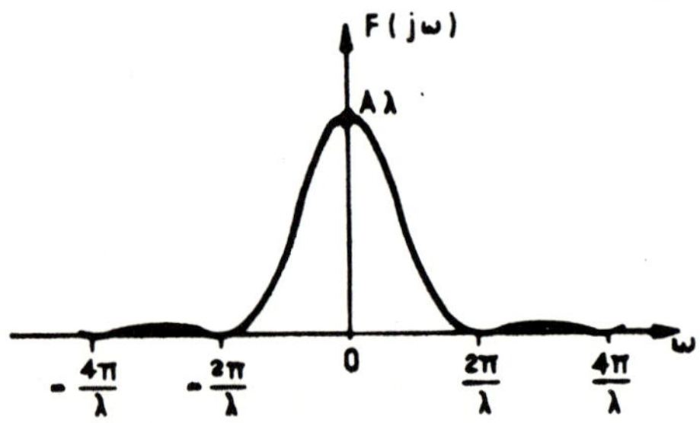

Figure 5

$$F(0) = 2 \int_0^{\infty} f(t) \, dt = A\lambda \; .$$

(The relationship between the Fourier Transform of this example and that of Example 3 will be discussed in a later section).

3. THE DISCRETE FOURIER TRANSFORM[2]

When we wish to compute a Fourier transform on a digital computer, it is necessary to use the Discrete Fourier Transform (DFT).

Let a sequence of samples be represented by:

$$\{f(nT)\} = f(0),\ f(T),\ f(2T)\ \ldots\ f((N-1)T).$$

Then the DFT is a sequence of complex samples $\{F_D(k\Omega)\}$ defined by

$$F_D(k\Omega) = \sum_{n=0}^{N-1} f(nT)\exp(-jk\Omega nT)\ ,\qquad k = 0,1,\ldots,(N-1) \tag{6}$$

where $\Omega = \frac{2\pi}{NT}$ is the separation of the components in the frequency domain.

It is readily shown that $F_D(k\Omega)$ *is periodic with period* $N\Omega$:

$$F_D((N+k_1)\Omega) = \sum_{n=0}^{N-1} f(nT)\exp(-j(k_1+N)\Omega nT)$$

$$= \sum_{n=0}^{N-1} f(nT)\exp(-j.2\pi\frac{N+k_1}{N}n)$$

But $\exp(-j2\pi n) = \cos 2\pi n - j\sin 2\pi n = 1$ since n is an integer.

$$\therefore\quad F_D((N+k_1)\Omega) = \sum_{n=0}^{N-1} f(nT)\exp(-j2\pi\frac{k_1}{N}n) = F_D(k_1\Omega).$$

Similarly, it can be shown that

$$F_D(k\Omega) = F_D^*([N-k]\Omega), \tag{7}$$

where the star denotes the complex conjugate. Thus, only one half of the Fourier coefficients are determined independently by the discrete Fourier transformation. This phenomenon is sometimes referred to as "frequency folding", or "aliasing".

3.1 Interpretation of the DFT

Consider a continuous time function f(t) that exists only in the time interval $0 \leq t \leq NT$. A convenient approximation can be found by representing the function as the limit of a sequence of equally-spaced impulses, where the strength of the impulse at t = nT is the area under the function f(t) in the interval $(n - \frac{1}{2})T \leq t < (n + \frac{1}{2})T$.

If the function does not change appreciably over one interval, then the strength of the δ-function can be taken as Tf(nT) and our approximation, $f_1(t)$ to the true function f(t) as

$$f_1(t) = T \sum_{n=0}^{N-1} f(nT).\ \delta(t-nT), \text{ where the total record length} = NT.$$

This can be used for computing the spectrum up to a frequency f = 1/(2T). By making T smaller, the fidelity of the representation can be extended to higher frequencies.

The (continuous) Fourier transform of $f_1(t)$ is given by

$$F_1(j\omega) = T \int_{-\infty}^{\infty} e^{-j\omega t} \sum_{n=0}^{N-1} f(nT).\ \delta(t-nT)dt$$

$$= T \sum_{n=0}^{N-1} f(nT) \exp(-j\omega nT)$$

When $\omega = k\Omega = \frac{2\pi k}{NT}$, this is related to the DFT of the original sequence {f(nT)}, i.e.

$$F_1(j\omega)\Big|_{\omega=k\Omega} = T.\ F_D\{f(nT)\},$$

so the DFT is $\frac{1}{T}$ times the sequence of values taken at intervals of Ω from the Fourier transform of the delta-function approximation to the continuous time function.

EXAMPLE 6

This example illustrates some elementary pitfalls that can occur in the interpretation of numerically computed Fourier transforms. Suppose we wish to compute an approximation to the continuous Fourier transform of the rectangular pulse u(t)-u(t-1) by means of the DFT. The sampled time function is represented in Fig. 6.

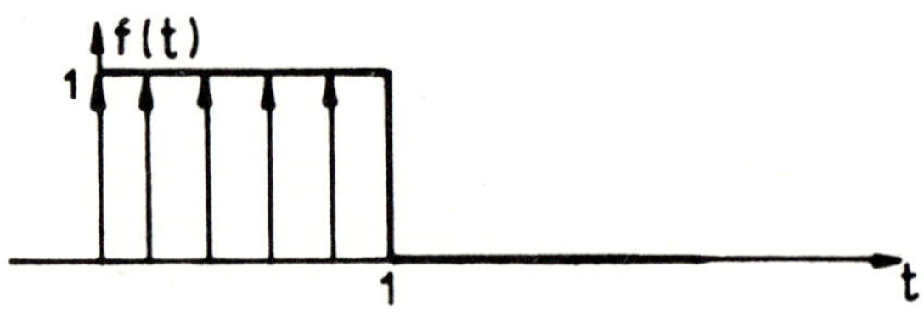

Figure 6

Let $T = \frac{1}{5}$, so {f(nT)} = {1, 1, 1, 1, 1.}

$$F_D\{f(nT)\} = \sum_{n=0}^{N-1} f(nT) \exp(-j.2\pi n\, k/N)$$

$$= \sum_{n=0}^{4} \exp(-j.2\pi nk/5)$$

$$= 5 \text{ for } k = 0$$

$$= 0, \text{ for } k = 1,2,3,4.$$

Samples of the continuous F.T. can now be computed:

$$F_1(j\omega)\Big|_{\omega=2\pi k} = T.F_D\{f(nT)\}$$

$$= 1,\ k = 0$$

$$= 0,\ k = 1,2,3,4.$$

But from Example 4, the exact Fourier Transform amplitude is $|F(j\omega)| = \frac{\sin(\omega/2)}{\omega/2}$ which appears at first sight to bear no resemblance to $F_1(j\omega)$. But

$$\Omega = \frac{2\pi}{NT} = \frac{2\pi}{5.\frac{1}{5}} = 2\pi, \text{ so the DFT is at intervals of } 2\pi.$$

Let us place some augmenting zeros at the end of the sequence. For example, placing five zeros at the end of the five ones,

$$g(nT) = \{1,1,1,1,1,0,0,0,0,0\}.$$

Keeping T the same, $F_D\{g(nT)\} = \sum_{n=0}^{9} g(nT) \exp(-j.2\pi nk/10)$, and the space is now $\Omega = \frac{2\pi}{NT} = \pi$.

For k = 1, $F_D\{g(nT)\} = \exp(0) + \exp(-j\frac{2\pi}{10}) + \exp(-j\frac{4\pi}{10})$

$$+ \exp(-j\tfrac{6\pi}{10}) + \exp(-j\tfrac{8\pi}{10})$$

$$= 1.00000 - j(3.07768)$$

so $|F_1(\pi)| = T.|F_D|_{k=1} = 0.64721.$

The true value is $F(\pi) = -j.0.63662$.

The error is due to *aliasing* - from the contributions from replications of the spectrum at all multiples of $N = \frac{2\pi}{T}$. There are significant spectral components in the signal at frequencies in excess of $\frac{1}{2T}$, i.e. greater than half the sampling frequency.

Aliasing error can be reduced by using more closely spaced samples. With $T = \frac{1}{10}$, there will be ten ones and ten zeros, but Ω will still be the same. $T|F_D|_{k=1} = 0.63925$ - an appreciable improvement over the previous case.

It is not usually possible to compute the aliasing error analytically, and the usual technique is to keep reducing sampling interval and re-computing the transform until no change in desired number of significant figures occurs.

If data has values for negative time, use can be made of the fact that the DFT *and its inverse* are periodic with periods $\frac{2\pi}{T}$ rad/sec., and NT sec. respectively, e.g. with reference to Fig. 7, the DFT of the waveform (a) would be computed by feeding in data corresponding to waveform (b).

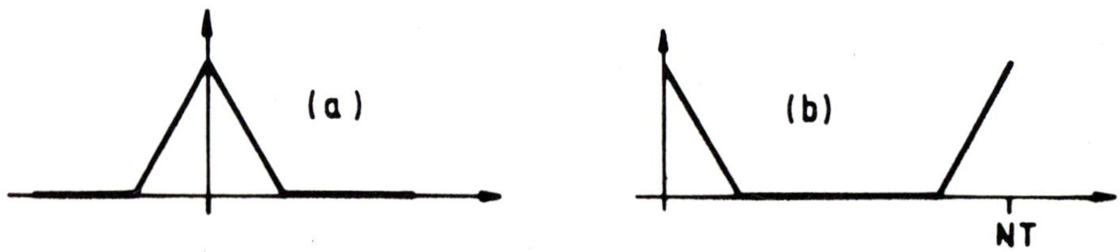

Figure 7

3.2 Fast Fourier Transform[2]

A substantial reduction in computing time can be achieved by use of *Fast Fourier Transform* algorithms, the most time-saving of which require that the total number of data samples be a power of 2. For a 4096 point data set, for example, the FFT requires 98,304 multiplications whereas direct computation of the DFT requires 16,777,216 multiplications.

EXAMPLE 7

Prepare a data sequence for input to a computer to compute using a radix 2 FFT algorithm the spectrum of the rectangular pulse shown in Fig.8. A frequency resolution of 100 Hz is required, with a maximum frequency of 25 KHz.

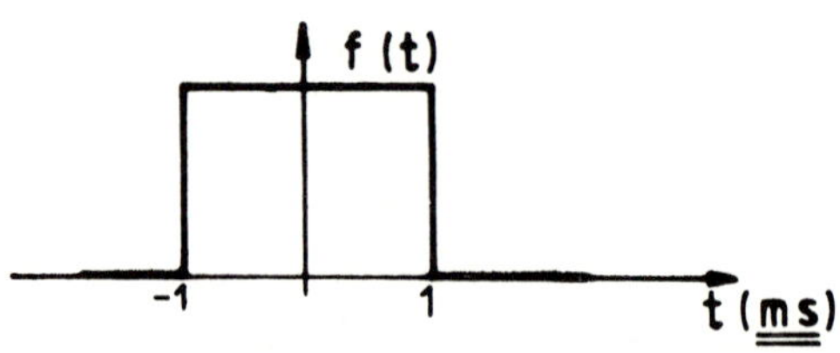

Figure 8

For DFT, $f_{max} = \frac{1}{2T}$

$$\therefore \quad \frac{1}{2T} \geqq 25.10^3 \text{ whence } T \leqq 2.10^{-5} \text{ sec.}$$

For frequency resolution of 100 Hz,

$$\frac{1}{NT} = 100. \quad \therefore \quad N = \frac{1}{100T} \geqq \frac{1}{(100)\ (2.10^{-5})} \geqq 500.$$

$$\therefore \quad \text{Take } N = 2^9 = 512.$$

$$\frac{1}{100T} = 512. \quad \therefore \quad T = \frac{1}{51200} = 19.53125\ \mu s.$$

Total duration, NT, of augmented signal $= \frac{512}{51200}$ sec. = 10 ms. The signal to be transformed is thus as shown in Fig. 9.

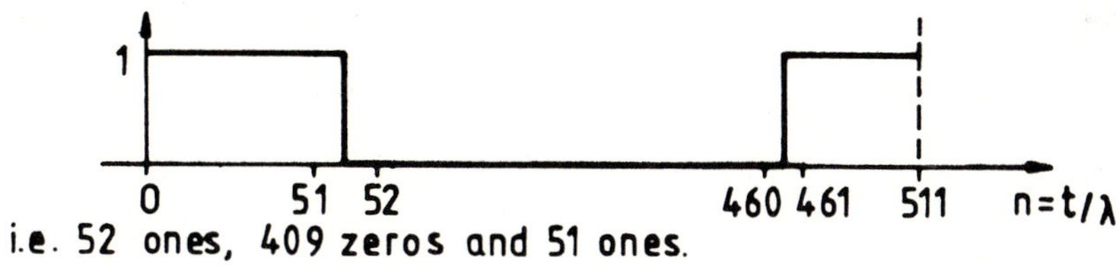

Figure 9

4. THE LAPLACE TRANSFORM

For a full discussion of the relationship between the Fourier and Laplace transformations, see for example Section 6.15 of the book, Ref. (3), 'Continuous and Discrete Signal and System Analysis', by McGillem & Cooper.

If f(t) is a known function of t for values of t ≧ 0, then the Laplace transform of f(t) is defined by

$$F(s) = \mathcal{L}[f(t)] = \int_0^\infty f(t)e^{-st}\, dt,$$

where s is the Laplace operator (p in many text-books).

The corresponding inverse transform is

$$f(t) = \mathcal{L}^{-1}[F(s)] = \frac{1}{2\pi j} \oint_{c-j\infty}^{c+j\infty} F(s)e^{st}\, ds,$$

where the contour of integration encloses all the poles of F(s). In general, tables of transforms are used. A selection of useful transforms is included at the end of these lecture notes (Table 1).

Note the close link between the Laplace and Fourier transforms. It is usual for Laplace transforms to be defined as one-sided (from zero to infinity) and Fourier transforms to be defined as two-sided. In general the Fourier transform cannot handle functions having finite average power without using special limiting operations. The Laplace transform is relatively simpler to use in such cases. The general utility of the Laplace transformation depends largely on the following properties, which are stated here without proof or further discussion. Any student who is unfamiliar with these concepts and their uses should refer to one of the many standard texts which deal with the subject (e.g. the book by Bajpai, Ref. (1)). In the following we denote by $\mathcal{L}\{f(t)\} = F(s)$ the Laplace transformation of the function $f(t)$, and take α,β to be arbitrary constants.

(i) Linearity:

$$\mathcal{L}\{\alpha f_1(t) + \beta f_2(t)\} = \alpha\, F_1(s) + \beta F_2(s) \tag{8}$$

(ii) Differentiation:

$$\mathcal{L}\left\{\frac{df}{dt}\right\} = s.F(s) - f(0^+) \tag{9}$$

where $f(0^+) = \lim_{\substack{\varepsilon\to 0 \\ \text{from above}}} f(\varepsilon)$

(iii) Integration:

$$\mathcal{L}\left\{\int_0^t f(u)du\right\} = \frac{F(s)}{s} \tag{10}$$

(iv) Shifting:

$$\mathcal{L}\{\exp(-\alpha t).\ f(t)\} = F(s + \alpha) \tag{11}$$

$$\mathcal{L}\ \{f(t - \beta)\} = \exp(-s\beta).\ F(s) \tag{12}$$

(v) Convolution:

$$\mathcal{L}\left\{\int_0^t f_1(t-\tau)\ f_2(\tau)d\tau\right\} = F_1(s).\ F_2(s) \tag{13}$$

(vi) Initial and Final Values:

$$\lim_{t\to 0} f(t) = \lim_{s\to\infty} s.F(s) \tag{14}$$

$$\lim_{t\to\infty} f(t) = \lim_{s\to 0} s.F(s) \tag{15}$$

(vii) Inverse Transformation:

If the Laplace transform F(s) can be expressed as a rational function:

$$F(s) = \frac{N(s)}{D(s)} = \frac{b_m s^m + b_{m-1}s^{m-1} + \ldots + b_1 s + b_o}{(s-p_1)(s-p_2) \quad \ldots\ldots \quad (s-p_n)}, \tag{16}$$

with $m \leq n$, and if the constants $\{p_i\}$, $i = 1,2,\ldots,n$ (i.e. the *poles* of F(s)) are all different, then the inverse transform of F(s) is

$$f(t) = \sum_{k=1}^{n} \left[(s-p_k).\frac{N(s)}{D(s)} e^{st}\right]_{s=p_K} \tag{17}$$

If F(s) has a pole of order r (i.e. a factor of the form $(s-p_i)^r$ in the denominator) then the contribution of that pole to f(t) is given by

$$\frac{1}{(r-1)!} \frac{d^{r-1}}{ds^{r-1}} \left[(s-p_i)^r . \frac{N(s)}{D(s)} e^{st}\right]_{s=p_i} \tag{18}$$

The remaining terms in f(t) are given by Eq. (17), for $k \neq i$.

EXAMPLE 8

Determine the time function which corresponds to the Laplace transform

$$F(s) = \frac{(s+2)}{s(s+1)^2 (s+3)} \quad .$$

Using Eqs. (17) and (18), we have

$$f(t) = \frac{d}{ds}\left[\frac{(s+2)}{s(s+3)} e^{st}\right]_{s=-1} + \left[\frac{(s+2)e^{st}}{(s+1)^2(s+3)}\right]_{s=0} + \left[\frac{(s+2)e^{st}}{s(s+1)^2}\right]_{s=-3}$$

$$= -\frac{1}{2}\left[\frac{3}{2} + t\right] e^{-t} + \frac{2}{3} + \frac{1}{12} e^{-3t}$$

5. SAMPLED DATA AND THE z-TRANSFORM[4]

When dealing with quantities which are defined, or sampled, only at discrete instants of time (rather than continuously), it is often convenient to employ a transform which plays a similar role to that of the Laplace transform for continuous data.

We consider a function of time, f(t), to be sampled at regular instants, which may be regarded as multiples of a *sampling interval*, T. The sampled sequence may be represented by a series of impulses occurring at sampling instants 0,T, 2T etc, of strength equal to the amplitude of the continuous time function at these instants. Using { } to represent the sampled sequence, we may write

$$\{f(nT)\} = \{f(0).\delta(t) + f(T).\delta(t-T) + \ldots.\} .$$

The Laplace transform of this sequence is

$$F(s) = \sum_{n=0}^{\infty} f(nT)e^{-snT}$$

Defining a new variable $z = e^{sT}$ gives the z transform of the time sequence

$$F(z) = Z\{f(nT)\} = \sum_{n=0}^{\infty} f(nT)z^{-n} \tag{19}$$

The corresponding inverse transformation is

$$f(nT) = \frac{1}{2\pi J} \oint_{\Gamma} F(z).z^{n-1}dz . \tag{20}$$

The path of integration, Γ, is in the region of convergence of $F(z)$, usually bounded by the unit circle.

z-transforms of various functions of time are extensively tabulated in the literature. A short table of transforms for selected functions is presented at the end of these lecture notes (Table 1).

Useful properties of the z-transform are set out below, with the z-transform of the sequence $f(nT)$ represented as

$$Z\{f(nT)\} = F(z).$$

Shifting Forward

$$Z\{f(t + T)\} = z[F(z) - f(0)] \tag{21}$$

Shifting Back

For any positive integer k,

$$Z\{f(t - kT).\ u(t - kT)\} = z^{-k}.F(z), \tag{22}$$

where u(..) represents the unit step sequence (zero for negative argument; a sequence of 'ones' for positive argument).

From this, the property of the variable z as a "shifting operator" may be clearly seen.

Exponential Factor

$$Z\{\exp(-\alpha nT).f(nT)\} = F[z \exp(\alpha T)] \tag{23}$$

Initial and Final Values

$$\lim_{n\to 0} f(nT) = \lim_{z\to\infty} z\,F(z) \tag{24}$$

$$\lim_{n\to\infty} f(nT) = \lim_{z\to 1} \frac{z-1}{z}\,F(z) \tag{25}$$

EXAMPLE 9

Find the inverse z-transform of the function

$$F(z) = \frac{z}{(z-1)^2}$$

(a) as a numerical sequence

(b) in closed form.

(a) $$\frac{z}{(z-1)^2} = \frac{z}{z^2 - 2z + 1} \; .$$

Using long division,

$$\begin{array}{r l} & \underline{0 + z^{-1} + 2z^{-2} + 3z^{-3} + \ldots} \\ z^2 - 2z + 1 \;) & z \\ & \underline{z - 2 + z^{-1}} \\ & 2 - z^{-1} \\ & \underline{2 - 4z^{-1} + 2z^{-2}} \\ & 3z^{-1} - 2z^{-2} \\ & \underline{3z^{-1} - 6z^{-2} + 3z^{-3}} \\ & \cdot \quad \cdot \quad \cdot \qquad \text{etc.} \end{array}$$

Thus, $F(z) = z^{-1} + 2z^{-2} + 3z^{-3} + \text{---}$,

$$\therefore \quad f(nT) = \{0, 1, 2, 3, \ldots\} \; .$$

(b) From Table 1 at the end of these notes,

$$Z[nT] = \frac{Tz}{(z-1)^2} \; .$$

Thus, with $T = 1$, the inverse transform is $f(nT) = n = \{0, 1, 2, 3, \text{---},\}$ as before.

Finally, the time sequence can be evaluated using a recurrence relationship. Letting the time sequence be $\{y(nT)\}$ the given transform can be written, after division by z^2,

$$y(nT) - 2y(\overline{n-1}\,T) + y(\overline{n-2}\,T) = 1(T)$$

or
$$y(nT) = 1(T) + 2y(\overline{n-1}\,T) - y(\overline{n-2}\,T).$$

Substituting n = 0, 1, 2, 3 gives

$$y(0) = 0,\ y(T) = 1;\quad y(2T) = 2;\ y(3T) = 3$$

6. RANDOM SIGNAL ANALYSIS

The range and scope of procedures for the dynamic analysis of random data is very wide in general, but it is found that a number of simple basic procedures occur frequently in a large class of applications. These basic procedures may be conveniently classified as follows:

(a) In the analysis of a record of a single variable $x(t)$, the following quantities often provide useful information. The circumflex is used to denote estimated quantities, and the superior bar ($\bar{\ }$) denotes averaging over a finite sample.

  (i) the *mean value* $\hat{\mu} = \bar{x}$

  (ii) the *mean square value* $\overline{x^2}$

  (iii) the *amplitude probability density function* $\hat{f}(X)$

  (iv) the *autocorrelation function* $\hat{R}_{xx}(\tau) = \overline{x(t).x(t+\tau)}$

  (v) the *power spectral density function* $\hat{S}_{xx}(\omega)$, which is the Fourier transform of $\hat{R}_{xx}(\tau)$.

(b) In the study of relationships between records of *two variables* $x(t)$ and $y(t)$, the following quantities are frequently employed:

  (iv) the *joint probability density function* $\hat{f}(X,Y)$,

  (vii) the *cross-correlation function* $\hat{R}_{xy}(\tau) = \overline{x(t-\tau).y(t)}$

  (viii) the *cross-spectral density function* $\hat{S}_{xy}(j\omega)$, the Fourier transform of $\hat{R}_{xy}(\tau)$.

  (ix) the *coherency function* $\hat{\gamma}^2_{xy}(\omega). = \dfrac{|\hat{S}_{xy}(j\omega)|^2}{\hat{S}_{xx}(\omega).\hat{S}_{yy}(\omega)}$

It is useful at this stage to note that data may be analysed using either analogue or digital techniques. As both methods may be of interest from time to time it is worth noting the principal differences and the relationships between them.

Analogue techniques assume continuous records, whereas digital techniques assume the data to be sampled at regular time intervals, each sample being converted into a digital representation for subsequent analysis. Decreasing the sample interval and hence increasing the number of samples tends to decrease the loss of information due to the sampling process, but it also increases the time and/or machine capacity required to perform the analysis. It can also introduce errors due to 'roundoff' effects in numerical processing. Thus, the choice of a suitable sampling interval for digital processing often requires careful consideration.

Analogue methods require special-purpose equipment for each procedure (e.g. square-law elements for measuring mean square value, time delays and multipliers for correlation analysis, tuned filters for frequency-domain analysis). Digital techniques use, in the main, general-purpose logical elements or computers, augmented by analogue-to-digital converters, with either software or hard-wired programs for analysis. It is thus more convenient to change parameters of the analytical process and to implement novel procedures digitally.

Digital apparatus is less prone to inaccurate operation due to drift, etc. than most analogue equipment.

Each of the quantities (i) to (ix) could be estimated using various analogue or digital techniques. Figs. 10 and 11 show typical examples of available analogue and digital techniques for measuring each of the indicated quantities.

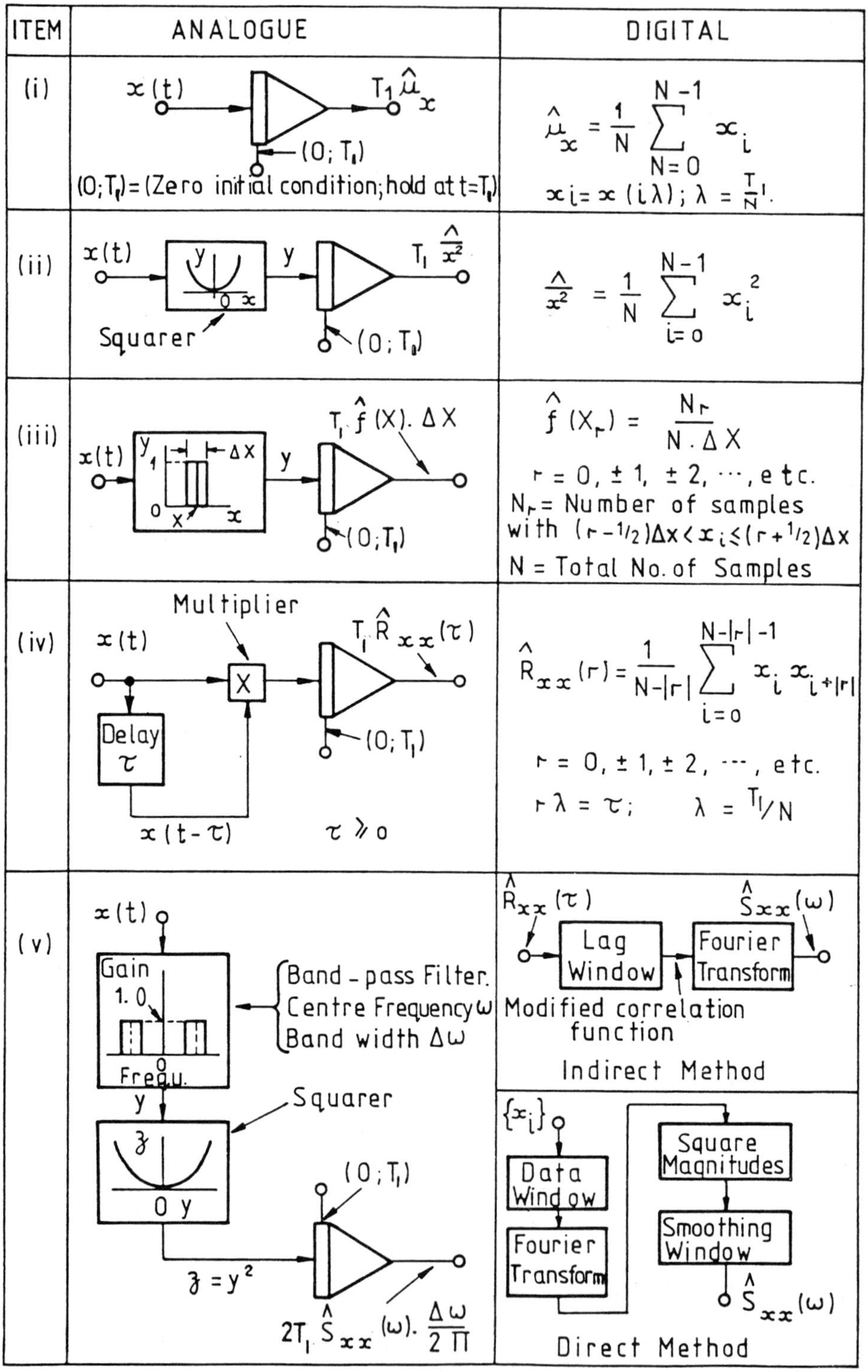

Figure 10. Basic Data Processing for Single Records

ITEM	ANALOGUE	DIGITAL
(vi)	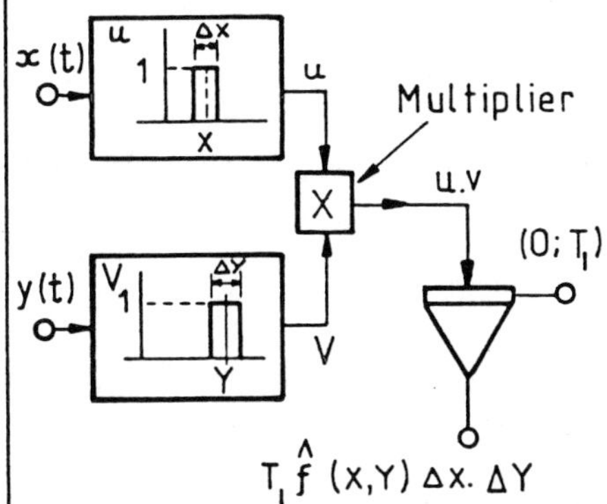	$\hat{f}(X_r, Y_s) = \frac{N_{rs}}{N.\Delta X.\Delta Y}$ $r, s = 0, \pm 1, \pm 2$, etc. N_{rs} = No. of occurrences of the joint event $\left. \begin{array}{l} (r-1/2)\Delta X < x_i \leqslant (r+1/2)\Delta X \\ (s-1/2)\Delta Y < y_i \leqslant (s+1/2)\Delta Y \end{array} \right\}$
(vii)	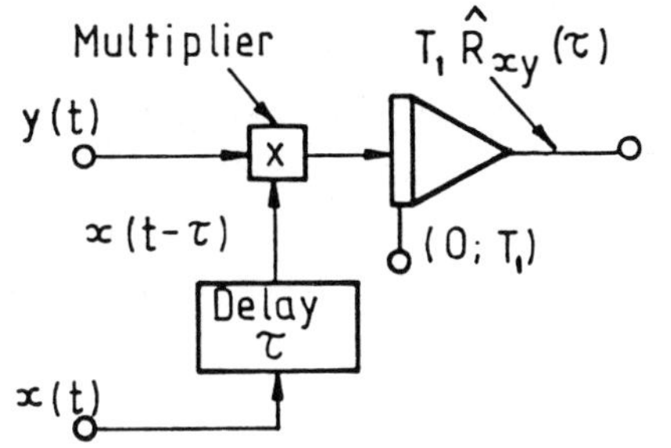	$\hat{R}_{xy}(r) = \frac{1}{N-\lvert r \rvert} \sum_{i=0}^{N-\lvert r \rvert-1} y_i x_{i-r}$ $r = 0, \pm 1, \pm 2, \cdots$, etc. $\tau = r\lambda; \quad \lambda = T_1/N$
(viii)	Cross-spectral Density Function Not Conveniently Measurable By Analogue Means. $S_{xy}(j\omega) = C_{xy}(\omega) + jQ_{xy}(\omega)$.	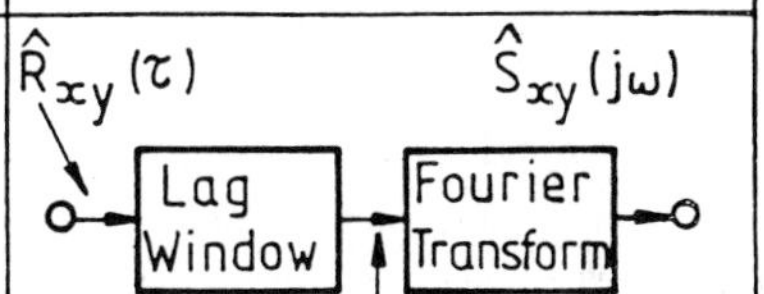Indirect Method 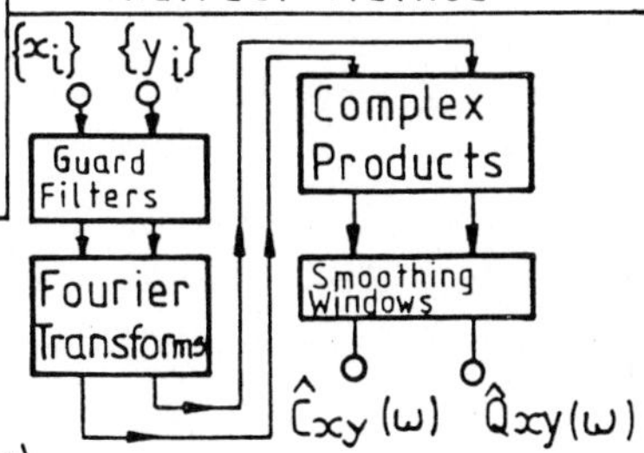Direct Method
(ix)	Coherency Function not Conveniently Measurable By Analogue Means.	$\hat{\gamma}^2_{xy}(\omega) = \frac{\lvert \hat{S}_{xy}(j\omega) \rvert^2}{\hat{S}_{xx}(\omega).\hat{S}_{yy}(\omega)}$

Figure 11. Basic Data Processing Operations for Related Records

7. ESTIMATION ERRORS

Whenever the characteristics of a random signal are estimated from a finite sample of data, errors will arise. These errors may be considered as two forms:

(i) *Random dispersion of the estimates about a mean value.*
The most commonly used measure of such errors is the *variance* of the estimate, which is defined as the mean square value of deviations from the mean value of the estimates.

(ii) *Deviation of the mean value of estimates from the true values of the quantity being estimated.*
The value of this deviation is referred to as the *bias* of the estimate.

An important problem which often arises in the planning of experiments is that of deciding in advance how much data must be collected so as to achieve a given accuracy in the computed results. The objective of an experiment should not be thwarted by large errors due to insufficient data, nor should time and effort be wasted by collecting and analysing more data than is necessary to achieve the desired confidence in the result. In order to resolve such considerations it is desirable to establish relationships between the probable levels of errors for different quantities of data. Such relationships will be established in subsequent lectures.

8. SUGGESTIONS FOR FURTHER READING

1. Bajpai, A.C., Mustoe, L.R., and Walker, D. "Advanced Engineering Mathematics", John Wiley & Sons, 1977.
2. Gold, B., and Rader, C.M., "Digital Processing of Signals", McGraw-Hill, 1969.
3. McGillem, C.D., and Cooper, G.R., "Continuous and Discrete Signal and System Analysis", Holt, Rinehart and Winston, 1974.
4. Jury, E.I., "Sampled-data Control Systems", Wiley, 1958.
5. Bendat, J.S., and Piersol, A.G., "Measurement and Analysis of Random Data", Wiley, 1966.
6. Blackman, R.B., and Tukey, J.W., "The Measurement of Power Spectra", Dover, 1958.
7. Special issue "Spectral Estimation" Proc. IEEE, September, 1982, Vol. 70.
8. Papoulis, A., "Circuits and Systems, a modern approach", Holt, Rinehart and Winston, Inc. 1980.
9. Jones, N.B. (ed.) "Digital Signal Processing", Peter Peregrinus Ltd. 1982.

TABLE 1 : A Selection of Laplace and z-Transforms

Laplace transform $F(s)$	Time function, $f(t)$, $t \geq 0$	z-transform, $F(z)$
e^{-kTs}	$\delta(t - kT)$	z^{-k}
$\frac{1}{s}$	$u(t)$	$\frac{z}{z - 1}$
$\frac{1}{s-a}$	e^{at}	$\frac{z}{z - e^{aT}}$
$\frac{\omega}{s^2+\omega^2}$	$\sin \omega t$	$\frac{z \sin \omega T}{z^2 - 2z \cos \omega T + 1}$
$\frac{s}{s^2+\omega^2}$	$\cos \omega t$	$\frac{z(z - \cos \omega T)}{z^2 - 2z \cos \omega T + 1}$
$\frac{b}{s^2-b^2}$	$\sinh bt$	$\frac{z \sinh bT}{z^2 - 2z \cosh bT + 1}$
$\frac{s}{s^2-b^2}$	$\cosh bt$	$\frac{z(z - \cosh bT)}{z^2 - 2z \cosh bT + 1}$
$\frac{1}{s^2}$	t	$\frac{zT}{(z - 1)^2}$
$\frac{2!}{s^3}$	t^2	$\frac{T^2 z(z + 1)}{(z - 1)^3}$
$F(s+a)$	$e^{-at}f(t)$	$F(e^{aT}z)$
$\frac{1}{(s+a)(s+b)}$	$\frac{1}{b-a}(e^{-at}-e^{-bt})$	$\frac{1}{b-a}(\frac{z}{z-e^{-aT}} - \frac{z}{z-e^{-bT}})$
$\frac{1}{(s+a)^2}$	te^{-at}	$\frac{Tze^{-aT}}{(z-e^{-aT})^2}$
$e^{-kTs}.F(s)$	$f(t - kT)u(t - kT)$	$z^{-k}F(z)$

Except for the first entry in the Table, the z-transforms shown are of the sequence $f(nT)$ where $n = 0,1,2,\ldots$ and T is the sampling interval. For the first entry, the sequence $f(nT) = 0$, $n \neq k$ and $= 1$, $n = k$. When taking inverse transforms, note that $\mathcal{Z}^{-1}[F(z)]$ does not equal $f(t)$ but does equal $f(t).\delta_T(t)$, where $\delta_T(t)$ is a periodic sequence of unit impulses occurring with period T.

SYSTEMS ANALYSIS I

Dr. K.R. Godfrey

1. INTRODUCTION

Much of the material in later lectures will be concerned with the characterization of response of dynamic systems to random excitation, with the processing of experimental input/output data to estimate unknown system parameters, and with the derivation of control strategies for systems which are subjected to random inputs and disturbances.

The material to be developed throughout this vacation school is based largely on techniques of linear systems analysis which will be well known to many graduate engineers. The aim of this introductory material is to outline the basic notation and prerequisite knowledge which will be assumed in the sequel, and to suggest fruitful areas for further reading, for the benefit of those students who recognise a need for such preparation. This introduction is necessarily brief, and it is inevitable that much important material must be omitted. Those requiring access to more comprehensive expositions of the subject of dynamic systems analysis are advised to consult the sources cited in the Recommendations for Further Reading at the end of these notes.

A wide class of dynamic systems with single input u(t) and output y(t) can be represented in terms of differential equations of the form

$$\frac{d^n y}{dt^n} = f\left(u, y, \frac{dy}{dt} \;\; \cdots\cdots \;\; \frac{d^{n-1} y}{dt^{n-1}};\, t\right) \qquad (1)$$

In this expression, n is the order of the system and the symbol t is included explicitly on the right-hand side to indicate that the dynamics of the system may change with time. An example of this may be found in a rocket, where the mass decreases as fuel is consumed. In an important sub-class of systems where the dynamics do not change with time (this is the most common situation which we shall consider), the function f(··) is independent of time, and t does not appear explicitly.

In this lecture, we shall further restrict attention to *linear* systems, so that the function f(··) is a linear function, and y and u are related by

$$a_n \frac{d^n y}{dt^n} + a_{n-1} \frac{d^{n-1} y}{dt^{n-1}} + \dots + a_1 \frac{dy}{dt} + a_o y(t)$$

$$= b_m \frac{d^m u}{dt^m} + b_{m-1} \frac{d^{m-1} u}{dt^{m-1}} + \dots + b_1 \frac{du}{dt} + b_o u(t) \qquad (2)$$

Powerful analytical techniques are available for this class of linear, time-invariant systems to express relationships between the system input and output.

A fundamental property of such systems is reflected in the so-called *'principle of superposition'*. This states that if an input $u_1(t)$ produces an output $y_1(t)$ and a second input $u_2(t)$ produces $y_2(t)$ then the combined input $[u_1(t) + u_2(t)]$ produces the response $[y_1(t) + y_2(t)]$. For linear systems this procedure can be applied repeatedly, and this can enable the response to quite complicated input signals to be found either analytically or numerically, if the system response to an appropriate simple input signal is known.

2. THE IMPULSE RESPONSE

One of the most useful simple input signals is the Dirac delta function or unit impulse written $\delta(t)$.

This may be defined as a function such that

$$\left.\begin{aligned} &\delta(t) = 0 && \text{for } t \neq 0 \\ &\int_a^b \delta(t)\, dt = 1 && \text{for } a < 0 < b \end{aligned}\right\} \qquad (3)$$

Thus, the unit impulse may be envisaged as a 'signal' which is zero except over a vanishingly small time interval around $t = 0$, such that the area under the signal, with respect to time, has unit value. Note that a unit impulse occurring at time $t = t_1$ is written $\delta(t - t_1)$.

Although the unit impulse can only be approximated in practice, it is an extremely useful concept.

In particular the response of a system to the unit impulse applied at time $t = 0$ is termed the impulse response of the system, written $h(t)$.

Some special cases of note are:

(a) The first-order system, obeying the differential equation

$$T \frac{dh}{dt} + h = \delta(t),$$

The impulse response is

$$h(t) = 0 \text{ for } t < 0$$

$$h(t) = \frac{1}{T} e^{-t/T} \text{ for } t > 0. \qquad \text{(see Fig. 1).}$$

Note the dimensions of $h(t)$ are $(\text{time})^{-1}$.

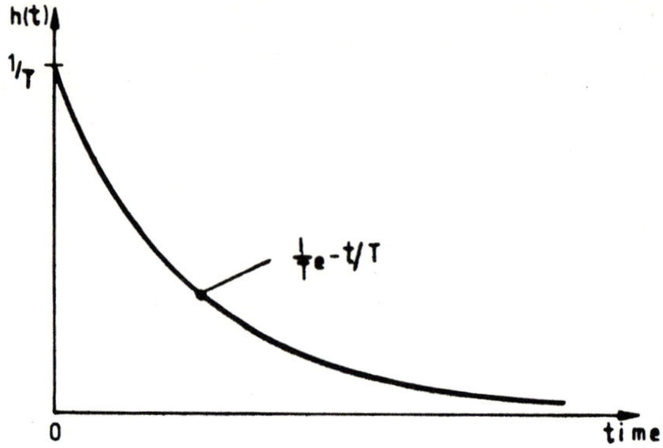

Figure 1. Impulse response of first order linear system

(b) For the second order system

$$\frac{1}{\omega_o^2}\,\frac{d^2h}{dt^2} + \frac{2\zeta}{\omega_o}\frac{dh}{dt} + h = \delta(t)$$

With ω_o and $\zeta > 0$, the response is zero for $t < 0$ and tends to zero as $t \to \infty$. For $0 < \zeta < 1$ the response is oscillatory whilst for $\zeta > 1$ the response is non-negative (Fig. 2).

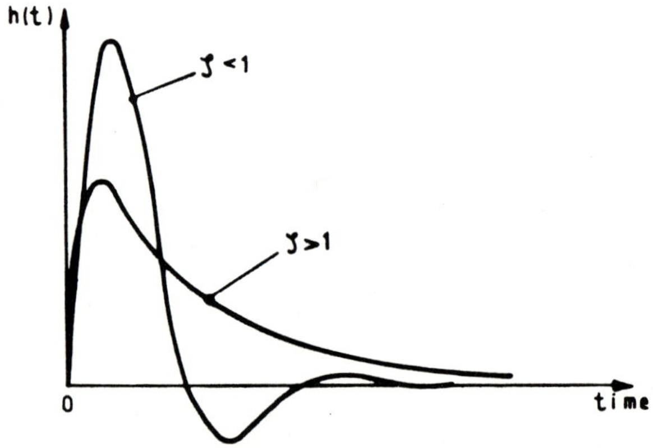

Figure 2. Impulse response of second order linear system for two values of damping parameter ζ.

All physically realisable systems must possess an impulse response which is zero for $t \leqq 0$. This simple observation has some important implications, as we shall see.

3. CONVOLUTION AND APPLICATIONS

All signals of engineering interest may be approximated as closely as desired by a train of closely spaced impulses of appropriate amplitude.*

Figure 3(a) and 3(b) demonstrate this representation for a particular example. The basic idea is that over each (vanishingly small) time interval, say t_1 to $(t_1 + \Delta t)$, the continuous signal is represented by an impulse of area equal to the area $\int_{t_1}^{t_1+\Delta t} u(t)dt$ which is approximated by $u(t_1).\Delta t$.

If we know the impulse response of the system to which the input u(t) is applied, then we can derive the response of u(t) as follows, using superposition. Referring to Figure 3(c), the system response at time T depends on the input up to time T. This is decomposed into the sum of the responses to all impulses representing the input signal up to time T.

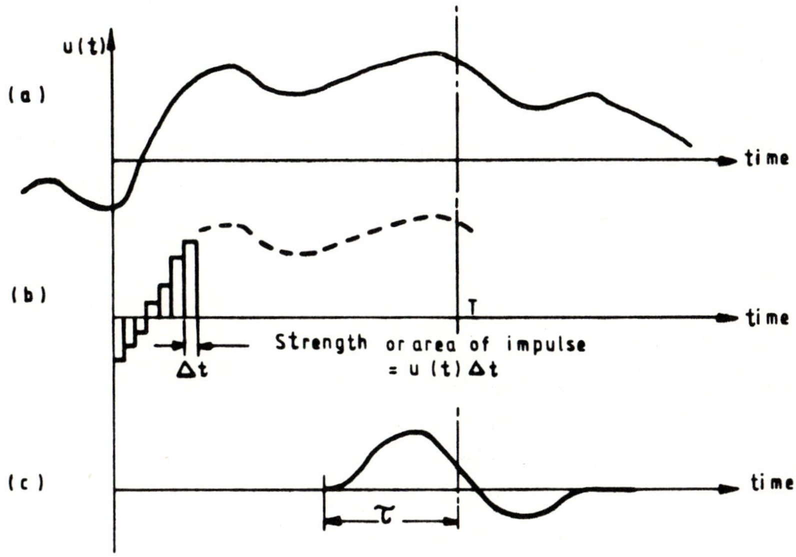

Figure 3. The Convolution Integral

(a) a continuous signal
(b) impulse representation
(c) response to input at $(T-\tau)$

Consider the influence of the signal $u(T-\tau)$, that is the input applied a time τ prior to the instant of interest. This signal is modelled over the time duration $\Delta\tau$ by an impulse of strength $u(T-\tau). \Delta\tau$. This excites a response at a time τ later equal to $h(\tau).\ u(T-\tau).\Delta\tau$. Summing or superimposing the response to all impulses for $\tau \geqq 0$ gives

* This does not apply strictly to all functions of time, e.g. $u(t) = t \sin t^3$ cannot be so represented as $t \to \infty$.

$$y(T) = [h(0).u(T) + h(\Delta\tau).u(T-\Delta\tau) + h(2\Delta\tau).u(T-2\Delta\tau) + \ldots]\Delta\tau.$$

Letting $\Delta\tau \to 0$ gives the Convolution integral:

$$y(T) = \int_0^T h(\tau)\, u(T - \tau)d\tau$$

This is more usually written

$$y(t) = \int_0^\infty h(\tau)\, u(t - \tau)d\tau, \quad (4)$$

where we assume the input u(t) to have commenced in the remote past.

The lower limit of integration may be changed to $-\infty$, since $h(\tau) = 0$ for $\tau < 0$ for a physically realisable system.

Applying the principle of superposition we may readily deduce that the step response function of a system, that is the response to the input

$$u(t) = 0 \qquad t \leqq 0$$
$$= 1 \qquad t > 0$$

is the time integral of the impulse response, given by $y(t) = \int_0^t h(\tau)d\tau$.

Similarly, the system response to a unit ramp input $u(t) = t$ is the time integral of the step response. Conversely, the impulse and step responses are the time derivatives of the step and unit ramp response respectively.

4. FREQUENCY RESPONSE

When a sinusoidal signal is applied to a linear time-invariant system, the response of the system can be considered as the sum of two components. There is a transient term due to the response of the system to the initial conditions and to any discontinuity at the instant at which the sinusoidal input is initially applied. If the system is stable, this transient term tends to zero with increasing time. The second component is the steady state response to the sinusoid, and is of the same frequency as the input signal. The frequency-response function relates the steady-state output signal to the input sinusoid. Letting $u(t) = Ae^{j\omega t}$ and the steady-state component of the output be $y(t) = Be^{j(\omega t+\phi)}$, the frequency-response function is defined as $H(j\omega) = \frac{B}{A}e^{j\phi}$. It is essential that the physical significance of this function be fully appreciated, and the following properties of $H(j\omega) \equiv |H|e^{j\phi} \equiv X+jY$ be thoroughly understood.

1. $|H|$ is the ratio (output amplitude) ÷ (input amplitude) and ϕ is the phase angle between output and input. If ϕ is in the range 0 to π the output is normally considered to lead the input. Note that measurement, (and certain analytical

results) cannot differentiate between a lagging phase angle of θ and a lead of $(2\pi - \theta)$, and ambiguities can easily arise if the value of θ exceeds π.

2. X and Y give the components of the output signal which are respectively in phase and in quadrature with the input sinusoid. A positive value for Y is associated with an output leading the input.

3. Transformation from Cartesian to polar co-ordinates and vice versa is apparently trivial, using

$$X = |H| \cos \phi \;; \qquad Y = |H| \sin \phi$$

$$|H| = \sqrt{X^2 + Y^2} \;; \qquad \tan \phi = Y/X \,.$$

Note however the mechanical application of the last expression gives only the principal value of ϕ and this is often not the appropriate value over the whole range of interest.

5. DETERMINATION OF THE FREQUENCY RESPONSE

Four methods of determining the frequency-response function of the type of system considered may be noted.

(i) The convolution integral gives the response of the system to an arbitrary input $u(t)$. Setting $u(t) = e^{j\omega t}$ gives (neglecting transients):

$$y(t) = \int_0^\infty h(\tau) e^{j\omega(t-\tau)} d\tau$$

$$= e^{j\omega t} \int_0^\infty h(\tau) e^{-j\omega\tau}\, d\tau.$$

Hence $H(j\omega)$ is the Fourier transform of the impulse response. It may be verified readily that the frequency response of a first-order system, with

$$h(t) = \frac{1}{T} e^{-t/T} \text{ is given by } H(j\omega) = \frac{1}{1 + j\omega T} \,.$$

(ii) The general differential equation describing the behaviour of the class of system considered is of the form

$$a_n \frac{d^n y}{dt^n} + a_{n-1} \frac{d^{n-1} y}{dt^{n-1}} + \ldots + a_1 \frac{dy}{dt} + a_o y(t)$$

$$= b_m \frac{du^m}{dt^m} + \ldots + b_1 \frac{du}{dt} + b_o\, u(t) \qquad (5)$$

Again, consider $u(t) = e^{j\omega t}$. Substituting for u,y and their derivatives gives

$$[a_n(j\omega)^n + a_{n-1}(j\omega)^{n-1} + \ldots + a_1(j\omega) + a_o].\ H(j\omega)$$

$$= [b_m(j\omega)^m + \ldots + b_1(j\omega) + b_o]$$

giving $H(j\omega)$ as a complex number in terms of ω and the coefficients of the differential equation.

(iii) The transfer function $H(s)$ of the system, introduced below, gives the frequency response directly by the substitution $s = j\omega$.

(iv) The frequency-response function $H(j\omega)$ may be determined experimentally by perturbing the input sinusoidally and cross-correlating the response respectively with in-phase and quadrature-related signals at the same frequency as the input.

This technique is of considerable practical importance, since it possesses inherently powerful noise-reduction properties. To see this suppose the input to the system contains a deliberately injected component of the form $V \sin \omega t$. The system response to this will have the form

$$y(t) = V[a\sin\omega t - b\cos\omega t] + n(t)$$

where V is the input amplitude, a is the real component of the complex system gain $H(j\omega)$, and b is the imaginary component. The quantity $n(t)$ is taken to represent the aggregated effects of random noise and other inputs to the system.

If the measured response $y(t)$ is multiplied by a sinusoidal reference signal, and then averaged over an integral number of cycles of the waveform, we get

$$\overline{y(t) \sin \omega t} = \frac{aV}{2} + \overline{n(t) \sin \omega t}$$

and similarly, correlating with respect to a cosine wave,

$$\overline{y(t) \cos \omega t} = \frac{-bV}{2} + \overline{n(t) \cos \omega t}.$$

The noise components occurring in these expressions can be made as small as desired by choosing a sufficiently long averaging period, provided that the noise is not correlated in any way with the input signal. To make these ideas more precise, statistical concepts must be applied. These will be developed and discussed in later lectures in the vacation school.

6. THE TRANSFER FUNCTION

In Revision Lecture R1, the Laplace transform of a time function $f(t)$ is defined as

$$F(s) = \int_0^\infty f(t)\, e^{-st}\, dt.$$

The relationship between the transforms of the input and output time signals for a linear system will now be derived.

Note firstly that the transform of the time derivative of a time function is intimately related to the transform of the original function since if these transforms are denoted as $F_1(s)$ and $F(s)$ respectively

$$F_1(s) = \int_0^\infty \frac{df(t)}{dt} e^{-st} dt$$

$$= [e^{-st} f(t)]_0^\infty + s \int_0^\infty f(t)e^{-st} dt$$

$$= -f(0^+) + sF(s).$$

The first term is the value of the function at time $t = 0^+$ i.e. just after $t = 0$. In the particular case we shall consider, all initial conditions will be taken as zero, and we can derive as above the general result for the transform $F_n(s)$ of the nth derivative of $f(t)$:

$$F_n(s) = s^n F(s).$$

Given the differential equation relating input and output of a linear system

$$a_n \frac{d^n y}{dt^n} + \dots + a_1 \frac{dy}{dt} + a_o y = b_m \frac{d^m u}{dt^m} + \dots + b_1 \frac{du}{dt} + b_o u$$

we take the transform of both sides, assume zero initial conditions, and use the above relationship to give

$$[a_n s^n + \dots + a_1 s + a_o] \; Y(s) = [b_m s^m + \dots + b_1 s + b_o] U(s)$$

in which $Y(s)$ and $U(s)$ are respectively the transforms of $y(t)$ and $u(t)$.

Hence we may write

$$\frac{Y(s)}{U(s)} \equiv H(s) = \frac{b_m s^m + \dots + b_1 s + b_o}{a_n s^n + \dots + a_1 s + a_o} \tag{6}$$

where $H(s)$, termed the *transfer function* of the system, is the ratio of the polynomials occurring in the previous equation. In general, for physical systems, the indices above must satisfy $m < n$.

Noting that the transform of a unit impulse is unity, it follows that the transfer function of a system is the transform of the impulse response,

$$H(s) = \int_0^\infty h(t) \; e^{-st} \; dt.$$

In summary, we note that the impulse response and the system transfer function contain the same information, in different forms, so that either permits the response of a system with zero initial conditions to be found for a given input signal.

7. SAMPLED-DATA SYSTEMS

When digital devices are employed for data analysis or control, certain system inputs and/or outputs will be constrained so that they may change only at certain time instants, or 'sampling' instants. If the sampling instants are uniformly spaced in time, the Z-transformation, introduced in Revision Lecture R1, may be used to characterise the system.

Sampling may be introduced into a system in many ways; for a comprehensive treatment of the subject, the student is referred to the suggestions for further reading at the end of these notes. Here we simply introduce some concepts which will be employed in subsequent lectures.

7.1 Difference Equations

A general form of linear difference equation of order n may be written

$$\alpha_n y_{t-n} + \alpha_{n-1} y_{t-n+1} + \dots + \alpha_1 y_{t-1} + \alpha_o y_t$$

$$= \beta_m u_{t-m} + \dots + \beta_1 u_{t-1} + \beta_o u_t \tag{7}$$

Here, and in what follows, the symbol t, when used as a subscript, will denote values of variables at discrete sampling instants, for example

$$x_t, \text{ for } t = 0, \pm 1, \pm 2, \dots \text{ etc.}$$

will denote values of $x(t)$ at the discrete time instants $0, \pm T, \pm 2T, \dots$ etc. where T is the sampling interval.

Using the Z-transformation, Eq. (7) may be rewritten as

$$(\alpha_n z^{-n} + \alpha_{n-1} z^{-n+1} + \dots + \alpha_1 z^{-1} + \alpha_o) Y(z)$$

$$= (\beta_m z^{-m} + \dots + \beta_1 z^{-1} + \beta_o) U(z). \tag{8}$$

Thus, we may invoke the idea of a *pulse transfer function* to represent the linear relationship between the discrete time sequences

$$\{y_t\} \text{ and } \{u_t\} \text{ for } t = 0,1,2,\dots, \text{ etc.}$$

$$H(z) \equiv \frac{Y(z)}{U(z)} = \frac{\beta_m z^{-m} + \dots + \beta_1 z^{-1} + \beta_o}{\alpha_n z^{-n} + \dots + \alpha_1 z^{-1} + \alpha_o} \tag{9}$$

7.2 'Zero-order' Hold Element

Many digital devices such as analogue/digital and digital/analogue converters operate in such a way that the converted quantity remains constant between sampling intervals.

This action may be represented by means of a sampler and a 'zero-order hold' element, as shown in Fig. 4.

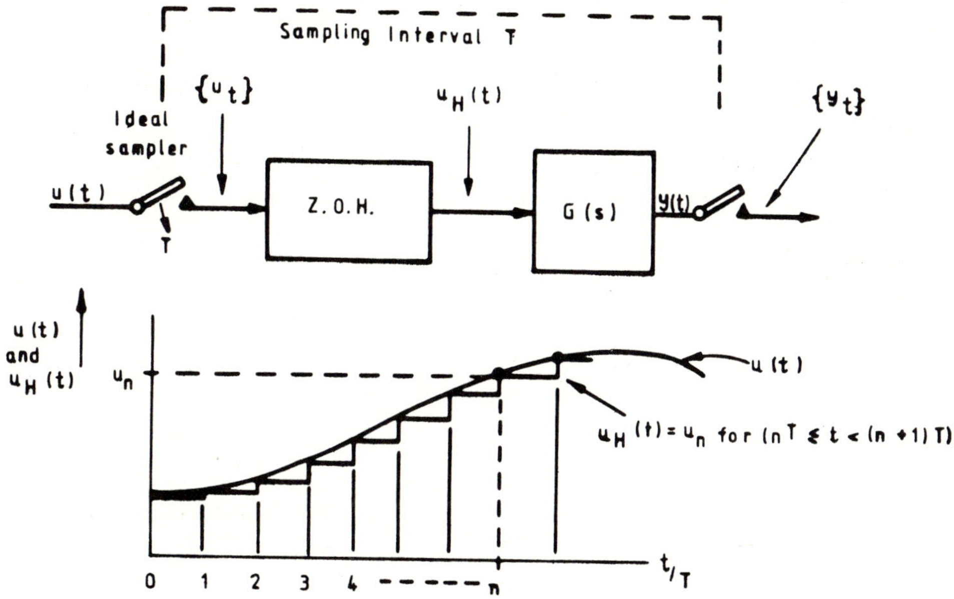

Figure 4. Action of 'Zero Order Hold' Element

The 'transfer function' of such an element has the form

$$\frac{U_H(s)}{U(s)} = \frac{1-e^{-sT}}{s} \tag{10}$$

and the pulse transfer of a linear system preceded by a sampler and zero-order hold element (with synchronously sampled output) is

$$G_H(z) = (1 - z^{-1})\mathcal{Z}\{\frac{H(s)}{s}\} \tag{11}$$

where H(s) is the transfer function of the system whose input is derived from the zero-order hold, and $\mathcal{Z}\{H(s)/s\}$ means 'Take the z-transform of the time function whose Laplace transform is H(s)/s'.

Example

Find the pulse transfer function of the system shown in Fig. 4, if the continuous transfer function H(s) has the form

$$H(s) = \frac{K}{1 + s\tau}$$

From Eq. (11),

$$G_H(z) = (1 - z^{-1})\mathfrak{z}\{\frac{K}{s(1 + s\tau)}\}$$

$$= (1 - z^{-1})\mathfrak{z}\{K[\frac{1}{s} - \frac{\tau}{1 + s\tau}]\}$$

From the table of z-transforms at the end of Revision Lecture R1,

$$G_H(z) = (1 - z^{-1})\ K\{\frac{z}{z - 1} - \frac{z}{z - e^{-T/\tau}}\}$$

$$= \frac{K(1 - e^{-T/\tau})}{z - e^{-T/\tau}}\ .$$

7.3 Convolution Sum

A useful modification to the *convolution integral* in Eq. (4) permits the output of a linear system to be calculated at regularly spaced sampling instants by means of a weighted sum of input values, when the input is applied via a sampler and a Zero-Order Hold element:

$$y_t = \sum_{i=1}^{\infty} w_i\ u_{t-i} \qquad (12)$$

where
$$w_i = \int_{(i-1)T}^{iT} h(\tau)d\tau \qquad (13)$$

The sequence of numbers $\{w_i\}$ for i = 0,1,2,..., etc. is called the 'weighting sequence' of the system. It represents the sequence of values (at the sampling instants) of the response of the system to a pulse input of unit height and duration equal to the sampling interval.

The derivation of Eqs.(12) and (13) follows from Eq. (4), with the time set equal to its value at (say) the k'th sampling instant, and with the input u(t) modified by the sample-and-hold system as shown in Fig. 4:

$$y(kT) = \int_0^{\infty} h(\tau)\ U(kT - \tau)d\tau \quad \text{for } k = 0,1,2,\ldots,\ \text{etc.}$$

But since $u(kT - \tau) = u_{k-i}$ for $(i-1)T \leq \tau < iT$, we have

$$y(kT) = \sum_{i=0}^{\infty} u_{k-i} \int_{(i-1)T}^{iT} h(\tau)d\tau = \sum_{i=0}^{\infty} w_i u_{k-i}.$$

Noting that $w_0 = 0$, Eq.(12) follows.

Example

Estimation of Unit Pulse Response using Cross-correlation

Consider the technique of estimating the weighting sequence (unit pulse response) of a linear system when its response to an input sequence $\{u_t\}$ is corrupted by a random noise sequence $\{n_t\}$, as in Fig. 5.

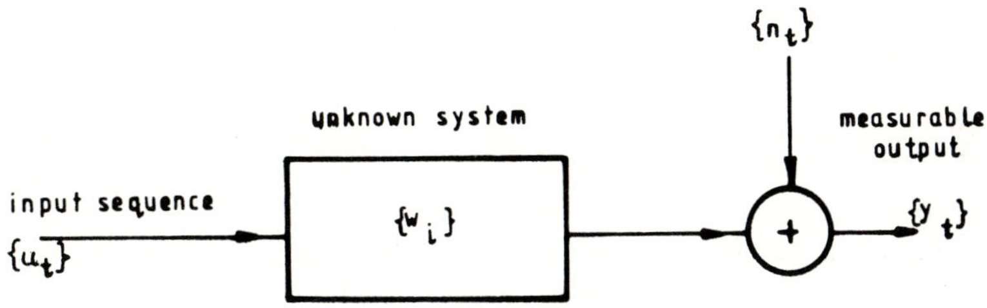

Figure 5

This is a typical example of a problem in system modelling or identification, and may be approached statistically by considering the *cross-correlation* of the measured output with delayed versions of the input:

$$R_{uy}(r) = \overline{y_t u_{t-r}} = \frac{1}{N} \sum_{t=1}^{N} y_t u_{t-r}.$$

From Eq. (12), we have

$$y_t = n_t + \sum_{i=1}^{\infty} w_i u_{t-i}.$$

Thus,

$$R_{uy}(r) = \overline{n_t u_{t-r}} + \sum_{i=1}^{\infty} w_i \cdot \overline{u_{t-i} \cdot u_{t-r}}.$$

Examining this expression, we note that if the noise sequence $\{n_t\}$ is uncorrelated with the input sequence $\{u_t\}$, and if the averaging process is taken over a large number of samples (N large), the first term on the right may be expected to have a very small value (vanishingly small for $N \to \infty$). Furthermore, it is possible to choose the input sequence u_t in such a way that the quantity $\overline{u_{t-i} \cdot u_{t-r}}$ satisfies

$$\overline{u_{t-i} \cdot u_{t-r}} = 0 \text{ for } i \neq r$$
$$= \overline{u^2} \text{ for } i = r.$$

Thus, the cross-correlation becomes

$$R_{uy}(r) = \overline{u^2} . w_r$$
$$\text{for } r = 1,2,\ldots, \text{ etc.}$$

The statistical implications of this type of procedure, and the choice of input sequences having the requisite characteristics, will be the subjects of lectures occurring later in the vacation school.

Concluding Comments

In this introductory review we have attempted to outline the most basic concepts which will be assumed to be familiar to the course participants at the outset. Those who are new to the subject are strongly urged to consult the extensive literature, a small selection of which is referenced here.

SUGGESTIONS FOR FURTHER READING

On Dynamic Systems Concepts Generally:

R.J. Richards, "An Introduction to Dynamics and Control". (Longman, 1979).
J. Gary Reid, "Linear System Fundamentals". (McGraw Hill, 1983).
C.D. McGillem and G.R. Cooper, "Continuous and Discrete Signal and System Analysis" (Holt, Rinehart and Winston, 1974).
T. Kailath, "Linear Systems". (Prentice Hall, 1980).

On Frequency Domain Concepts:

See the notes for Lecture L8. The following texts are also recommended:

J.S. Bendat and A.G. Piersol, "Random Data: Analysis and Measurement Procedures". (Wiley, 1971).
J.S. Bendat and A.G. Piersol, "Engineering Applications of Correlation and Spectral Analysis". (Wiley, 1980).

<u>On Sampled-data Control Systems</u>:

G.F. Franklin and J.D. Powell, "Digital Control of Dynamic Systems". (Addison-Wesley, 1980).

C.L. Phillips and H.T. Nagle, "Digital Control System Analysis and Design". (Prentice-Hall, 1984).

J.R. Leigh, "Applied Digital Control". (Prentice-Hall, 1984).

<u>On System Identification (including many practical aspects)</u>:

J.P. Norton, "An Introduction to Identification". (Academic Press, 1986).

MATRIX TECHNIQUES

Dr. M.T.G. Hughes

1. INTRODUCTION

This section outlines some basic concepts and techniques of matrix analysis. Some of these will be employed in later lectures. The intention here is to provide a guide to further study for those who are unfamiliar with the subject, and to introduce the notation which will be employed subsequently in the vacation school.

2. ELEMENTARY DEFINITIONS [1,4]

A $m \times n$ matrix is defined here as an array of numbers arranged in m rows and n columns, thus:

$$A = \begin{bmatrix} a_{11} & a_{12} \cdots\cdots & a_{1n} \\ a_{21} & a_{22} \cdots\cdots & a_{2n} \\ \vdots & & \vdots \\ a_{m1} & a_{m2} & a_{mn} \end{bmatrix} \tag{1}$$

We note that individual elements are identified by ordered subscripts, thus a_{ij} is the element in the i'th row and j'th column.

Occasionally, the notation $[a_{ij}]$ will be found convenient in order to specify something about the typical element of the matrix.

Illustrative Example:

Suppose we have a set of m variables $\{y_i\}$, $i = 1,2,\ldots,m$, and suppose that each member of this set is a function of the n variables $\{x_j\}$, $j = 1,2,\ldots,n$.

This may be written in full as

$$\left.\begin{aligned} y_1 &= y_1(x_1,x_2,\ldots,x_n) \\ y_2 &= y_2(x_1,x_2,\ldots,x_n) \\ &\cdots\cdots\cdots\cdots \\ y_m &= y_m(x_1,x_2,\ldots,x_n) \end{aligned}\right\} \tag{2}$$

or, more concisely as

$$y = y(x) \tag{3}$$

where the quantities y and x are special kinds of matrices, referred to as *column vectors*:

$$y = \begin{bmatrix} y_1 \\ y_2 \\ \vdots \\ y_m \end{bmatrix} \qquad x = \begin{bmatrix} x_1 \\ x_2 \\ \vdots \\ x_n \end{bmatrix} \tag{4}$$

Here, y is a m × 1 matrix, referred to as an m-dimensional column vector, or simply as a m-vector. Similarly, the n × 1 matrix (or column vector) x is a n-vector.

Equations (2) and (3) can be regarded as alternative ways of representing a *transformation* of the n variables $\{x_1, x_2, \ldots, x_n\}$ into a set of m variables $\{y_1, y_2, \ldots, y_m\}$ or more concisely as a transformation of the n-vector x into the m-vector y. In analytical geometry an important quantity associated with this transformation is the so-called *Jacobian matrix*[(2)], defined as

$$J = \begin{bmatrix} \frac{\partial y_1}{\partial x_1} & \frac{\partial y_1}{\partial x_2} & \cdots & \frac{\partial y_1}{\partial x_n} \\ \vdots & & & \\ \frac{\partial y_m}{\partial x_1} & \frac{\partial y_m}{\partial x_2} & \cdots & \frac{\partial y_m}{\partial x_n} \end{bmatrix} \tag{5}$$

To represent this quantity more concisely, we may either use the $[a_{ij}]$ notation:

$$J = \left[\frac{\partial y_i}{\partial x_j}\right] \tag{6}$$

or we may regard the matrix J as being (formally) the 'derivative' of the vector y with respect to the vector x. Thus:

$$J = \frac{dy}{dx} \tag{7}$$

This concludes the illustration, but the uses of such contracted notation will be demonstrated later in the lecture.

Transposition

This operation is defined simply by the interchange of rows and columns of a matrix, and is denoted by the superscript $(\ldots)^T$, thus:

$$\left.\begin{aligned} \text{If} \quad A &= [a_{ij}] \\ \text{Then} \quad A^T &= [a_{ji}] \end{aligned}\right\} \tag{8}$$

A simple special case of transposition is one which converts a n × 1 matrix (or *column vector*) into a 1 × n matrix (or *row vector*). For example, if x is defined as in Eq. (4), then

$$x^T = (x_1\ x_2\ \dots\ x_n) \tag{9}$$

a *row vector*. This notation is often used simply to save page space when defining column vectors though, of course, it has other more significant uses.

Some Special Types of Matrix

The zero matrix, denoted as 0, is a matrix whose elements are all zero. *A diagonal matrix* is a square matrix (m = n) whose elements are zero, except for those elements on the *principal diagonal* (where i = j).

A special case of a diagonal matrix is the *unit matrix:*

$$I = \begin{bmatrix} 1 & 0 & \dots\dots & 0 \\ 0 & 1 & 0 \dots & 0 \\ \hline 0 & 0 & \dots 0 & 1 \end{bmatrix} \tag{10}$$

Sometimes, the order of I is indicated by a subscript (e.g. I_n, to denote a n × n unit matrix).

The *trace* of a square matrix A, denoted as Tr(A), is simply the sum of all elements on the principal diagonal of A.

A symmetric matrix is one which is unaltered by transposition, i.e. $A^T \equiv A$, or $[a_{ij}] \equiv [a_{ji}]$.

3. ELEMENTARY OPERATIONS AND RELATIONS

Partitioning. It is sometimes helpful to divide a matrix into convenient sub-matrices as follows, for example:

$$A = \left[\begin{array}{cc|c} a_{11} & a_{12} & a_{13} \\ a_{21} & a_{22} & a_{23} \\ \hline a_{31} & a_{32} & a_{33} \end{array}\right] = \left[\begin{array}{c|c} A_{11} & A_{12} \\ \hline A_{21} & A_{22} \end{array}\right] \tag{11}$$

where

$$A_{11} = \begin{bmatrix} a_{11} & a_{12} \\ & \\ a_{21} & a_{22} \end{bmatrix}, \qquad A_{12} = \begin{bmatrix} a_{13} \\ \\ a_{23} \end{bmatrix}, \qquad (12)$$

$$A_{21} = [a_{31} \ \ a_{32}], \qquad A_{22} = a_{33}$$

Equality: we say that A = B if $[a_{ij}] = [b_{ij}]$
for all i and j. (13)

Addition/Subtraction: $C = A \pm B$ if $[c_{ij}] = [a_{ij} \pm b_{ij}]$
for all i and j. (14)

Multiplication: C = A B if $[c_{ij}] = \left[\sum_{k=1}^{n} a_{ik}\, b_{kj}\right]$
for all i and j, with n = number of columns in A
= number of rows in B. (15)

In general, it should be fairly clear that matrix addition is both commutative and associative, i.e. A + B = B + A, and (A + B) + C = A + (B + C), whereas matrix multiplication is associative, but not commutative: (A.B).C = A.(B.C), but A B $\neq$ B A in general. Combined multiplication and addition/subtraction with matrices possesses the distributive property:

$$A.(B \pm C) = A.B \pm A.C$$

Operations with scalars. Multiplication of a matrix by a scalar has the effect of multiplying all elements of the matrix. Thus, if α is a scalar,

$$\alpha A = [\alpha a_{ij}] \qquad (16)$$

Addition/subtraction of scalar quantities with a matrix is not, however, defined (see Eq (14)).

If the elements of a matrix X are functions of some scalar quantity (say t), then X(t) may be differentiated or integrated with respect to t, thus:

$$\dot{X} = \frac{dX}{dt} = [\dot{x}_{ij}] = \left[\frac{dx_{ij}}{dt}\right], \qquad (17)$$

$$\int X dt = \left[\int x_{ij} dt\right] \qquad (18)$$

We have already indicated the way in which this concept would require extension to deal with the case of differentiation with respect to a vector (Eqs. (5) to (7)).

Clearly the property of Eq. (18) may be extended to the case of any linear transform of a matrix. Thus, if the symbol $\mathcal{L}(\ldots)$ denotes a Laplace transform, for example, we have:

$$\mathcal{L}(X(t)) = \int_0^\infty X(t)\exp(-st)dt$$

$$= \left[\int_0^\infty x_{ij}(t)\exp(-st)dt\right] = \left[X_{ij}(s)\right] = X(s) \tag{19}$$

Determinant and Matrix Inverse

The *determinant* of a square $(n \times n)$ matrix A may be defined formally as: "The sum of the signed products of all possible combinations of n elements, where each element is taken from a different row and column".

More conveniently, the determinant of A (written as det A or $|A|$) may be evaluated by repeated use of the relation

$$|A| = \sum_{j=1}^{n} a_{ij}\gamma_{ij} \tag{20}$$

for any fixed i in the range $i = 1,2,\ldots,n$, with

$$\gamma_{ij} = (-1)^{i+j}\mu_{ij} \ . \tag{21}$$

Here, μ_{ij} is the determinant of the $(n-1) \times (n-1)$ matrix formed by deleting the row and column through the element a_{ij}, and is called the *minor* of element a_{ij}. The signed quantity γ_{ij} is called the *cofactor* of a_{ij}.

The transposed matrix of cofactors of a square matrix A is called the *Adjoint* or *Adjugate*, denoted as Adj A, thus,

$$\text{Adj } A = [\gamma_{ij}]^T = [\gamma_{ji}] \tag{22}$$

Matrix Inverse

The 'inverse' of a matrix can be defined in a number of ways. For instance, if A is a non-square $m \times n$ matrix $(m \neq n)$, a *generalised inverse* of A, viz. A^I, may be defined[5] such that

$$A\,A^I\,A = A, \text{ and } A^I\,A\,A^I = A^I \tag{23}$$

Such generalised inverses are associated with solutions, or approximate solutions, of equations of the form $Ax = y$, in which the number of independent equations differs from the number of unknowns.

In the more familiar case, where the matrix A is square $(m = n)$, a unique inverse A^{-1} will exist such that

$$A^{-1} A = A A^{-1} = I \qquad (24)$$

provided the matrix A is *nonsingular*, that is, $\det A \neq 0$.
The elements of the inverse matrix A^{-1} are defined by the relation

$$A^{-1} = \frac{\text{Adj } A}{\det A} \qquad (25)$$

Linear Independence and Rank

A set of n vectors $x_1, x_2, \ldots, x_n$ is said to be *linearly dependent* if there exists a set of scalar constants $\alpha_1, \alpha_2, \ldots, \alpha_n$, at least one of which is nonzero, such that

$$\sum_{i=1}^{n} \alpha_i x_i = 0. \qquad (26)$$

If such a set of constants does *not* exist, the vectors $\{x_j\}$ are said to be *linearly independent.*

The *rank* of any $m \times n$ matrix is defined as the order of the largest nonsingular square matrix which can be formed simply by deleting rows and/or columns in the original matrix.

Consider the matrix equation

$$A x = y$$

Given the $m \times n$ matrix A and the m-vector y, the problem is to find the n-vector x. Two principal cases may be discerned:

Inhomogeneous Case ($y \neq 0$)

We consider here the rank of the matrix A (say, r) and that of the so-called *Augmented matrix* A^1, formed by appending the column vector y to the A matrix. Let the rank of A^1 be r^1.

The inhomogeneous equations will be consistent (i.e. at least one value of x will satisfy them) if $r = r^1$. They will be inconsistent if $r < r^1$. Note that r cannot exceed r^1.

If $n > (r = r^1)$, then (n-r) of the elements of x may be given arbitrary values and the remaining r unknowns found uniquely in terms of them.

If $n = r = r^1$, the original equations may be solved uniquely for x.

Homogeneous Case ($y = 0$)

If the rank $r = n$, the equation $Ax = 0$ will have the unique (and trivial) solution $x = 0$.

If $r < n$, then as before, $(n - r)$ of the elements of x may be assigned arbitrary values, and the remaining r unknowns found uniquely in terms of them.

4. LINEAR DIFFERENTIAL EQUATIONS

The study of dynamic systems described by linear differential equations may be greatly facilitated through the use of vector-matrix concepts.

In the case of a linear, constant-coefficient system, the general nth order differential equation describing the response y(t) to input u(t) may be written as

$$\frac{d^n y}{dt^n} + a_{n-1}\frac{d^{n-1}y}{dt^{n-1}} + \dots + a_1\frac{dy}{dt} + a_0 y(t)$$

$$= b_m\frac{du^m}{dt^m} + b_{m-1}\frac{du^{m-1}}{dt^{m-1}} + \dots + b_1\frac{du}{dt} + b_0 u(t) \tag{27}$$

where, for a physical system, $m < n$.

This nth-order equation may be expressed as a vector-matrix differential equation of *first order* in a variety of ways. For example, we could consider initially a reduced equation of the form

$$\frac{d^n x}{dt^n} + a_{n-1}\frac{d^{n-1}x}{dt^{n-1}} + \dots + a_1\frac{dx}{dt} + a_0 x(t) = u(t) \tag{28}$$

with the new variable x(t) satisfying the same initial conditions as y(t) in Eq. (27).

Now we could introduce a new set of variables $\{x_1(t), x_2(t), \dots, x_n(t)\}$, called *state variables*, which may be defined as follows:

$$x_1(t) = x(t);\quad x_2(t) = \frac{dx_1}{dt};\ \dots\ x_n(t) = \frac{dx_{n-1}}{dt} \tag{29}$$

In terms of these new variables, Eq. (28) could be re-written (using dot notation for derivatives) as

$$\left.\begin{aligned} \dot{x}_1 &= x_2 \\ \dot{x}_2 &= x_3 \\ &\cdots\cdots \\ &\cdots\cdots \\ \dot{x}_{n-1} &= x_n \\ \dot{x}_n &= -a_0x_1 - a_1x_2 - \dots - a_{n-1}x_n + u(t) \end{aligned}\right\} \tag{30}$$

or in vector-matrix form,

$$\dot{x}(t) = A\,x(t) + B\,u(t) \tag{31}$$

where

$$x(t) = (x_1(t),\ x_2(t), \dots, x_n(t))^T \tag{32}$$

$$A = \begin{bmatrix} 0 & 1 & 0 & \cdots & & 0 \\ 0 & 0 & 1 & 0 & \cdots & 0 \\ \cdot & \cdot & \cdot & \cdot & \cdot & \cdot \\ \cdot & \cdot & \cdot & \cdot & \cdot & \cdot \\ 0 & 0 & \cdots & & 0 & 1 \\ -a_0 & -a_1 & & & & -a_{n-1} \end{bmatrix} \tag{33}$$

$$B = (0 \quad 0 \; . \; . \; . \; . \; . \; . \; . \; . \; . 1)^T, \tag{34}$$

$$u(t) = \text{scalar} \tag{35}$$

Equation (31) represents the general form of the *state equations* for an nth-order dynamic system. The system equations could, of course, be stated differently from those of Eq. (27), and the state variables could be defined in a different way from Eq. (30). In such a case it would merely be necessary to redefine the matrices A, B, x(t) and u(t) accordingly.

Assuming that a solution of Eq. (31) is obtained for x(t), for given initial conditions x(0), the final solution for the required system output y(t) may be obtained by applying the principle of superposition, to obtain:

$$y(t) = b_0x_1(t) + b_1x_2(t) + \ldots + b_mx_{m-1}(t) \tag{36}$$

The structural form of the equations represented by Eqs. (30) to (36) is illustrated schematically in Fig. 1.

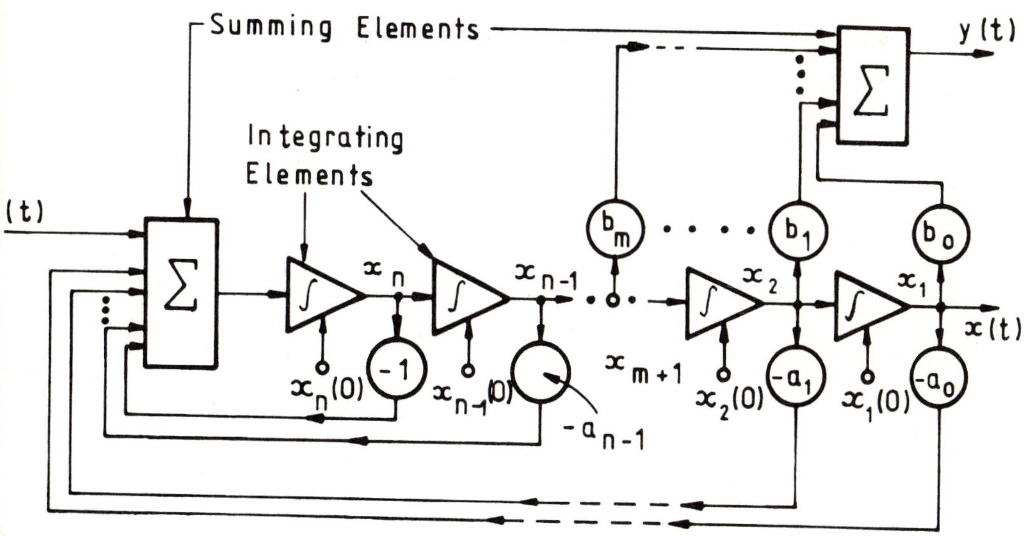

Figure 1. State-variable representation of linear differential equation

In a more general case, we might have several outputs and inputs, and the vector-matrix form of the differential equations will be

$$\left.\begin{aligned} \dot{x} &= A\,x(t) + B\,u(t), \\ x(t) &= x_0 \text{ at } t = t_0, \\ y(t) &= C\,x(t) \end{aligned}\right\} \tag{37}$$

where A, B, C are matrices, which may in general be functions of time, x(t), y(t), u(t) are vectors having appropriate dimensions, and t_0 represents the 'initial time', at which the input is presumed to start. A schematic representation of Eq. (37) is shown in Fig. 2.

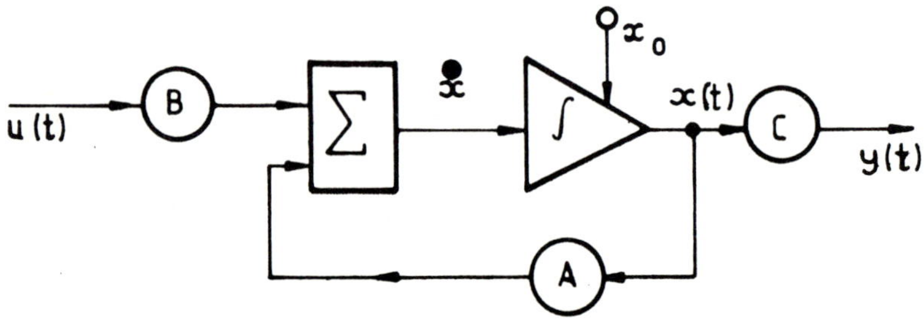

Figure 2. Matrix schematic diagram of linear system

An important special case of Eq. (37) is the *homogeneous* case, in which the system input vector u(t) is absent. Here, we have the *homogeneous equation*

$$\left.\begin{aligned} \dot{x}(t) &= A\,x(t) \\ x(t) &= x_0 \text{ at } t = t_0. \end{aligned}\right\} \tag{38}$$

Even in the case where the matrix A is a function of time, the solution of Eq. (38) can be shown to have the form

$$x(t) = \Phi(t,t_0)x_0 \tag{39}$$

The quantity $\Phi(t,t_0)$ is known as the *state transition matrix* of the system described by Eq. (37), and it plays an important part in linear system theory, possessing as it does characteristics which are analogous to those of the scalar exponential function.

Once the state transition matrix is found by solving Eq. (38), the solution of the inhomogeneous equation (37) may be written as[1]:

$$x(t) = \Phi(t,t_0)x_0 + \int_{t_0}^{t} \Phi(t,\tau).B.u(\tau)d\tau \qquad (40)$$

This is an important result, as it separates clearly the transient components of system response (representing the recovery from initial conditions) from the components due to input fluctuations.

The actual calculation of system responses from Eq. (40) can be quite difficult in the general case of a time-varying A matrix. In the case where A is a constant matrix, however, relatively simple general methods of solution are available. One of these is the method of Laplace Transformation, which is outlined here:

Choosing a time origin such that $t_0 = 0$, Laplace transformation of both sides of Eq. (37) yields

$$sX(s) - x_0 = AX(s) + BU(s) \qquad (41)$$

where X(s) is the Laplace transform of the state vector x(t), and A, B are assumed to be constant matrices. The quantity s is used here to denote the (scalar) Laplace transform variable.

By algebraic manipulation of Eq. (41), we obtain

$$(sI - A)X(s) = x_0 + BU(s),$$

from which

$$X(s) = (sI - A)^{-1}x_0 + (sI - A)^{-1}BU(s) \qquad (42)$$

The solution for x(t) follows, as usual, by finding the inverse Laplace transformation of the elements of the vector X(s).

The state transition matrix in this case is seen to be a function of only one variable - the elapsed time - and may be found from

$$\Phi(t) = \mathcal{L}^{-1}\{(sI - A)^{-1}\} \qquad (43)$$

Example

A certain dynamic system is described by the differential equation

$$\ddot{y} + 3\dot{y} + 2y = u + 2\dot{u}.$$

Put this equation in state variable form, and find the state transition matrix of the system.

First, let x(t) be a new variable satisfying the equation

$$\ddot{x} + 3\dot{x} + 2x = u \; .$$

Then $\quad y = x + 2\dot{x}$.

Now let $x_1 = x$, $x_2 = \dot{x}_1$,

then we have the system equations

$$\dot{x}_1 = x_2$$

$$\dot{x}_2 = -2x_1 - 3x_2 + u(t)$$

or, in vector-matrix form,

$$\dot{x}(t) = Ax(t) + bu(t),$$

$$y(t) = c^T x(t)$$

where $x = (x_1\ x_2)^T$, $\qquad u(t) = \text{scalar},$

$$A = \begin{bmatrix} 0 & 1 \\ -2 & -3 \end{bmatrix}, \quad b = \begin{bmatrix} 0 \\ 1 \end{bmatrix},$$

$$c^T = (1 \quad 2).$$

The Laplace transform of the state transition matrix is

$$(sI - A)^{-1} = \begin{bmatrix} s & -1 \\ 2 & s+3 \end{bmatrix}^{-1}$$

$$= \begin{bmatrix} \frac{s+3}{(s+1)(s+2)} & \frac{1}{(s+1)(s+2)} \\ \frac{-2}{(s+1)(s+2)} & \frac{s}{(s+1)(s+2)} \end{bmatrix}$$

Thus, by inverse Laplace transformation we obtain finally

$$\Phi(t) = \begin{bmatrix} 2e^{-t} - e^{-2t} & e^{-t} - e^{-2t} \\ 2e^{-2t} - 2e^{-t} & 2e^{-2t} - e^{-t} \end{bmatrix}$$

As a check on this result, we may observe that $\Phi(0) = I$, as is obviously required in general, and that $\lim_{t \to \infty} \Phi(t) = 0$ for a stable system.

5. EIGENVECTORS AND EIGENVALUES

Many physical problems can be greatly simplified through the use of eigenvalue analysis. The literature on this subject is very extensive, and all of the references listed at the end of this lecture employ it in various ways. At this point, it is only possible to present a brief outline of the main ideas and uses of eigenvalue analysis. For a more complete treatment, the reader could consult any of refs. 1, 4, 5.

All eigenvalue problems can be reduced to the problem of finding a scalar λ and a vector v to satisfy an equation of the form

$$(A - \lambda I)v = 0 \tag{44}$$

or a vector w such that

$$w(A - \lambda I) = 0 \tag{45}$$

In either case, the matrix A is square ($n \times n$), and λ is a scalar. The quantity w is a row vector while v is a column vector.

The values of λ which satisfy Eqs. (44) and (45) are called the *Eigenvalues* of the matrix A. The corresponding values of vector v are the *column eigenvectors*, and the values of vector w are the *row eigenvectors*.

A necessary and sufficient condition for Eqs. (44) and (45) to have nontrivial solutions is that the $n \times n$ matrix $(A - \lambda I)$ has rank n-1. This requires that

$$\det(A - \lambda I) = 0 \tag{46}$$

This constitutes a polynomial equation of the nth degree in the scalar quantity λ, which yields exactly n (not necessarily distinct) *characteristic values* or *eigenvalues* $\{\lambda_1, \lambda_2, \ldots, \lambda_n\}$.

Corresponding to each distinct eigenvalue λ_i, there will be a row eigenvector w_i and a column eigenvector v_i, as defined by Eqs. (44) and (45) respectively.

Example

For the matrix

$$A = \begin{bmatrix} 0 & 1 \\ -2 & -3 \end{bmatrix} \tag{47}$$

the column eigenvectors are defined by Eq. (44):

$$\begin{bmatrix} -\lambda & 1 \\ -2 & -3-\lambda \end{bmatrix} \begin{bmatrix} v_1 \\ v_2 \end{bmatrix} = \begin{bmatrix} 0 \\ 0 \end{bmatrix}$$

For a nontrivial solution, we require

$$\det \begin{bmatrix} -\lambda & 1 \\ -2 & -3-\lambda \end{bmatrix} = 0,$$

that is $\quad \lambda(3 + \lambda) + 2 = 0,$

or $\quad \lambda^2 + 3\lambda + 2 = 0.$

This is satisfied by two values of λ, (the eigenvalues of A):

$$\lambda_1 = -1, \lambda_2 = -2 \tag{48}$$

Thus, since the eigenvalues are distinct, it is possible to find *two* distinct eigenvectors v_1 and v_2

with $\lambda = \lambda_1 = -1$:

$$\begin{bmatrix} 1 & 1 \\ -2 & -2 \end{bmatrix} \begin{bmatrix} v_{11} \\ v_{21} \end{bmatrix} = \begin{bmatrix} 0 \\ 0 \end{bmatrix}$$

Clearly, there are infinitely many solutions to this equation. Thus, we may assign an arbitrary value to any one element of v_1, and evaluate the remaining element accordingly.

Choosing $v_{11} = 1$ (arbitrarily), we obtain

$$v_1 = \begin{bmatrix} v_{11} \\ v_{21} \end{bmatrix} = \begin{bmatrix} 1 \\ -1 \end{bmatrix} \tag{49}$$

Similarly, with $\lambda = \lambda_2 = -2$:

$$\begin{bmatrix} 2 & 1 \\ -2 & -1 \end{bmatrix} \begin{bmatrix} v_{12} \\ v_{22} \end{bmatrix} \begin{bmatrix} 0 \\ 0 \end{bmatrix}$$

from which, choosing $v_{12} = 1$, we obtain

$$v_2 = \begin{bmatrix} v_{12} \\ v_{22} \end{bmatrix} = \begin{bmatrix} 1 \\ -2 \end{bmatrix} \tag{50}$$

The row eigenvectors, similarly, are defined by Eq. (45):

$$(w_1 \;\; w_2) \begin{bmatrix} -\lambda & 1 \\ -2 & -3-\lambda \end{bmatrix} = (0 \quad 0)$$

which yields the same values of λ as previously, for a nontrivial solution. Thus, similarly to before,

with $\lambda = \lambda_1 = -1$:

$$(w_{11} \;\; w_{12}) \begin{bmatrix} 1 & 1 \\ -2 & -2 \end{bmatrix} = (0 \quad 0)$$

If we choose $w_{11} = 1$ (arbitrarily), we obtain

$$(w_{11} \;\; w_{12}) = (1 \quad \tfrac{1}{2}) \tag{51}$$

with $\lambda = \lambda_2 = -2$:

$$(w_{21}\ w_{22}) \begin{bmatrix} 1 & 1 \\ -2 & -1 \end{bmatrix} = (0\ \ 0)$$

from which, setting $w_{21} = 1$,

$$(w_{21}\ w_{22}) = (1\ \ 1) \qquad (52)$$

It is found that the row and column eigenvectors corresponding to distinct eigenvalues possess some remarkable and convenient properties. These are discussed below:

Orthogonality

When the eigenvalues of A are distinct, it can be shown that row and column eigenvectors corresponding to different eigenvalues are orthogonal. That is,

$$w_i v_j = 0,\ i \neq j \qquad (53)$$

This follows from the fact that, for $i,j = 1,2,\ldots,n$,

$$Av_j = \lambda_j v_j \qquad (54)$$

$$w_i A = \lambda_i w_i \qquad (55)$$

Premultiplying Eq. (54) by w_i, and postmultiplying Eq. (55) by v_j, we get from Eq. (54),

$$w_i A v_j = \lambda_j w_i v_j$$

and from Eq. (55),

$$w_i A v_j = \lambda_i w_i v_j$$

Thus, $(\lambda_i - \lambda_j) w_i v_j = 0$,

and if $\lambda_i \neq \lambda_j$, we have

$$w_i v_j = 0$$

thus confirming Eq. (53).

Referring again to the example, since the absolute scaling of the vectors defined by Eqs. (49) - (52) is arbitrary, we may adjust the scaling factors in such a way that the products $w_i v_i$ all have unit value. When this is done, if the rescaled column eigenvectors v_i are arranged alongside one another to form a square matrix V and the rescaled row eigenvectors w_j are arranged beneath one another to form a square matrix W, we have

$$W.V = \begin{bmatrix} 2 & 1 \\ 1 & 1 \end{bmatrix} \begin{bmatrix} 1 & -1 \\ -1 & 2 \end{bmatrix} = \begin{bmatrix} 1 & 0 \\ 0 & 1 \end{bmatrix} \tag{56}$$

This example illustrates a very convenient property of the scaled row - and column - eigenvector matrices which generally holds only in the case of distinct eigenvalues*:

$$W.V = V.W = I \tag{57}$$

*The situation in the case of repeated eigenvalues is more complicated than this, but a full discussion of that case would be beyond the scope of these introductory notes.

Spectral Resolution of a Square Matrix

By building up Eqs. (54) and (55) to include all the column and row eigenvectors for the full respective ranges of indices i and j, it is possible to write

$$AV = V\Lambda \tag{58}$$

and

$$WA = \Lambda W \tag{59}$$

where

$$\Lambda = \begin{bmatrix} \lambda_1 & 0 & \cdots\cdots\cdots & 0 \\ 0 & \lambda_2 & 0 \cdots\cdots & 0 \\ \hline 0 & 0 & 0 & \lambda_n \end{bmatrix} \tag{60}$$

is a diagonal matrix in which the (distinct) eigenvalues of matrix A appear along the principal diagonal, and 'zeros' appear elsewhere.

Eqs. (57) to (60) may be employed to advantage in several ways. For instance, noting that $W = V^{-1}$, it is possible to perform a 'diagonalising transformation' on matrix A, as follows:

$$WAV = V^{-1}AV = \Lambda \tag{61}$$

Alternatively, it is often helpful to resolve the matrix A into a product of three component matrices, as follows:
From Eqs. (59) and (57):

$$VWA = A = V\Lambda W \tag{62}$$

<u>Example</u>

Here we illustrate the use of Eq. (62) in the solution of the differential equation

$$\begin{bmatrix} \dot{x}_1 \\ \dot{x}_2 \end{bmatrix} = \begin{bmatrix} 0 & 1 \\ -2 & -3 \end{bmatrix} \begin{bmatrix} x_1(t) \\ x_2(t) \end{bmatrix}$$

with initial conditions

$$x_1(0) = 1, \quad x_2(0) = 0.$$

This equation has the form

$$\dot{x} = Ax$$

and by Eq. (62), could be rewritten as

$$\dot{x} = V\Lambda Wx$$

Premultiplying both sides of this equation by W, noting that WV = I, and introducing a change of variable:

$$Wx = y$$

we obtain

$$\dot{y} = \Lambda y$$

This is a set of uncoupled differential equations of first order, the solution of which can be written by inspection:

$$y_1(t) = y_1(0)\exp(\lambda_1 t)$$

$$y_2(t) = y_2(0)\exp(\lambda_2 t)$$

We have already established (in Eq. (48)) that the eigenvalues of the matrix A used in this example are $\lambda_1 = -1$, $\lambda_2 = -2$, and we know the elements of matrices W and V from Eq. (56). Thus we have, since $x(0) = [1 \ \ 0]^T$,

$$\begin{bmatrix} y_1(0) \\ y_2(0) \end{bmatrix} = \begin{bmatrix} 2 & 1 \\ 1 & 1 \end{bmatrix} \begin{bmatrix} 1 \\ 0 \end{bmatrix} = \begin{bmatrix} 2 \\ 1 \end{bmatrix}.$$

$$y_1(t) = 2\exp(-t),$$

$$y_2(t) = \exp(-2t),$$

and finally, since $x(t) = Vy(t)$,

$$\begin{bmatrix} x_1(t) \\ x_2(t) \end{bmatrix} = \begin{bmatrix} 1 & -1 \\ -1 & 2 \end{bmatrix} \begin{bmatrix} y_1(t) \\ y_2(t) \end{bmatrix}$$

Thus, $x_1(t) = 2\exp(-t) - \exp(-2t)$

$$x_2(t) = -2\exp(-t) + 2\exp(-2t).$$

This concludes the example.

The main benefit of eigenvalue analysis lies in its property of isolating, or uncoupling, the fundamental modes or interconnections of a system. With large complex systems, this has both conceptual and computational advantages, and eigenvalue analysis can often be used to good effect in clarifying otherwise obscure problems.

Example

Consider the *controllability* of a constant-coefficient linear system with a single input u(t). The state equations of such a system may be written in the form

$$\dot{x} = Ax + bu,$$

where b is a constant vector of suitable dimension. The fundamental issue of controllability of the state of such a system is concerned with the question of whether any particular state can be reached from any other given state (which may be taken to be the origin) for some choice of control input u(t). Eigenvalue analysis can provide a useful insight into this problem, as follows:

Resolving the matrix A into the spectral form $V\Lambda W$, premultiplying the state equation by W, and changing the state variable x to z = Wx, we obtain

$$\dot{z} = \Lambda z + Wbu.$$

It is fairly clear from this expression that if any element in the vector Wb is zero, then the corresponding element of the state vector z will be effectively disconnected from the control. Consequently, any elements of x made up of linear combinations of these z's will be uncontrollable. Thus, if the system is to be totally controllable, all the elements of the vector Wb must be nonzero.

This is of course a very simplified instance of the general problem of controllability. For a more extensive treatment of the subject, the reader is referred to the suggested further reading[1].

6. DISCRETE-TIME SYSTEMS

By analogy to the continuous-time case, a natural mode of representation for discrete-time (or *sampled-data*) systems is through vector-matrix *difference* equations such as

$$\left.\begin{aligned} x_{k+1} &= Fx_k + Gu_k, \\ y_k &= Hx_k \end{aligned}\right\} \qquad (63)$$

Here, as before, F,G,H are matrices which may in general change as functions of the time index k. The vectors u_k, x_k, y_k are respectively the values of the input, state, and output vectors at the discrete time instant $t = k\lambda$, where λ is the sampling interval.

This mode of representation tends to be a natural choice when digital computation is involved, and questions of controllability and observability may be dealt with relatively straighforwardly compared with the continuous-time case.

Controllability Condition

For initial simplicity, consider a system with a single input u, such that $x_{k+1} = Fx_k + qu_k$ where q is a constant vector. For a given x_0, we seek conditions under which the control necessary to drive the system to some arbitrary state x_n may be determined.

From the given initial state, we have

$$x_1 = Fx_0 + qu_0$$

$$x_2 = Fx_1 + qu_1 = F^2x_0 + Fqu_0 + qu_1$$

$$\vdots$$

$$x_n = F^nx_0 + F^{n-1}qu_0 + F^{n-2}qu_1 + \ldots + qu_{n-1}$$

From this, we find

$$x_n - F^nx_0 = F^{n-1}qu_0 + F^{n-2}qu_1 + \ldots + qu_{n-1}$$

$$= [q \quad Fq \text{ ----- } F^{n-1}q] \begin{bmatrix} u_{n-1} \\ u_{n-2} \\ \vdots \\ u_1 \\ u_0 \end{bmatrix}$$

Since x_n, F^n, and x_0 are given, the condition for a unique solution to exist for the u's is that the matrix

$$M_1 = [q \; Fq \text{ ----- } F^{n-1}q] \qquad (64)$$

should have full rank (n).

Where this condition is satisfied, then F, q are referred to as a *controllable pair*.

Observability Condition

Again, for simplicity, consider a system having a single output y_k, and assume the system equations to have the form

$$x_{k+1} = Fx_k,$$
$$y_k = h^T x_k$$

where h is a constant column vector.

We may now seek the condition under which the unknown state x_0 may be determined from observations of the y's.

We have, starting with the unknown initial state,

$$y_0 = h^T x_0$$
$$y_1 = h^T F x_0$$
$$\vdots$$
$$y_{n-1} = h^T F^{n-1} x_0$$

or

$$\begin{bmatrix} y_0 \\ y_1 \\ \vdots \\ y_{n-1} \end{bmatrix} = \begin{bmatrix} h^T \\ h^T F \\ \vdots \\ h^T F^{n-1} \end{bmatrix} x_0$$

If x_0 is to be determined uniquely from this, the matrix

$$M_2 = \begin{bmatrix} h^T \\ h^T F \\ \vdots \\ h^T F^{n-1} \end{bmatrix} \tag{65}$$

must have full rank (i.e. must be nonsingular).

7. QUADRATIC FORMS

A quadratic form is defined by the expression

$$Q(x) = x^T A x \tag{66}$$

Here, the quantity Q is a scalar, x is a n-vector, and A is a $n \times n$ matrix $[a_{ij}]$.

Expansion of the terms in Eq. (66) shows the structure of Q to be a weighted sum of all pairwise products of the elements of x (including the squares of the elements). Thus

$$Q(x) = \sum_{i=1}^{n} \sum_{j=1}^{n} a_{ij} x_i x_j \qquad (67)$$

A convenient feature of all quadratic forms is that the total coefficient of the product $x_i x_j$ in Eq. (67) is the sum of the matrix elements a_{ij} and a_{ji}. *Thus it is always possible to treat the matrix associated with a quadratic form as though it were symmetric. If it is not so, the matrix can be replaced by a symmetric one with elements equal to* $(a_{ij} + a_{ji})/2$ *without affecting the value of* Q.

Quadratic forms occur widely in problems involving maxima or minima of functions of several variables. They are used to define measures of cost or of error in optimal control problems, and in the fitting of system models to experimental data. It is thus worth examining a few typical problem areas in outline before proceeding to the relatively detailed material to be presented in subsequent lectures.

Diagonalisation

If the matrix associated with a quadratic form is diagonal, then Q(x) will consist of a weighted sum of squares of the elements of x.

Diagonalisation of symmetric matrices is particularly simple provided the eigenvalues are distinct, for it can be shown that

(i) The eigenvalues of a symmetric matrix are always real.

(ii) The matrix of column eigenvectors (V) of a real symmetric matrix is merely the transpose of the matrix of row eigenvectors (W).

$$\text{Thus, } W = V^{-1} = V^T \qquad (68)$$

provided A is symmetric.

Consider the quadratic form

$$Q = x^T A x.$$

This can be written as

$$Q = x^T V \Lambda W x$$

where, as usual, Λ is the diagonal matrix of eigenvalues (all real numbers when A is symmetric).

Now note that $V = W^T$, in the case considered, so that if we set

$$y = Wx,$$

we obtain

$$Q = y^T \Lambda y = \sum_{i=1}^{n} \lambda_i y_i^2 \tag{69}$$

That is, we have reduced the quadratic form to a sum of squares.

Sign Definiteness

A quadratic form is said to be *positive definite* if it is positive for all non-zero values of the vector x.

Negative definiteness is obviously defined in a similar way, and various degrees of *semi-definiteness* can be defined to cover cases where the values of Q may actually reach zero.

Since the sign-definiteness of a quadratic form depends entirely on the coefficients of the matrix which is involved, the qualities of definiteness are naturally ascribed to the matrix itself. Such qualities are of importance in many situations, a well-known one being associated with the occurrence of maxima or minima. We shall consider such problems presently.

The sign-definiteness of a matrix may be determined in a number of ways. We mention two below:

One straightforward, but laborious, test is to examine the determinant of A, and all of the principal minors thereof. If all of these are positive (negative), then A is positive (negative) definite.

An alternative test, which is more convenient in many ways, is to examine the eigenvalues of A. For a symmetric matrix, these will always be real; and if they are all positive (negative), then A will be positive (negative) definite. This may be deduced from Eq. (69).

8. TAYLOR'S SERIES, MAXIMA AND MINIMA

The use of Taylor's series for extrapolation of a function of a single variable is well known, but the extension to functions of several variables is less familiar. In fact, the use of matrix notation, and the notion of differentiation with respect to a vector (Eqs. (2) to (7)) makes possible a concise statement of procedures which are closely analogous to the single-variable case.

Consider the state equations of a nonlinear dynamic system of nth order. In terms of a state vector x(t) and an input vector u(t), the state equations may be written as

$$\dot{x} = f(x,u;t), \tag{70}$$

where f is a vector-valued function of dimension n.

If the variables x and u are changed to (X + x), (U + u), where X, U are now 'reference' vectors, and x,u are 'small' deviations, we have

$$\dot{X} + \dot{x} = f(X + x, U + u;t)$$

and expanding this in a Taylor's series, we may write

$$\dot{X} + \dot{x} = f(X,U;t) + Ax + Bu + O(\|x\|^2, \|u\|^2) \tag{71}$$

where (cf. Eq. (6):

$$A = [a_{ij}] = \left[\frac{\partial f_i}{\partial x_j}\right]_{X,U} \tag{72}$$

$$B = [b_{ij}] = \left[\frac{\partial f_i}{\partial u_j}\right]_{X,U} \tag{73}$$

$$O(\|x\|^2, \|u\|^2) = (\text{"Terms of order } x^2 \text{ and } u^2\text{"}) \tag{74}$$

Thus, discarding terms of higher order than the first, and noting Eq. (70), we obtain

$$\dot{x} = Ax + Bu + \text{"small" errors} \tag{75}$$

provided the conditions necessary for good approximation have been satisfied.

In maxima/minima problems, of course, the second order terms are very important, so they need to be retained in the expansion. For notational simplicity here, it is convenient to deal with such functions one at a time rather than vector-valued functions. Thus we might often be concerned with the location of an extremum of a scalar function of n variables:

$$f(x) = f(x_1 \dots x_n) \tag{76}$$

It is known[2] that the partial derivatives of such a function with respect to the elements of x must all vanish at an extremum. This is equivalent to

$$\frac{df}{dx} = \left(\frac{df}{dx_1} \;.\;.\;.\;.\; \frac{df}{dx_n}\right)^T = 0. \tag{77}$$

The nature of f(x) in the region of the point defined by Eq. (77) may be examined by considering the so-called *curvature matrix*:

$$\frac{d^2f}{dx^2} = \left[\frac{\partial^2 f}{\partial x_i \partial x_j}\right] = C \tag{78}$$

If this matrix is negative definite, then the point concerned will be a maximum. If it is positive definite, the point will be a minimum. If it is not sign definite, the point will not be either a true maximum or a true minimum, but might for example be a 'saddle point'.

Quadratic Functions

In the region of an extremum, a suitably 'smooth' function may be expected to exhibit approximately quadratic behaviour. This may be described by an expression of the form

$$f(x) = \tfrac{1}{2}x^T Ax + b^T x + c \tag{79}$$

If an extremum of this function exists, it will be at a point x_0 defined by

$$\frac{df(x_0)}{dx} = x_0^T A + b^T = 0$$

or

$$x_0 = -A^{-1}b \tag{80}$$

The curvature matrix is given by

$$\frac{d^2 f}{dx^2} = A, \tag{81}$$

so if matrix A is negative definite, the function f will possess a unique maximum at the point x_0, etc.

Example

Consider the following problem of *multiple linear regression*, which we shall consider in greater detail in later lectures.

A set of N observations, regarded as elements of a vector y is believed to be linearly related to a set of p unknown parameters (elements of a vector θ), but is also subject to random errors of measurement.

This situation may be represented as follows:

$$y_1 = x_{11}\theta_1 + x_{12}\theta_2 + \dots + x_{1p}\theta_p + \varepsilon_1$$

$$y_2 = x_{21}\theta_1 + x_{22}\theta_2 + \dots + x_{2p}\theta_p + \varepsilon_2$$

$$\cdots\cdots\cdots\cdots\cdots\cdots\cdots\cdots$$

$$\cdots\cdots\cdots\cdots\cdots\cdots\cdots\cdots$$

$$y_N = x_{N1}\theta_1 + x_{N2}\theta_2 + \dots + x_{Np}\theta_p + \varepsilon_N$$

Here, the quantities $\{x_{ij}\}$ are assumed known, and the random errors are represented by the $\{\varepsilon_i\}$. The above set of equations can be condensed to vector-matrix form as:

$$y = X\theta + \varepsilon, \tag{82}$$

which is the standard linear model for a linear regression problem. The approach taken here is to seek that value of θ which minimises the sum of squares of the errors, i.e.

$$\min_{\theta} \sum_{i=1}^{n} \varepsilon_i^2 = \min_{\theta} \varepsilon^T \varepsilon \qquad (83)$$

Thus, the quantity to be minimised is (using Eqs. (82) and (83)):

$$s = (y - X\theta)^T (y - X\theta) \qquad (84)$$

It can be shown that the generalised derivative satisfies the 'chain rule', viz.

$$\frac{ds}{d\theta} = \frac{ds}{d\varepsilon} \cdot \frac{d\varepsilon}{d\theta}, \qquad (85)$$

provided the correct order of multiplication is observed.

Furthermore, the derivative of a quadratic form can be shown to be

$$\frac{d}{d\varepsilon}(\varepsilon^T A \varepsilon) = 2\varepsilon^T A \qquad (86)$$

with A in our case being a unit matrix.

Since $\varepsilon = (y - X\theta)$, we have

$$\frac{d\varepsilon}{d\theta} = -X \qquad (87)$$

Thus, the quantity $\varepsilon^T\varepsilon$ will have an extremum at the point $\theta = \hat{\theta}$, where

$$-2(y - X\hat{\theta})^T X = 0 \qquad (88)$$

The solution of Eq. (88) is obtained by multiplying out the terms in the bracket (noting that X is not necessarily square in general):

$$y^T X - \hat{\theta}^T X^T X = 0$$

$$\therefore X^T y = X^T X \hat{\theta}$$

$$\therefore \hat{\theta} = (X^T X)^{-1} X^T y \qquad (89)$$

This result is the matrix form of the well-known *normal equations of least squares*, and it will be encountered frequently in connection with subsequent developments.

9. CONCLUDING COMMENTS

In this introductory review we have attempted to outline the most basic concepts which will be assumed to be familiar to the course participants at the outset. Those who are new to the subject are strongly urged to consult the extensive literature, a small selection of which is referenced here.

SUGGESTIONS FOR FURTHER READING

1. *For a condensed, but clear development of matrix concepts applied to linear system theory, see Chapter 2 of the book:* 'Stochastic Optimal Linear Estimation and Control' by J.S. Meditch, McGraw-Hill, 1969.
2. *For a fundamental text on matrix concepts applied to functions of several variables,:* 'Calculus of Several Variables' by Serge Lang, Addison Wesley, 1973.
3. *For a very condensed but authoritative development of matrix theory relevant to stochastic modelling:* "Dynamic Stochastic Models from Empirical Data" by R.L. Kashyap and A.R. Rao, Academic Press, 1976.
4. *For a useful self-instruction text on state-space concepts and techniques,* 'Schaum's Outline on State Space and Linear Systems', by D.M. Wiberg, McGraw-Hill.
5. *For a fundamental mathematical text on matrix theory,* 'Theory of Matrices', by P. Lancaster, Academic Press, 1969.

MAIN LECTURES

RELEVANT PROBABILITY THEORY

Dr. R.P. Jones

1. INTRODUCTION

The aim of this lecture is to introduce the essential ideas of probability theory as background to the analysis and understanding of random signals and their properties. Note that probability theory can be presented in a precise and mathematically rigorous manner but that this approach is beyond the intended scope of this vacation school. An alternative, less rigorous approach is adopted here, based on intuitive considerations closely allied to experimental observation.

2. BASIC CONCEPTS

2.1 Probability

Probability theory is concerned with providing a mathematical description of random phenomena in which there is always uncertainty as to whether a particular event will or will not occur. For such phenomena, individual events occur in a haphazard manner and it is not possible to predict, in advance, the occurrence of a particular event. However, over a large number of occurrences of events an average pattern or characteristic emerges and it is this average characteristic which forms the basis of the concept of probability.

To illustrate this, consider the phenomenon of the tossing of a perfectly balanced coin. In this case, two possible events may occur, viz. a head or a tail. We know that we cannot predict, with certainty, the outcome in advance of tossing the coin. However, we know from experience that if we toss the same coin a large number of times we will obtain approximately an equal number of heads and tails, i.e. a definite 'average' pattern emerges.

As a measure of the chance or *probability* with which we expect an event A to occur we assign a number P(A), with $0 \leq P(A) \leq 1$, termed the probability of the event A. If the event A is certain to occur then P(A) = 1, and if it is certain that A will not occur, then P(A) = 0.

The probability P(A) of an event A occurring may be interpreted, intuitively, as the relative frequency with which A occurs in the outcome of a large number of events. In the case of the tossing of the coin, it is clear that P (Head) = P (Tail) = 0.5.

2.2 Joint Probability

If A and B are any two events, then

$$P\,[A \text{ or } B] = P\,[A] + P[B] - P\,[A \text{ and } B]$$

where the compound event [A *or* B] denotes the occurrences of A or B or both, and the notation [A *and* B] denotes the joint occurrence of both A and B.

2.3 Conditional Probability

We shall denote by P[A|B] and the probability of event A given that event B has occurred, i.e. the *conditional probability* of A given B.

$$P[A|B] = \frac{P[A \text{ and } B]}{P[B]} \quad .$$

This relationship is valid, provided $P\,[B] \neq 0$.

2.4 Independent Events

If P[A|B] = P[A], i.e. the probability of event A occurring is not affected by the occurrence or non-occurrence of event B, then A and B are said to be independent events. Then P[A and B] = P[A].P[B].

2.5 Bayes' Theorem

Suppose that $A_1, A_2, \ldots, A_n$ are mutually exclusive events such that $P[A_1] + P[A_2] + \ldots + P[A_n] = 1$. Then if A is any event,

$$P[A_k|A] = \frac{P[A_k]\cdot P[A|A_k]}{\sum_{j=1}^{n} P[A_j]\cdot P[A|A_j]} \quad .$$

This theorem forms the basis of several useful 'Bayesian' concepts in statistical inference.

2.6 Example

A new X-ray test for the detection of small fractures in concrete members is to be evaluated. From a large number of tests in the laboratory, it was ascertained that 98% of concrete members having small fractures reacted positively to the test but that 4% of those not having such fractures also did so. If this test is applied in the field to a large number of concrete members containing 3% with small fractures show that:

(i) 43.1% of members which react positively to the test actually have small fractures.

(ii) 0.0644% of members which react negatively to the test will have small fractures.

Define the events: T = positive result from test
$\bar{T}$ = negative result from test
F = Fracture present
$\bar{F}$ = Fracture not present

We are given $P[T|F] = 0.98$ and $P[T|\bar{F}] = 0.04$. Therefore $P[\bar{T}|F] = 0.02$ and $P[\bar{T}|\bar{F}] = 0.96$.

For the field trials, $P[F] = 0.03$, therefore $P[\bar{F}] = 0.97$. We require $P[F|T]$ and $P[F|\bar{T}]$.

Using Bayes' Theorem with n = 2 (there are just two possible outcomes, viz. F and $\bar{F}$).

$$P[F|T] = \frac{P[F] \cdot P[T|F]}{P[F] \cdot P[T|F] + P[\bar{F}] \cdot P[T|\bar{F}]}$$

$$= \frac{(0.03) \cdot (0.98)}{(0.03)\,(0.98) + (0.97)\,(0.04)}$$

$$= 0.431$$

and

$$P[F|\bar{T}] = \frac{P[F] \cdot P[\bar{T}|F]}{P[F] \cdot P[\bar{T}|F] + P[\bar{F}] \cdot P[\bar{T}|\bar{F}]}$$

$$= \frac{(0.03)\,(0.02)}{(0.03)\,(0.02) + (0.97)\,(0.96)}$$

$$= 0.000644.$$

3. DISCRETE RANDOM VARIABLES

3.1 Single variable

Consider a discrete random variable x with possible values $X_1, X_2, X_3 \ldots$ arranged in increasing order of magnitude. The *probability function* P(X) defines the probability that the random variable x takes the value X. We note that P(X) is always non-negative, and that

$$\sum_i P(X_i) = 1$$

The *cumulative distribution function* F(X) defines the probability that the random variable x takes any value less than or equal to X and is given by

$$F(X) = \sum_{\substack{i \\ X_i \leq X}} P(X_i)$$

The *expected value* (or *mean value*) of x, written E[x] (sometimes $\bar{x}$ or μ_x), is defined by

$$E[x] = \sum_i X_i \; P(X_i).$$

The *variance* of x, which is sometimes written as σ_x^2 is defined by

$$\mathrm{Var}[x] = E[x - E[x]]^2 = \sum_i (X_i - \mu_x)^2 \; P(X_i).$$

Note that the *standard deviation* σ_x of x is the positive square root of the variance of x. Also, note that it can easily be shown that

$$\mathrm{Var}[x] = E[x^2] - (E[x])^2,$$

a result which can be used to simplify the evaluation of Var[x].

3.2 Example

Consider the situation in which we roll two dice and count the dots on the upper two faces. There are 36 possible combinations, and the dice are considered fair if each combination has an equal probability. If a random variable x is defined as the *sum* of the two upper faces, and X represents the *possible* values of x, we have:

X	2	3	4	5	6	7	8	9	10	11	12
P(X)	$\frac{1}{36}$	$\frac{2}{36}$	$\frac{3}{36}$	$\frac{4}{36}$	$\frac{5}{36}$	$\frac{6}{36}$	$\frac{5}{36}$	$\frac{4}{36}$	$\frac{3}{36}$	$\frac{2}{36}$	$\frac{1}{36}$
F(X)	$\frac{1}{36}$	$\frac{3}{36}$	$\frac{6}{36}$	$\frac{10}{36}$	$\frac{15}{36}$	$\frac{21}{36}$	$\frac{26}{36}$	$\frac{30}{36}$	$\frac{33}{36}$	$\frac{35}{36}$	$\frac{36}{36}$

Mean value of x $= 2 \times \frac{1}{36} + 3 \times \frac{2}{36} + \;\ldots\; + 12 \times \frac{1}{36} = 7.$

Variance, $\sigma^2 = (2-7)^2 \times \frac{1}{36} + (3-7)^2 \times \frac{2}{36} + \ldots + (12-7)^2 \times \frac{1}{36} = \frac{210}{36}$.

3.3 Two variables: joint probability distributions

Consider now a pair x,y of discrete random variables with possible values $X_1, X_2, X_3, \ldots$ and $Y_1, Y_2, Y_3, \ldots$, respectively. The *joint probability distribution* P(X,Y) defines the probability that the random variable x takes the value X and the random variable y takes the value Y, where X and Y represent possible values X_i, Y_j, $i,j = 1,2,\ldots$, respectively.

Note that

$$\sum_{\text{all } i} P(X_i, Y) = P(Y)$$

and

$$\sum_{\text{all } j} P(X, Y_j) = P(X).$$

The random variables x and y are said to be *independent* if

$$P(X,Y) = P(X)P(Y)$$

for all possible values X and Y.

Finally, the *conditional probability distribution* $P(X|Y)$ is defined by

$$P(X|Y) = \frac{P(X,Y)}{P(Y)}$$

for $P(Y) \neq 0$.

3.4 Example

The joint probability distribution function for two discrete random variables is given by $P(X,Y) = k(2X + Y)$ where x and y can assume integer values defined by $0 \leq x \leq 2$, $0 \leq y \leq 3$.

(a) Find k.
(b) Find P(2,1).
(c) Find $P[x \geq 1, y \leq 2]$.
(d) Find $P(Y|2)$
(e) Find $P(y = 1 \mid x = 2)$.

x \ y	0	1	2	3
0	0	k	2k	3k
1	2k	3k	4k	5k
2	4k	5k	6k	7k

(a) Total = 42k. Therefore $k = \frac{1}{42}$

(b) $P(2,1) = \frac{5}{42}$

(c) $P[x \geq 1, y \leq 2] = \frac{1}{42}(2 + 3 + 4 + 4 + 5 + 6) = \frac{24}{42} = \frac{4}{7}$

(d) $P(Y|X) = \frac{P(X,Y)}{P(X)}$ so $P(Y|2) = \frac{(4 + Y)/42}{22/42} = \frac{4 + Y}{22}$

(e) $P[y = 1 \mid x = 2] = \frac{4 + 1}{22} = \frac{5}{22}$

4. CONTINUOUS RANDOM VARIABLES

4.1 Single Variable

If x is a continuous random variable, the probability that x takes on any one particular value X is generally zero. Therefore we cannot define a probability function in the same way as for a discrete random variable. We note, however, that the probability that x lies between two distinct values X_1, X_2 is meaningful and this motivates the introduction of a continuous *probability density function* f(x) (p.d.f.) with the properties:

(i) If the continuous random variable x has a minimum value of x_{min} and a maximum value of x_{max}, then

$$\int_{x_{min}}^{x_{max}} f(X)dX = 1$$

(ii) The integral $\int_a^b f(X)dX$ is the probability that the variable x lies between the limits a and b.

The expected value E(x) (or mean) of a continuous random variable (sometimes written as $\bar{x}$ or μ_x) is defined by

$$E[x] = \int_{x_{min}}^{x_{max}} Xf(X)dx \ .$$

The variance Var[x] (sometimes written as σ_x^2) is defined by

$$\text{Var}[x] = E[x - E[x]]^2 = \int_{x_{min}}^{x_{max}} (X-\mu_x)^2 \ f(X)dX \ .$$

Note the result that

$$\text{Var}[x] = E[x^2] - (E[x])^2$$

$$= \int_{x_{min}}^{x_{max}} X^2 f(X)dX - \mu_x^2 \ .$$

As in the case of a discrete random variable, we can define a *cumulative distribution function* F(X) by

$$F(X) = \int_{X_{min}}^{X} f(u)du$$

4.2 Example

A non-negative continuous random variable x has a p.d.f. $f(X) = kX.\exp(-X)(X \geq 0)$. Show that the probability that x lies between 0 and 1 is 0.264. Determine the variance of x. Show that $F(X) = 1 - (X + 1)\exp(-X)$. Since $x_{min} = 0$ and $x_{max} = \infty$,

$$k \int_0^\infty X.\exp(-X)dX = 1, \text{ giving } k = 1.$$

$$P[0 \leq x \leq 1] = \int_0^1 X.\exp(-X)dX = 1 - 2\exp(-1) = 0.264.$$

$$E[x] = \int_0^\infty X.f(X)dX = \int_0^\infty X^2 \exp(-X)dX = 2.$$

$$\begin{aligned} Var[x] &= \int_0^\infty X^2 f(X)dX - \mu_x^2 \\ &= \int_0^\infty X^3 \exp(-X)dX - 4 \\ &= 6 - 4 = 2 \end{aligned}$$

$$\begin{aligned} F(X) = \int_0^X f(u)du &= \int_0^X u.\exp(-u)du \\ &= 1 - (X + 1)\exp(-X). \end{aligned}$$

4.3 Two variables: joint probability density functions

Consider now a pair x, y of continuous random variables with associated *joint probability density function* $f(X,Y)$ satisfying the properties:

(i) $\int_{Y_{min}}^{Y_{max}} \int_{X_{min}}^{X_{max}} f(X,Y)dXdY = 1$

(ii) $\int_{Y_1}^{Y_2} \int_{X_1}^{X_2} f(X,Y)\,dXdY = P(X_1 \leq x < X_2 \text{ and } Y_1 \leq y < Y_2)$

(iii) $\int_{X_{min}}^{X_{max}} f(X,Y)dX = f(Y)$

(iv) $\int_{Y_{min}}^{Y_{max}} f(X,Y)dY = f(X)$

The random variables x and y are said to be *independent* if $f(X,Y) = f(X)f(Y)$ for all possible values X and Y.

Finally, we introduce the *conditional probability density function* $f(X|Y)$ with properties:

(i) $\int_{X_{min}}^{X_{max}} f(X|Y)dX = 1.$

(ii) $\int_{X_1}^{X_2} f(X|Y)dX$ is the probability that $X_1 \leq x < X_2$, given Y

(iii) $$f(X|Y) = \frac{f(X,Y)}{f(Y)}$$

for $f(Y) \neq 0$.

4.4 Example

The joint probability density function of two continuous random variables x and y is given by $f(X,Y) = 8XY,\ 0 \leqq X \leqq 1,\ 0 \leqq Y \leqq X$

$= 0$, otherwise.

Find (i) f(X), (ii) f(Y), (iii) f(X|Y), (iv) f(Y|X).

(i) $f(X) = \int_0^X 8XY\, dY = 4X^3$ for $0 < X < 1$

$= 0$, otherwise.

(ii) $f(Y) = \int_Y^1 8XY\, dX = 4Y(1 - Y^2),\ 0 < Y < 1$

$= 0$, otherwise.

(iii) $f(X|Y) = \dfrac{f(X,Y)}{f(Y)} = \dfrac{8XY}{4Y(1 - Y^2)} = \dfrac{2X}{1 - Y^2},\ Y \leqq X \leqq 1$

$= 0$ for other values of X.

(f(X|Y) is not defined when f(Y) = 0).

(iv) $f(Y|X) = \dfrac{f(X,Y)}{f(X)} = \dfrac{8XY}{4X^3} = \dfrac{2Y}{X^2},\ 0 \leqq Y \leqq X$

$= 0$ for other values of Y.

f(Y|X) is not defined when f(X) = 0).

Note that $f(X,Y) \neq f(X).\ f(Y)$, so X and Y are not independent in the region of interest on the (X,Y) plane.

5. EXPECTED VALUES AND MOMENTS

5.1 Moments

The *k-th moment* m_k of a continuous random variable x is defined by

$$m_k = \int_{X_{min}}^{X_{max}} X^k f(X)dX = E[x^k]$$

and is a straight forward generalisation of the expected value, with $E[x] = m_1$ and $Var[x] = m_2 - (m_1)^2$

The use of moments results largely from the linearity of the "expected value" operation; i.e. if x and y are any random variables, and α and β are any (non-random) coefficients

$$E[\alpha x + \beta y] = \alpha E[x] + \beta E[y].$$

This simplifies the manipulation of expressions involving expected values of random variables to an extent which is not possible with probability distributions as they stand.

The concept of a moment generalises to a function g(x) of a continuous random variable x where the expected value is now given by

$$E[g(x)] = \int_{X_{min}}^{X_{max}} g(X)f(X)dX.$$

The *central moments* of a continuous random variable x are simply the moments of the p.d.f. about the mean, μ_x, rather than the origin, i.e.

$$C_k = E[(x - \mu_x)^k] = \int_{X_{min}}^{X_{max}} (X - \mu_x)^k.F(X)dX$$

Note that $C_1 = 0$

$$C_2 = \sigma^2_x$$

The *joint moment* of order (k + r) is given by

$$m_{kr} = E[x^k.y^r] = \int_{Y_{min}}^{Y_{max}} \int_{X_{min}}^{X_{max}} X^k.\ Y^r.f(X,Y)dXdY.$$

The quantity m_{11} is of considerable importance. It is called the *correlation* between x and y and is often written R_{xy},

$$m_{11} = R_{xy} = E[x.y].$$

When $E[xy] = 0$, the random variables x and y are said to be *orthogonal*. The central moment C_{11} is called the *covariance* between x and y, usually written Cov(x,y). Clearly

$$Cov(x,y) = E[(x - E[x])\ (y - E[y])]$$

Note that $Cov(x,y) = R_{xy} - E[x]\cdot E[y]$

The *correlation coefficient* $\rho_{xy} = \dfrac{Cov(x,y)}{\sigma_x \cdot \sigma_y}$.

We also note that

$$|\rho_{xy}| \leq 1.$$

If x and y are independent random variables,

$$R_{xy} = E(x)\cdot E(y)$$

and so $Cov(x,y) = 0$ and $\rho_{xy} = 0$.

Note that these concepts are defined in a similar manner for discrete random variables.

5.2 Example

For the joint probability distribution function of Example 3.4, find $E(x)$, $E(y)$, $E(xy)$, $E(x^2)$, $E(y^2)$, σ_x^2, σ_y^2, $Cov(x,y)$ and ρ_{xy}.

(i) $$E(x) = \sum_X X \sum_Y P(X,Y)$$
$$= 1(2 + 3 + 4 + 5)\cdot \frac{1}{42} + 2(4 + 5 + 6 + 7)\cdot \frac{1}{42} = \frac{58}{42}$$

(ii) $$E(y) = \sum_Y Y\cdot \sum_X P(X,Y)$$
$$= 1(1 + 3 + 5)\frac{1}{42} + 2(2 + 4 + 6)\frac{1}{42} + 3(3 + 5 + 7)\cdot \frac{1}{42} = \frac{78}{42}$$

(iii) $$E(xy) = \sum_X \sum_Y XYP(X,Y)$$
$$= (1)(1)\cdot\frac{3}{42} + (1)\cdot(2)\cdot \frac{4}{42} + (1)\cdot(3)\cdot \frac{5}{42}$$
$$+ (2)(1)\frac{5}{42} + (2)(2)\frac{6}{42} + (2)(3)\frac{7}{42} = \frac{102}{42}$$

(iv) $$E(x^2) = \sum_X X^2 \sum_Y P(X,Y)$$
$$= 1^2\cdot \frac{14}{42} + 2^2\cdot\frac{22}{42} = \frac{102}{42}$$

(v) $$E(y^2) = \sum_Y Y^2\cdot \sum_X P(X,Y)$$
$$= 1^2\cdot \frac{9}{42} + 2^2 \cdot \frac{12}{42} + 3^2 \cdot \frac{15}{42} = \frac{192}{42}$$

(vi) $\sigma_x^2 = \frac{102}{42} - \left(\frac{58}{42}\right)^2 = \frac{230}{441}$

(vii) $\sigma_y^2 = \frac{192}{42} - \left(\frac{78}{42}\right)^2 = \frac{55}{49}$

(viii) $\mathrm{Cov}(x,y) = E(xy) - E(x)\,E(y) = \frac{102}{42} - \frac{58}{42}\frac{78}{42} = -\frac{20}{147}$

(ix) $\rho_{xy} = \frac{\mathrm{Cov}(x,y)}{\sigma_X\,\sigma_y} = -\frac{20/147}{\left[\frac{230}{441}\;\frac{55}{49}\right]^{\frac{1}{2}}} = -0.21$

5.3 Conditional Expectation

Similar relationships hold for the moments of a conditional density function. The *conditional expectation*, or conditional mean of y given x is given by

$$E[y|X] = \int_{-\infty}^{\infty} Y.f(Y|X)dY. \quad \text{Note that } E[y] = \int_{-\infty}^{\infty} E[y|X].f(X)dX.$$

Again, this is defined in a similar manner for a discrete random variable.

5.4 Example

Find the conditional expectation of y given x for the joint p.d.f. of Example 4.4. Find also the conditional variance of y given x

(i) $E[y|X] = \int_{-\infty}^{\infty} Y.f(Y|X)dY = \int_0^X Y.\,\frac{2Y}{X^2}\,dY = \frac{2X}{3}$

(ii) Conditional variance $= E[(y - \bar{y})^2|X] = \int_{-\infty}^{\infty} (Y - \bar{y})^2 f(Y|X)dY$

$$= \int_0^X \left(Y - \frac{2X^2}{3}\,\frac{2Y}{X^2}\right)dY = \frac{X^2}{18}\,.$$

6. SOME USEFUL DISTRIBUTIONS

6.1 Uniform Distribution

A continuous random variable x is said to be uniformly distributed between a and b if its p.d.f. is given by

$$f(X) = \frac{1}{b-a}\,, \quad a \leq X < b$$

$$= 0 \quad , \text{ otherwise.}$$

The cumulative distribution function is given by

$$F(X) = 0, \qquad X < a$$
$$= \frac{X-a}{b-a}, \qquad a \leqq X < b$$
$$= 1, \qquad X \geqq b.$$

The mean, μ_X is $\frac{1}{2}(a+b)$, obviously, and the second moment is

$$E[x^2] = \int_a^b \frac{X^2 dX}{b-a} = \frac{b^3 - a^3}{3(b-a)} = \frac{b^2 + ab + a^2}{3}$$, giving the variance

$$\sigma_X^2 = \frac{b^2 + ab + a^2}{3} - \frac{a^2 + 2ab + b^2}{4} = \frac{1}{12}(b-a)^2$$

6.2 Binomial Distribution

The Binomial Distribution is a discrete distribution concerned with results of trials in which only two outcomes are possible, usually denoted 'success' and 'failure'. The probabilities of success, p and of failure q are thus related by

$$p = 1 - q.$$

The probability function is given by:

$$P(X) = \frac{n!}{X!(n-X)!} p^X q^{n-X}$$

where the random variable x denotes the number of successes in n trials ($X = 0,1,\dots,n$).

$$E[x] = \sum_j X_j . P(X_j). = np,$$

$$Var[x] = \sum_j X_j^2 P(X_j) - (E[x])^2 = npq.$$

To see this, consider the expansion of:

$$(q + pt)^n = q^n + {}^nC_1 q^{n-1} pt + {}^nC_2 q^{n-2} p^2 t^2 + \dots + p^n t^n$$

$$= P(0) + t.P(1) + t^2.P(2) + \dots + t^n.P(n).$$

Differentiating both sides with respect to t,

$$np(q + pt)^{n-1} = 0 + P(1) + 2t P(2) + \dots + nt^{n-1} P(n) \qquad (A)$$

If $t = 1$, $np = P(1) + 2P(2) + \dots + nP(n) = \sum_{X=0}^{n} XP(X) = E(x)$, i.e. the mean of the binomial distribution is np.

To find the variance, multiply equation (A) by t:

$$npt(q + pt)^{n-1} = t.P(1) + 2t^2P(2) + \dots + nt^n P(n)$$

Differentiating with respect to t,

$$np(q + pt)^{n-1} + npt(n-1)p(q + pt)^{n-2}$$

$$= P(1) + 2^2 t\, P(2) + \dots + n^2 t^{n-1} P(n).$$

Putting t = 1, $np + n(n-1)p^2 = 1^2P(1) + 2^2P(2) + \dots + n^2P(n)$

i.e. $np(1 - p) + n^2p^2 = E[x^2]$

giving $\sigma^2 = E[x^2] - (E[x])^2 = np(1 - p) = npq.$

6.3 Poisson Distribution

The Poisson distribution is a discrete distribution relating to a random variable which may take values 0,1,2,...,r, r+1,... . The probability function is given by

$$P(r) = \frac{\mu^r}{r!} e^{-\mu}$$

The mean $E[r] = \sum_{r=1}^{\infty} r.\frac{\mu^r}{r!} e^{-\mu} = \mu e^{-\mu} \sum_{r=1}^{\infty} \frac{\mu^{r-1}}{(r-1)!}$

$$= \mu e^{-\mu}.\, e^{\mu} = \mu.$$

The second moment $E[r^2] = \sum_{r=1}^{\infty} r^2 \frac{\mu^r}{r!} e^{-\mu} = \sum_{r=1}^{\infty} r(r-1) + r \frac{\mu^r}{r!} e^{-\mu}$

$$= \mu + \mu^2 e^{-\mu} \sum_{r=2}^{\infty} \frac{\mu^{r-2}}{(r-2)!}$$

$$= \mu + \mu^2 e^{-\mu}.e^{\mu} = \mu + \mu^2$$

so the variance

$$\sigma^2 = \mu + \mu^2 - \mu^2 = \mu.$$

The Poisson distribution is used to describe the situation in which the probability of an event occurring in a short time interval λ is proporitional to λ and is independent of the probability in any other time interval. The probability of two events is an order higher in λ and is neglected. If the constant of proportionality is ν, then the probability of r events in a time interval T is given by

$$P(r) = \frac{(\nu T)^r}{r!} e^{-\nu T} .$$

The Poisson distribution is also used to approximate the binomial distribution with mean $\mu = np$ when p (or q) < 0.1 and $np < 6$.

6.4 Normal Distribution

The normal (or Gaussian) distribution is one of the most important continuous probability distributions and is frequently used in practical situations. The probability density function takes the form

$$f(X) = \frac{1}{(2\pi)^{\frac{1}{2}}\sigma} \exp\{-\frac{(X-\mu)^2}{2\sigma^2}\}$$

where the random variable x is such that $-\infty < X < \infty$. The normal distribution has mean $E[x] = \mu$ and variance $Var[x] = \sigma^2$.

For finding probabilities, we use a standard normalised variate u given by

$$u = \frac{X - \mu}{\sigma}$$

which has zero mean, a variance of one and p.d.f.

$$f(U) = \frac{1}{\sqrt{2\pi}} \exp(-\frac{U^2}{2}) .$$

There are many tabulations of the normal curve, see for example Ref. 5, Tables 3 and 4.

The normal distribution can be used as an approximation to the binomial distribution for large n and not too small p ($np \geq 5$).

The normal distribution can also be used as an approximation to the Poisson distribution, provided μ is not too small.

6.5 Example

The average proportion of defectives in a product is 5%. Find the probability that a batch of size 100 will contain at least 9 defective. Applying the binomial distribution

$$p = 0.05, \quad q = 0.95, \quad np = 5, \quad npq = 4.75.$$

$$x = 5, \quad \sigma = \sqrt{4.75} = 2.18,$$

so $$u = \frac{x - 5}{2.18}$$, in the 'normal' approximation.

To allow for change from discrete to continuous variable, the requirement for the discrete variable to be 9 or more corresponds to the continuous variable more than 8.5.

Therefore the probability of at least nine defective = $P\left[u > \frac{8.5 - 5}{2.18}\right]$

From Tables, we find this to be 0.5000 - 0.4463

= 0.0537.

6.6 Example

A radioactive disintegration emits a mean number of 69 particles per sec. What is the probability of 60 particles or less in a 1 sec. interval?

Applying the Poisson distribution

$$\mu = 69, \quad \sigma = \sqrt{69} = 8.3.$$

Upper limit this time is 60.5, so required probability is, from 'normal' approximation,

$$P\left[u < \frac{60.5 - 69}{8.3}\right] = 0.5000 - 0.3473 = 0.1527, \text{ from Tables.}$$

6.7 Sampling Distribution of Sums and Differences

If two independent continuous random variables, normally distributed with means μ_1 and μ_2 and variances σ_1^2 and σ_2^2 are added, the resulting sum will also be normally distributed. The mean and variance of the new variable will be given by:

$$\mu = \mu_1 + \mu_2,$$
$$\sigma^2 = \sigma_1^2 + \sigma_2^2.$$

If the second random variable is *subtracted* from the first, the result again will be a normally distributed random variable with,

$$\mu = \mu_1 - \mu_2$$
$$\sigma^2 = \sigma_1^2 + \sigma_2^2$$

Note that the variance still add.

7. CHARACTERISTIC FUNCTION

The characteristic function (c.f.) of a continuous random variable x is defined by

$$\Phi(\omega) = E[\exp(j\omega x)]$$
$$= \int_{X_{min}}^{X_{max}} f(X).\exp(j\omega X)dX$$

Note that this is the same as the Fourier transform of f(X), except for the reversal in sign of the argument ω.

The characteristic function has the following useful properties:

(i) The c.f. of a sum of independent random variables is the product of the individual c.f.'s.

(ii) The derivatives of the c.f. are related to the moments of the random variable:

$$\frac{d^n\phi(0)}{d\omega^n} = j^n \, E[x^n] \quad \text{where } j = \sqrt{(-1)}$$

(iii) $|\phi(\omega)| \leq 1$ for all ω. (Obviously $\phi(0) = 1$).

The c.f. of an infinite sum of independent random variables tends to the limiting form

$$\phi(\omega) = \exp(j\omega\mu - \frac{\omega^2\sigma^2}{2})$$

under relatively general conditions concerning the c.f.'s of the individual components of the sum. The inverse Fourier transform of this is the normal probability density function:

$$f(X) = \frac{1}{\sigma\sqrt{2\pi}} \exp(- \frac{(X - \mu)^2}{2\sigma^2})$$

The above forms the basis of proof of most versions of the *central limit theorem*, which is now stated. Let x_1, x_2,... be independent random variables which are identically distributed (same probability function in the discrete case; same p.d.f. in the continuous case) with finite mean μ and variance σ^2. Then if $S_n = x_1 + x_2 + \ldots + x_n$,

$$\lim_{n\to\infty} P\,[a \leq \frac{S_n - n\mu}{\sigma\sqrt{n}} \leq b] = \frac{1}{\sqrt{2\pi}} \int_a^b e^{-u^2/2} \, du$$

i.e. the random variable $(S_n - n\mu)/\sigma\sqrt{n}$ is asymptotically normal. The theorem is also true under more general conditions, e.g. when x_1, x_2,... are independent random variables with the same mean and variance, but not necessarily identically distributed.

8. VECTOR RANDOM VARIABLES

8.1 Probability distributions

When considering the joint properties of a *set* of random variables $\{x_1, x_2, \ldots, x_n\}$, it is convenient to regard the individual variables x_i as the elements of a random vector

$$x = [x_1, x_2, \ldots, x_n]^T$$

This enables the use of techniques and concepts of matrix analysis to express the complex interrelationships among the elements of x with attractive economy and precision.

The probability density function f(X) of a continuous random vector x is defined as the joint probability density function of the elements of x. Thus

$$\int_{X_1^a}^{X_1^b} \cdots \int_{X_n^a}^{X_n^b} f(X)dX_1 \ldots dX_n = P(X_1^a \leq x_1 \leq X_1^b \text{ and } \ldots X_n^a \leq x_n \leq X_n^b)$$

The quantity f(X) is a scalar, is non-negative, and the n-fold integral of f(X) over the entire space of possible values of x has unit value, i.e.

$$\int_{-\infty}^{\infty} \cdots \int_{-\infty}^{\infty} f(X_1, X_2, \ldots, X_n)dX_1 dX_2 \ldots dX_n = 1.$$

If f(X) is integrated with respect to any subset of elements of X, the result is the joint density function of the remaining variables. Such density functions are termed *marginal densities.*

Note that, if the elements of a continuous random vector x are independent,

$$f(X) = f(X_1)\, f(X_2) \ldots f(X_n)$$

For two random vectors x and z the conditional density function $f(X|Z)$ is defined just as in the scalar case by

$$f(X|Z) = \frac{f(X,Z)}{f(Z)} \quad .$$

This now permits us to consider the effects of a whole *set* of observations z on the probabilities associated with some related random variable x.

Finally, it should be noted that the above generalisations apply to discrete random variables in an analogous manner.

The joint density function of a function of a random vector can be found, in certain simple cases, by the following method. If y and x are both n-vectors, such that y = y(x), i.e.

$$\begin{aligned} y_1 &= y_1(x_1, x_2, \ldots, x_n) \\ y_2 &= y_2(x_1, x_2, \ldots, x_n) \\ &\cdots\cdots \\ y_n &= y_n(x_1, x_2, \ldots, x_n) \end{aligned}$$

and the joint density function of x is known, then the joint density function for y can be found from

$$f^{(y)}(Y) = \frac{1}{|\det J|} f^{(x)}(X(Y))$$

where $f^{(y)}(Y)$ and $f^{(x)}(X)$ represent the joint density functions of x and y respectively, and det J is the Jacobian determinant

$$\det J = \begin{vmatrix} \frac{\partial y_1}{\partial x_1} & \cdots & \frac{\partial y_1}{\partial x_n} \\ \cdots & \cdots & \cdots \\ \frac{\partial y_n}{\partial x_n} & \cdots & \frac{\partial y_n}{\partial x_n} \end{vmatrix}$$

Note that the equation $y = y(x)$ is solved to obtain values $X = X(Y)$ for use in the above expression.

8.2 Example

As a simple illustration of the above concepts consider the problem of finding the joint, conditional, and marginal distributions of the sum and difference of two independent uniformly distributed random variables.

Given $x = [x_1 \; x_2]^T$

$y = [y_1 \; y_2]^T$

with $f^{(x)}(X) = 1$ for $0 \leqq X_1, X_2 < 1$,

$= 0$ otherwise,

$y_1 = x_1 + x_2$

$y_2 = x_1 - x_2$,

we require $f^{(y)}(Y)$, $f(Y_1)$, $f(Y_2)$, and $f(Y_2|Y_1)$.

The Jacobian in this case is

$$J = \begin{bmatrix} \frac{\partial y_1}{\partial x_1} & \frac{\partial y_1}{\partial x_2} \\ \frac{\partial y_2}{\partial x_1} & \frac{\partial y_2}{\partial x_2} \end{bmatrix} = \begin{bmatrix} 1 & 1 \\ 1 & -1 \end{bmatrix}$$

and thus $|\det J| = 2$,

and $f^{(y)}(X) = \frac{1}{2}$ for $0 \leqq X_1, X_2 < 1$

$= 0$ otherwise.

But to express this as a function of Y, we must express X_1, X_2 in terms of Y_1, Y_2:

$$X_1 = \tfrac{1}{2}(Y_1 + Y_2) \text{ and } X_2 = \tfrac{1}{2}(Y_1 - Y_2)$$

Thus, $\quad f^{(y)}(Y) = \tfrac{1}{2}$ for $0 \leqq Y_1 + Y_2 < 2,\ 0 \leqq Y_1 - Y_2 < 2$

$\quad = 0$ otherwise.

Thus, $f^{(y)}(Y)$ is a function which takes the value 0.5 within the shaded region shown in Fig. 1, and is zero elsewhere. Note that the total volume represented by this function has unit value

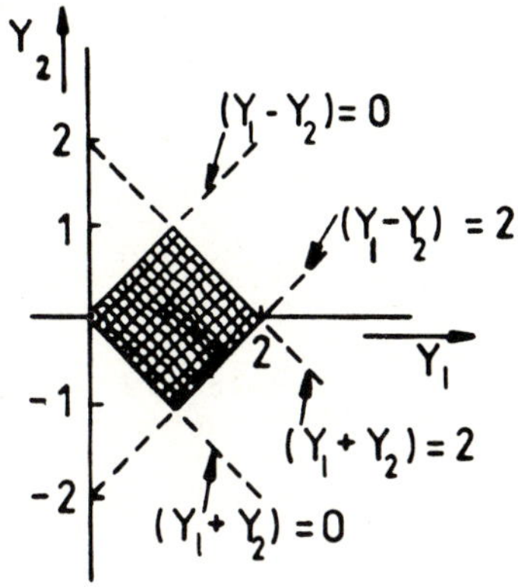

Fig. 1

The marginal distributions are obtained from

$$f(Y_1) = \int_{-\infty}^{\infty} f^{(y)}(Y)dY_2 = \tfrac{1}{2}\int_{-Y_1}^{Y_1} dY_2 \text{ for } 0 \leqq Y_1 < 1$$

$$= \tfrac{1}{2}\int_{Y_1-2}^{2-Y_1} dY_2 \text{ for } 1 \leqq Y_1 < 2.$$

Thus, $f(Y_1) = Y_1$ for $0 \leqq Y_1 < 1$

$\quad = (2 - Y_1)$ for $1 \leqq Y_1 < 2$.

Similarly,

$$f(Y_2) = \tfrac{1}{2}\int_{Y_2}^{2-Y_2} dY_1 = 1 - Y_2 \text{ for } 0 \leqq Y_2 < 1$$

$$= \tfrac{1}{2}\int_{-Y_2}^{2+Y_2} dY_1 = 1 + Y_2 \text{ for } -1 \leqq Y_2 < 0.$$

It may be noted further that since $f^{(y)}(Y) \neq f(Y_1)f(Y_2)$, the sum and difference variables y_1 and y_2 are *not* independent. The conditional density function shows this:

$$f(Y_2|Y_1) = \frac{f^{(y)}(Y)}{f(Y_1)} = \frac{1}{2Y_1} \text{ for } 0 \leq Y_1 \leq 1, \quad -Y_1 \leq Y_2 < Y_1$$

$$= \frac{1}{2(2-Y_1)} \text{ for } 1 \leq Y_1 < 2, \; (Y_1 - 2) \leq Y_2 < (2-Y_1)$$

8.3 Vector moments

The mean, or expected value, of a random vector x is simply a vector composed of the mean values of its elements. Thus,

$$E[x] = \begin{bmatrix} E[x_1] \\ \vdots \\ E[x_n] \end{bmatrix} = \begin{bmatrix} \mu_1 \\ \vdots \\ \mu_n \end{bmatrix} = \mu$$

The concept of scalar *variance* generalises to that of a *covariance* matrix in the vector case:

$$V = \text{Cov}\,[x] = E[(x - \mu)(x - \mu)^T]$$

$$V_{ij} = C_{ij} = E[(x_i - \mu_i)(x_j - \mu_j)]$$

$$= \text{Cov}\,[x_i, x_j]$$

$$\text{for } i, j = 1,2,\ldots,n$$

The matrix generated in this way, for a real random vector, will be square, symmetric, and positive definite.

The joint characteristic function of a random vector is defined in terms of a vector $[\omega_1,\ldots,\omega_n]^T$ as

$$\Phi(\omega_1,\omega_2,\ldots,\omega_n) = E[\exp(j\omega_1 x_1 + j\omega_2 x_2 + \ldots + j\omega_n x_n)]$$

$$= E[\exp\,(j\,\omega^T x)]$$

This function is related to the various mixed moments of the elements of x by the relation

$$\frac{\partial^r \Phi(0)}{\partial\omega_1^a\,\partial\omega_2^b \ldots \partial\omega_n^q} = (j^r)\,E[x_1^a\,x_2^b,\ldots,x_n^q]$$

where $r = a + b + \ldots + q$, for positive integers a, b, ...,q, $j = \sqrt{-1}$, and o represents the null vector.

8.4 Normal Random Vectors

Using the fact that the c.f. of a sum of independent random n-vectors is the product of the individual c.f's, it is possible to show that, under fairly general conditions on the individual c.f.'s, the sum of a large number of such vectors will possess a c.f. of the following form:

$$\Phi(\omega) = \exp(j\omega^T\mu - \tfrac{1}{2}\,\omega^T V\,\omega)$$

By inverse Fourier transformation, the corresponding p.d.f. can be shown to be given by

$$f(X) = (2\pi)^{-n/2}\,|\det V|^{-\frac{1}{2}}\,.\,\exp[\,-\tfrac{1}{2}\,(X - \mu)^T\,V^{-1}(X - \mu)]$$

which is the multidimensional form of the normal p.d.f. The normal random vector is thus seen to be completely characterised by the vector of mean values,

$$\mu = E[x]$$

and the covariance matrix

$$V = E[(x - \mu)\,(x - \mu)^T]$$

A process which generates random variables that are all jointly normal in this way is termed a *Gaussian process*. Such processes dominate modern developments in the fields of stochastic control and identification.

Suppose that x is a Gaussian random n-vector with mean μ_x and covariance matrix P_{xx}. We consider the formation of a new m-vector y by the linear transformation $y = Ax$ where A is a constant $m \times n$ matrix.

The c.f. of x is known to be

$$\Phi_x(\omega) = E[\exp\,(j\,\omega^T\,x)] = \exp\,\{j\,\omega^T\mu_x - \tfrac{1}{2}\,\omega^T\,P_{xx}\,\omega)$$

and the c.f. of y is, by definition,

$$\Phi_y(s) = E[\exp\,(j\,s^T\,y)]$$

where s is an m-vector. Thus,

$$\begin{aligned}\Phi_y(s) &= E[\exp\,(j\,s^T\,A\,x)] = \Phi_x\,(A^Ts)\\ &= \mathrm{Exp}\,\{j\,s^T\,A\,\mu_x - \tfrac{1}{2}\,s^T\,A\,P_{xx}\,A^T\,s\}\end{aligned}$$

and y is also Gaussian, with mean

$$E[y] = A\, E\, [x]$$

and covariance matrix

$$P_{yy} = A.\, P_{xx}.\, A^T$$

Finally, if x and z are jointly normal random vectors, the conditional p.d.f. of x for given observations Z, is of the form

$$f(X|Z) = (2\pi)^{-n/2}\, |\det P|^{-\frac{1}{2}}.\exp[-\tfrac{1}{2}\, (X - \mu)^T\, P^{-1}(X - \mu)]$$

where n is the dimension of x,

$$\mu = \mu_x + P_{xz}P_{zz}^{-1}\, (Z - \mu_z)$$

is the conditional mean $E[x|Z]$ and

$$P = P_{xx} - P_{xz}\, P_{zz}^{-1}\, P_{zx}$$

is the covariance matrix of the elements of x, conditional on the observations z.

The other terms in these expressions are given by

$$\mu_x = E[x],\ \mu_z = E[z],$$

$$P_{xx} = E[(x - \mu_x)\, (x - \mu_x)^T],\ P_{zz} = E[(z - \mu_z)\, (z - \mu_z)^T]$$

and $$P_{xz} = E[(x - \mu_x)\, (z - \mu_z)^T] = P_{zx}^T.$$

9. STOCHASTIC PROCESSES

The term 'stochastic process' refers to a quantity which evolves with time under the influence of a random variable. Other terms which may be used to denote the same quantity include, 'random process', 'random function' and 'random signal'.

A stochastic process may be visualised as a function of two variables, t and ω. The argument t could in principle represent any continuous variable such as distance, temperature, etc., but is usually taken to represent time in control system applications. The other argument, ω, may be taken loosely to represent some random event which is governed by probability laws. Then $x(t,\omega)$ defines a family of time functions, one for each value of the random variable ω.

In the literature on stochastic processes, the simpler notation x(t) is more commonly employed, the dependence of x(t) on the outcome of some chance mechanism being generally inferred from the context.

Examples of stochastic processes are given by

(i) $x(t) = A \ \text{Sin}\ \omega t$,

with ω a constant, and A a random variable.

(ii) $x(t) = A\ \text{Sin}\ (\omega t + \phi)$,

with A a constant, and ϕ a random variable.

In characterising the dynamic structure of a stochastic process, it is found that the process means and covariances provide virtually all the usefully available knowledge about the probability structure of the process. For this reason, most applications of stochastic process theory make extensive use of the *mean*

$$\mu(t) = E[x(t)],$$

the *covariance function*

$$C_{xx}(t_1,t_2) = E[(x(t_1) - \mu(t_1))(x(t_2) - \mu(t_2))],$$

the *correlation function*

$$R_{xx}(t_1,t_2) = E[x(t_1).x(t_2)] = C_{xx}(t_1,t_2) + \mu(t_1).\mu(t_2)$$

or various functions closely related to these.

By further use of vector-matrix notation, the foregoing discussion may be extended to cover vectors of stochastic processes,

$$x(t) = [x_1\ (t)\quad x_2(t)\ \dots\ x_n(t)]^T$$

the elements of which may be regarded as individual scalar stochastic processes.

In the case of stochastic vectors the most convenient probability functions for use in analysis are *the vector of mean values*

$$\begin{aligned}\mu(t) &= [\mu_1(t)\quad \mu_2(t)\dots \mu_n(t)]^T \\ &= [E[x_1(t)]\quad E[x_2(t)]\ \dots\ E[x_n(t)]]^T,\end{aligned}$$

and the *covariance matrix*

$$C_{xx}(t_1,t_2) = E[(x(t_1) - \mu(t_1))\ (x(t_2) - \mu(t_2))^T]$$

where $C_{ij}(t_1,t_2) = \text{Cov}\ (x_i(t_1),\ x_j(t_2))$, is the covariance function of $x_i(t_1)$ and $x_j(t_2)$.

A stochastic process is said to be *strictly stationary* when its probability characteristics remain invariant under a shift in the time origin.

Such a requirement can rarely be assured completely; however, it is possible to simplify the analysis of stochastic processes satisfying much weaker criteria of stationarity to a degree which is just as useful as if they were strictly stationary.

In particular, the property of *'wide-sense'* (or *'weak'*) stationarity merely requires that the mean value of the process is a constant, and that the autocorrelation $R_{xx}(t_1,t_2) = E[x(t_1)\ x(t_2)]$ depends only on the time shift $t_1 - t_2$.

Thus, setting $t_1 = t$ and $t_2 = t + \tau$, we see that the autocorrelation of a weakly stationary process is a function of the single variable τ

i.e. $R_{xx}(\tau) = E[x(t).\ x(t+\tau)]$.

This consequence of stationarity is one of the most important in the application of stochastic process theory.

BIBLIOGRAPHY

1. Spiegel, M.R.: "Probability and Statistics". Schaum Outline Series, McGraw-Hill, 1975.
2. Papoulis, A.: "Probability, Random Variable, and Stochastic Processes". McGraw-Hill, 1965.
3. Meditch, J.S.: "Stochastic Optimal Linear Estimation and Control", McGraw-Hill, 1969.
4. Melsa, J.L. and Sage, A.P.: "An Introduction to Probability and Stochastic Processes", Prentice-Hall, 1973.
5. Murdoch, J. and Barnes, J.A.: "Statistical Tables for Science, Engineering, Management and Business Studies", Macmillan, 1974.

TUTORIAL EXAMPLES - LECTURE L.1

1. The length x of the side of a square is uniformly distributed between 3 and 5. Show that the area A of the square is distributed with p.d.f. $\frac{1}{4}A^{-\frac{1}{2}}$, $9 \leq A \leq 25$ and find the mean area.

   (ANS. $16\frac{1}{3}$ square units).

2. Starting with the relationship

$$E[g(x,y)] = \int_{-\infty}^{\infty} dY \int_{-\infty}^{\infty} g(X,Y).f(X,Y)dX,$$

   verify the following statements concerning random variables x and y:

   (a) $E[\alpha x + \beta y] = \alpha E[x] + \beta E[y]$.

   (b) $E[x.y] = E[x].E[y]$ if x and y are independent.

   (c) $\Phi_z(\omega) = \Phi_x(\omega).\Phi_y(\omega)$ for $z = x + y$, x and y independent.

   (d) $\left.\dfrac{d^k\Phi(\omega)}{d\omega^k}\right|_{\omega=0} = (j)^k\ E[x^k]$.

3. Normally distributed random numbers x with zero mean and unit standard deviation are to be generated by taking the sum of n independent random numbers u which are uniformly distributed in the range (0,1). If n is large, the sum $S_n = \sum_{i=1}^{n} u_i$ is very nearly normal. Show that

   $x = \frac{\sqrt{3}}{n}(2S_n - n)$ has the required statistics.

4. A random 2-vector $y = [y_1\ y_2]^T$ is constructed from a linear combination of the elements of a random 3-vector $w = [w_1\ w_2\ w_3]^T$, by the transformation

   $y = A\ w$

   The elements of w are independent and each has a zero mean value, and the same standard deviation σ_w.

   (i) Find the mean and covariance matrix of the vector y,

   (a) for
$$A = \begin{bmatrix} 1 & 1 & 1 \\ 1 & 0 & 1 \end{bmatrix}$$

   (b) for
$$A = \begin{bmatrix} 1 & 1 & 1 \\ -1 & 2 & -1 \end{bmatrix}$$

(ii) If the elements of w are jointly normal, will the elements of y be independent in either case?

5. By differentiation of the joint characteristic function for a 4-dimensional normal random vector, show that the mixed moment of 4th order $E[x_1.x_2.x_3.x_4]$ is given by

$$E[x_1.\ x_2.\ x_3.\ x_4] = R_{12}.R_{34} + R_{13}.R_{24} + R_{14}.\ R_{23}$$

where $R_{12} = E[x_1.x_2]$ etc. and the expected value of each x_i is zero.

SOLUTIONS TO TUTORIAL EXAMPLES

1. $f(X) = \frac{1}{2},\ 3 \leq X \leq 5.$

$$1 = \int_3^5 \tfrac{1}{2}\ dX. \qquad (1)$$

Now, $X = A^{\frac{1}{2}}$, so $dX = \frac{1}{2}A^{-\frac{1}{2}}\ dA$.

Changing variable in (1), $1 = \int_9^{25} \frac{1}{2}.\ \frac{1}{2}A^{-\frac{1}{2}}\ dA$

$\therefore\ \underline{f(A) = \frac{1}{4}A^{-\frac{1}{2}}}$

$$\text{Mean} = \bar{A} = \frac{1}{4}\int_9^{25} A.\ A^{-\frac{1}{2}}\ dA = \frac{1}{4}\cdot\frac{2}{3}\,[A^{3/2}]_9^{25}$$

$$= \frac{1}{6}\,(125 - 27) = 16\tfrac{1}{3}$$

2. (a) $$E[\alpha x + \beta y] = \int_{-\infty}^{\infty} dY \int_{-\infty}^{\infty} dX\ (\alpha X + \beta Y)\ f(X,Y)$$

$$= \alpha\int_{-\infty}^{\infty} X\ dX \int_{-\infty} f(X,Y)dY + \beta\int_{-\infty}^{\infty} Y\ dY \int_{-\infty}^{\infty} f(X,Y)dX$$

$$= \alpha\int_{-\infty}^{\infty} X\ dX.f(X) + \beta\int_{-\infty}^{\infty} YdY.f(Y)$$

$$= \alpha\ .E[x] + \beta\ .E[y]$$

(b) $$E[x.y] = \int_{-\infty}^{\infty} dY \int_{-\infty}^{\infty} dX\ X.Y\ f(X,Y)$$

If independent, $f(X,Y) = f(X).\ f(Y)$

$$\therefore\ E[x.y] = \int_{-\infty}^{\infty} X.f(X)dX.\ \int_{-\infty}^{\infty} Y.f(Y)dY = E[x].\ E[y].$$

(c) $\Phi_z(\omega) = E[\exp j\omega(x + y)]$

$$= \int_{-\infty}^{\infty} dY \int_{-\infty}^{\infty} dX.\exp[j\omega(X + Y)].\ f(X).f(Y) \text{ (since x,y independent)}$$

$$= \int_{-\infty}^{\infty} \exp(j\omega X).f(X)dX.\ \int_{-\infty}^{\infty} \exp(j\omega Y).f(Y)dY$$

$$= \Phi_X(\omega).\ \ \Phi_Y(\omega)$$

(d) $\Phi(\omega) = E[\exp(j\omega x)] = E[1 + j\omega x + \frac{1}{2!}(j\omega x)^2 + \frac{1}{3!}(j\omega x)^3 + \ldots].$

Assuming term-by-term differentiation possible, and that we can interchange order of E[ ] and $\frac{d}{d\omega}$ [ ],

$$\frac{d\Phi}{d\omega} = E[jx(1 + j\omega x + \ldots)] = E[jx.\ \exp(j\omega x)].$$

$$\therefore \left.\frac{d\Phi}{d\omega}\right|_{\omega=0} = j\ E[x].$$

Similarly,

$$\frac{d^k\Phi}{d\omega^k} = E[(jx)^k \exp(j\omega x)]$$

$$= (j)^k\ E[x^k] \text{ at } \omega = 0.$$

3. $E[u_i] = \frac{1}{2}$, $E[u_i^2] = \int_0^1 u^2.du = \frac{1}{3}$

$$\therefore \sigma_u^2 = \frac{1}{12}.$$

Mean of $S_n = \frac{n}{2}$; variance of $S_n = \frac{n}{12}$.

To standardise, take $x = \dfrac{S_n - \frac{n}{2}}{\sqrt{\frac{n}{12}}} = \sqrt{\frac{3}{n}}\,(2\ S_n - n).$

4. (i) With y = Aw,

$$E[y] = A.E[w] = 0 \text{ in each case.}$$

$$\therefore \text{Cov}[y] = E[yy^T] = E[Aww^TA^T]$$

$$= A\ E[ww^T]A^T$$

$$\text{Now } E[ww^T] = \begin{bmatrix} \sigma_w^2 & 0 & 0 \\ 0 & \sigma_w^2 & 0 \\ 0 & 0 & \sigma_w^2 \end{bmatrix} = \sigma_w^2 I$$

Case (a)

$$A = \begin{bmatrix} 1 & 1 & 1 \\ 1 & 0 & 1 \end{bmatrix}$$

$$AA^T = \begin{bmatrix} 1 & 1 & 1 \\ 1 & 0 & 1 \end{bmatrix} \begin{bmatrix} 1 & 1 \\ 1 & 0 \\ 1 & 1 \end{bmatrix} = \left[\begin{array}{c|c} 3 & 2 \\ \hline 2 & 2 \end{array}\right]$$

$$\therefore \quad \text{Cov}[y] = \left[\begin{array}{cc} 3\sigma_w^2 & 2\sigma_w^2 \\ \hline 2\sigma_w^2 & 2\sigma_w^2 \end{array}\right]$$

Case (b)

$$AA^T = \begin{bmatrix} 1 & 1 & 1 \\ -1 & 2 & -1 \end{bmatrix} \begin{bmatrix} 1 & -1 \\ 1 & 2 \\ 1 & -1 \end{bmatrix} = \begin{bmatrix} 3 & 0 \\ 0 & 6 \end{bmatrix}$$

$$\therefore \quad \text{Cov}[y] = \begin{bmatrix} 3\sigma_w^2 & 0 \\ 0 & 6\sigma_w^2 \end{bmatrix}$$

(ii) y is now also a zero mean Gaussian process. For case (b) the covariance matrix of y is diagonal and hence the elements of y are independent.

5. Let $x = [x_1, x_2, x_3, x_4]^T$

$$V = \begin{bmatrix} R_{11} & R_{12} & R_{13} & R_{14} \\ R_{21} & R_{22} & R_{23} & R_{24} \\ R_{31} & R_{32} & R_{33} & R_{34} \\ R_{41} & R_{42} & R_{43} & R_{44} \end{bmatrix}$$

$\omega = [\omega_1, \omega_2, \omega_3, \omega_4]^T$

then for normal vector, with zero mean value,

$$\Phi(\omega) = \exp\left(-\frac{1}{2}\omega^T V \omega\right)$$

$$E[x_1 . x_2 . x_3 . x_4] = \frac{\partial^4 \Phi(0)}{\partial\omega_1 \partial\omega_2 \partial\omega_3 \partial\omega_4} .$$

Now

$$\omega^T V \omega = \sum_{i=1}^{4} \sum_{j=1}^{4} R_{ij}\, \omega_i\, \omega_j$$

and (after some labour!) we can obtain

$$\frac{\partial\Phi}{\partial\omega_1} = \left(-\sum_{i=1}^{4} R_{1i}\, \omega_i\right) \Phi(\omega)$$

$$\frac{\partial^2\Phi}{\partial\omega_1\, \partial\omega_2} = \left\{ \sum_{i=1}^{4} R_{1i}\, \omega_i \sum_{j=1}^{4} R_{2j}\, \omega_j - R_{12}\right\} \Phi(\omega)$$

$$\frac{\partial^3\omega}{\partial\omega_1\, \partial\omega_2\, \partial\omega_3} = \left\{R_{12} \sum_{i=1}^{4} R_{3i}\, \omega_i + R_{13} \sum_{i=1}^{4} R_{2i}\, \omega_i + R_{23} \sum_{i=1}^{4} R_{1i}\, \omega_i - \sum_{i=1}^{4} R_{1i}\, \omega_i \sum_{j=1}^{4} R_{2j}\, \omega_j \sum_{k=1}^{4} R_{3k}\, \omega_k\right\} \Phi(\omega)$$

$$\frac{\partial^4\Phi}{\partial\omega_1\, \partial\omega_2\, \partial\omega_3\, \partial\omega_4} = \left\{R_{12}R_{34} + R_{13}R_{24} + R_{14}R_{23} + (\text{summations of } \omega_1\omega_2\omega_3\omega_4)\right\} \times \Phi(\omega)$$

Setting finally $\omega_1 = \omega_2 = \omega_3 = \omega_4 = 0$, the value of Φ is unity and we get

$$E[x_1 . x_2 . x_3 . x_4] = R_{12}R_{34} + R_{13}R_{24} + R_{14}R_{23}$$

RELEVANT STATISTICAL THEORY

Dr. M.T.G. Hughes

1. INTRODUCTION

When we observe a random process over a finite time interval, we are effectively taking a finite sample of the infinite *population* of samples which may be generated by the process.

For example, taking values of random function $x(t)$ at the instants $t_1, t_2, \ldots, t_N$ yields a sample vector

$$z = (x_1, x_2, \ldots, x_N)^T$$

where $x_i = x(t_i)$ etc. The number N is called the *size* of the sample z. Any function, say $g(x_1, x_2, \ldots, x_N)$, of the sample z is called a *statistic*. Such quantities, like the samples from which they are computed, are random, and can be characterised by probability distributions. Normally, a statistic will be formulated in such a way that it estimates the value of some unknown parameter of the process generating the sample. In such cases, knowledge of certain probability distributions associated with the statistic will enable us to formulate useful confidence statements, or to test hypotheses concerning the unknown parameters of the process which is under observation.

2. ASSESSING THE QUALITIES OF AN ESTIMATE[1]

If $\hat{\theta}$ is an estimate of some scalar parameter θ, based on N samples of a random variable, we may assess the accuracy with which $\hat{\theta}$ may be expected to represent θ using the following criteria:

(1) *Bias*

The bias of $\hat{\theta}$, written as $b[\hat{\theta}]$, is defined as follows:

$$b[\hat{\theta}] = E[\hat{\theta}] - \theta \qquad (1)$$

We prefer estimates to be at least *asymptotically unbiased*, which requires:

$$\lim_{N \to \infty} b[\hat{\theta}] = 0.$$

(ii) *Variance*

The variance of $\hat{\theta}$ is written as $\text{Var}[\hat{\theta}]$, and is defined by the relation

$$\text{Var}[\hat{\theta}] = E[\{\hat{\theta}-E[\hat{\theta}]\}^2] = E[\hat{\theta}^2] - \{E[\hat{\theta}]\}^2 \tag{2}$$

(iii) *Consistency*

$\hat{\theta}$ is said to be consistent if

$$\lim_{\substack{N\to\infty \\ \varepsilon\to 0}} \text{Prob}\,[|\hat{\theta}-\theta| < |\varepsilon|] = 1 \tag{3}$$

This is a desirable property because it ensures 'convergence in probability' of the estimate towards the true value of the quantity being estimated, as the sample size increases.

(iv) *Efficiency*

Strictly, the term 'efficiency', as applied to estimates, is defined in a relative sense.

Thus if $\hat{\theta}_1$ and $\hat{\theta}_2$ are two estimates of the same quantity θ, and if $\text{Var}[\hat{\theta}_1] < \text{Var}[\hat{\theta}_2]$, then $\hat{\theta}_1$ is said to be *'more efficient'* than $\hat{\theta}_2$. However, in recent usage the term has come to be accepted in an absolute sense, such that an estimate θ is said to be '*efficient*' if it has smaller variance than *any other* estimate of the same quantity.

3. THE STATISTICS OF SAMPLES

Consider a sample N observations drawn from a large population with mean $\mu = E[x] = 0$ (this is without loss of generality) and variance $\sigma^2 = E[x^2]$.

The sample mean $\bar{x}_s = (x_1 + x_2 + \ldots + x_N)/N$.

The expected value of the sample mean is

$$\hat{\mu} = E[\bar{x}_s] = E\left[\frac{x_1}{N}\right] + E\left[\frac{x_2}{N}\right] + \ldots + E\left[\frac{x_N}{N}\right]$$

$$= \frac{\mu}{N} + \frac{\mu}{N} + \ldots + \frac{\mu}{N}$$

$$= \mu = 0.$$

The mean of the sample thus provides a correct estimate of the mean of the parent population. It is said to be an *unbiased estimator*. Variance of the sample mean is $\text{Var}[\hat{\mu}] = E[x_s^2]$

$$= E\left[\frac{x_1 + x_2 + \ldots + x_N}{N}\right]^2 = \sum_i E\left[\frac{x_i}{N}\right]^2 + \sum_{i\neq j}\sum E\left[\frac{x_i x_j}{N^2}\right]$$

If $E[x_i x_j] = E[x_i]\, E[x_j] \neq 0$, for i = j (assuming independent samples), this simplifies to:

$$\mathrm{Var}[\hat{\mu}] = N.\frac{\sigma^2}{N^2} + 0$$

$$= \frac{\sigma^2}{N}.$$

This highlights the great convenience of working with independent quantities in statistics.

The variance of the sample is not an unbiased estimator of the overall population variance, as we shall see. Let s^2 be the mean squared deviation measured from the sample mean $\bar{x}_s$. If it were measured from μ, it would be greater by $(\bar{x}_s - \mu)^2$, so that as an estimate of σ^2, it is biased on the low side, as we will now prove.

Say we take a sample of size N from a population with mean μ and variance σ^2. We seek the best estimate of σ^2 that we can obtain from the sample, (assuming, as before, independent samples). Remember that $E[\bar{x}] = \mu$ (unbiased) and the sample variance $s^2 = \frac{1}{N}\sum_{i=1}^{N}(x_i - \bar{x})^2$.

$$E(s^2) = E[\frac{1}{N}\Sigma\,(x_i - \bar{x})^2] = \frac{1}{N}E[\Sigma\,(x_i - \bar{x})^2]$$

$$= \frac{1}{N}E[\Sigma\,(x_i - \mu + \mu - \bar{x})^2]$$

$$= \frac{1}{N}E[\Sigma\,(x_i - \mu)^2 - 2(\bar{x} - \mu)\,\Sigma\,(x_i - \mu) + \Sigma(\bar{x} - \mu)^2]$$

$$= \frac{1}{N}E[\Sigma\,(x_i - \mu)^2 - 2(\bar{x} - \mu).\,N(\bar{x} - \mu) + N.(\bar{x} - \mu)^2]$$

$$= \frac{1}{N}E[\Sigma\,(x_i - \mu)^2 - N(\bar{x} - \mu)^2]$$

But we have already seen that

$$\sigma^2 = E[(x_i - \mu)^2], \text{ while } E[(\bar{x} - \mu)^2] = \frac{\sigma^2}{N}.$$

$$\therefore\quad E(s^2) = \frac{1}{N}(N\sigma^2 - \sigma^2) = \frac{N-1}{N}\sigma^2.$$

The sample variance is thus a *biased estimator* of population variance.

4. HYPOTHESIS TESTING

If we can establish the form of probability distributions governing our estimate of $\hat{\theta}$, it is possible to test between two alternatives:

(i) The so-called *NULL HYPOTHESIS:* that θ has some specified value θ_0.

(iii) The *ALTERNATIVE HYPOTHESIS:* that θ does *not* have the value θ_o. Even if θ does equal θ_o, the estimate $\hat{\theta}$ will almost certainly *not* take this value. How much error can be allowed before we are justified in rejecting the null hypothesis?

We could commit *two* types of errors:

Type I: Rejection of the hypothesis: $\hat{\theta} = \theta_o$ when it is in fact true.

Type II: Acceptance of the hypothesis when it is in fact false.

We may now consider the probabilities of these two types of errors. Suppose the sampling distribution of $\hat{\theta}$ is known, and that the null hypothesis is true. In this case, the probability density curve for $\hat{\theta}$ could be drawn as follows:

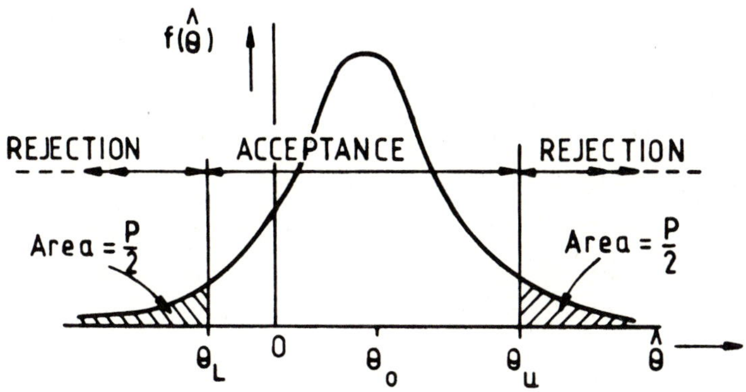

Figure 1. Probabilities relating to type I error.

For an unbiased estimate, if a 'small' value for P is chosen, it is unlikely that values of $\hat{\theta}$ outside the interval $\theta_L \leq \hat{\theta} \leq \theta_u$ would occur (the probability is just P). Hence, if the value of $\hat{\theta}$ turned out to lie outside this interval, there would be good reason to reject the hypothesis: $\hat{\theta} = \theta_o$ at the *100P% level of significance.* Bearing in mind the possibility of erroneous rejection of the hypothesis (a Type I error), we see that the probability of our committing such an error is just P, the level of significance for the test.

The procedure for determining the probability of a Type II error is rather more involved than the preceding one, and will not be considered further here. However, a good discussion of this subject will be found in the book by Bendat and Piersol[2] (Chapter 4).

5. CONFIDENCE INTERVALS

The preceding discussion forms the basis of a widely-used procedure for estimating parameters of random variables. It involves the determination of an interval which will include the parameter being estimated to a known degree of uncertainty, based on assumptions concerning the probability distributions of the observations. Consider the case where a sample provides an estimate $\hat{\theta}$ of a parameter θ of a random variable. $\hat{\theta}$ is estimated in terms of an interval, between θ_L and θ_u, say, where there is some measure of certainty that $\hat{\theta}$ lies within that interval. This is usually written

$$\text{Prob } [\theta_L \leqq \hat{\theta} \leqq \theta_u] = 1 - P$$

(see Fig. 1). Clearly the smaller we make P, the greater will be our certainty, but on the other hand, the width of the interval concerned will be wider. It is thus necessary to exercise judgement in the selection of a level of significance appropriate to the problem in hand. A commonly-used value of P is 0.05, and the corresponding confidence interval is called the 95% confidence interval. Section 6 will deal with how to determine the values θ_L and θ_u.

6. SAMPLING DISTRIBUTIONS

(i) Confidence Interval for Mean, with Population Variance Known

Suppose we wish to estimate the mean value μ of a normally-distributed random variable with *known* standard deviation σ. If our estimate is based on the mean, $\bar{x}$, of a sample of size N, then we know from Section 3 that

$$E[\bar{x}] = \mu \text{ and } \operatorname{Var}[\bar{x}] = \frac{\sigma^2}{N}$$

The quantity $u = \dfrac{\bar{x} - \mu}{\sigma/\sqrt{N}}$ (4)

is normally distributed with zero mean and unit standard deviation. If u_α is the value of u obtained from the Tables corresponding to a level of significance P (where $\alpha = \frac{1}{2}P$), we may write

$$\text{Prob } \left[-u_\alpha \leqq \frac{(\bar{x} - \mu)\sqrt{N}}{\sigma} \leqq u_\alpha\right] = 1 - P \tag{5}$$

so that the quantity (1 - P) is the probability of finding the value of μ somewhere in the interval

$$(\bar{x} - \frac{u_\alpha \sigma}{\sqrt{N}}) \leqq \mu \leqq (\bar{x} + \frac{u_\alpha \sigma}{\sqrt{N}})$$

The required probability thus depends on areas under a standardized (i.e. zero mean, unit standard deviation) normal curve. Values of such areas are extensively tabulated, see for example, ref. 8, Table 4. For 95% confidence limits, we are interested in $u_{0.025} = 1.96$, so the confidence interval is

$$(\bar{x} - \frac{1.96\sigma}{\sqrt{N}}) \leqq \mu \leqq (\bar{x} + \frac{1.96\sigma}{\sqrt{N}})$$

(ii) Confidence Interval for Mean, with Population Variance Unknown

It is rare for the population variance to be known, and it usually has to be estimated from a sample. The quantity

$$t = \frac{(\bar{x} - \mu)}{s/\sqrt{N-1}} \qquad (6)$$

(where s^2 is the sample variance, as defined in Section 3) is distributed as Student's 't' with $\nu = (n-1)$ degrees of freedom. The properties presented in Tables (e.g. ref. 8, Table 7) permit the formulation of confidence statements involving this distribution.

The shape of Students t-distribution appears to be quite similar to that of the normal distribution, and indeed t_∞ (i.e. t with $\nu \to \infty$) is the normal distribution. As N becomes small, the area in the tail of the 't'-distribution does differ quite significantly from that in the tail of the normal distribution, and erroneous confidence intervals result if the normal distribution is used when the population variance is not known.

Suppose that $\bar{x}$ and s are respectively the mean and standard deviation of a sample of N = 30 independent observations. Then $t = \frac{(\bar{x} - \mu)\sqrt{29}}{s}$ is distributed as Students 't' with 29 degrees of freedom. From the Tables, $t_{0.025;29} = 2.045$ so

$$\text{Prob}\left[-2.045 \leqq \frac{(\bar{x} - \mu)\sqrt{29}}{s} \leqq 2.045\right] = 0.95$$

from which the 95% confidence interval for μ is

$$\bar{x} - \frac{2.045s}{\sqrt{29}} \leqq \mu \leqq \bar{x} + \frac{2.045s}{\sqrt{29}}$$

(iii) Confidence Interval for Variance

The quantity:

$$\chi^2 = \frac{N.s^2}{\sigma^2} \qquad (7)$$

is distributed as chi-squared with

$\nu = (N-1)$ degrees of freedom (see, for example, ref. 8, Table 8).

Suppose that we require 95% confidence intervals for the population variance σ^2, given the variance s^2 of a sample of size N.

Then $\quad \text{Prob}\left[\chi^2_{0.975;29} \leqq \frac{30\ s^2}{\sigma^2} \leqq \chi^2_{0.025;29}\right] = 0.95.$

From Tables

$$\chi^2_{0.975;29} = 16.047$$

$$\chi^2_{0.025;29} = 45.722$$

so that the 95% confidence interval is

$$\frac{16.047}{30} \leq \frac{s^2}{\sigma^2} \leq \frac{45.722}{30}$$

i.e. $$\frac{30\ s^2}{45.722} \leq \sigma^2 \leq \frac{30\ s^2}{16.047}$$

$$0.656s^2 \leq \sigma^2 \leq 1.87\ s^2$$

(iv) Confidence Interval for Ratio of Variance

The 'F'-distribution (see, for example, ref. 8, pages 18 and 19) is concerned with ratios of variance. If two independent random samples of sizes M and N respectively having variances s_1^2 and s_2^2 are drawn from two normally-distributed populations of (unknown) variances σ_1^2 and σ_2^2, then the variable

$$F = \frac{\hat{\sigma}_1^2/\sigma_1^2}{\hat{\sigma}_2^2/\sigma_2^2} \tag{8}$$

(where $\hat{\sigma}_1^2$ is the best estimate of population variance from sample 1, i.e. $\frac{N}{N-1}\ s_1^2$, similarly for $\hat{\sigma}_2^2$) has an F-distribution with (M-1), (N-1) degrees of freedom. 95% confidence intervals are thus

$$F_{0.975;\nu_1,\nu_2} \leq \frac{\hat{\sigma}_1^2/\sigma_1^2}{\hat{\sigma}_2^2/\sigma_2^2} \leq F_{0.025;\nu_1,\nu_2}$$

giving

$$\frac{1}{F_{0.025;\nu_1,\nu_2}}\ \frac{\hat{\sigma}_1^2}{\hat{\sigma}_2^2} \leqslant \frac{\sigma_1^2}{\sigma_2^2} \leq \frac{1}{F_{0.975;\nu_1,\nu_2}}\ \frac{\hat{\sigma}_1^2}{\hat{\sigma}_2^2}$$

In ref. 8, $F_{\alpha;\nu_1,\nu_2}$ is tabulated for α = 0.05, 0.025, 0.01 and 0.001 and the lower percentage points of the distribution may be obtained from the relation

$$F_{(1-\alpha);\nu_1,\nu_2} = (F_{\alpha;\nu_2,\nu_1})^{-1}$$

Example

Two samples of sizes 25 and 9 respectively are drawn at random from two normal populations. The sample variances are found to be 24 and 16 respectively. Determine the 90% confidence interval for the ratio of variances of the two populations.

$$s_1^2 = 24, \text{ so } \hat{\sigma}_1^2 = \frac{25}{24} \cdot 24 = 25; \; \nu_1 = 24.$$

$$s_2^2 = 16, \text{ so } \hat{\sigma}_2^2 = \frac{9}{8} \cdot 16 = 18; \; \nu_2 = 8.$$

Directly from Tables, $F_{0.05;24,8} = 3.12$

$$F_{0.95;24,8} = (F_{0.05;8,24})^{-1} = \frac{1}{2.36}$$

so 90% confidence limits for $\frac{\sigma_1^2}{\sigma_2^2}$ are given by:

$$\frac{1}{3.12}\left(\frac{25}{18}\right) \leq \frac{\sigma_1^2}{\sigma_2^2} \leq 2.36\left(\frac{25}{18}\right)$$

i.e. $$0.445 \leq \frac{\sigma_1^2}{\sigma_2^2} \leq 3.28$$

7. THE CRAMÉR-RAO VARIANCE BOUND[7]

Consider a set of observations (generally regarded as an N-vector z) which are taken to constitute a particular realisation of a random N-vector x. The p.d.f. of x is assumed to be of known structural form, but depends on a set of unknown parameters (generally a p-vector θ) which we wish to estimate, i.e.

$f(X;\theta)$ is a function of known form, but θ is an unknown vector.

We perform an experiment in which the elements of x take particular values represented by vector z, and we wish to obtain the "best" estimate of θ, of the form $\hat{\theta} = \hat{\theta}(z)$.

It is helpful at this stage to consider the simpler case where x, z and θ are all scalar quantities and to consider the question: "What is the smallest possible variance with which an estimate of θ can be determined, by any method, from the observation z?". The resulting answer can be extended readily to the vector case.

Suppose $\hat{g}(z)$ is an unbiased estimate of some given function $g(\theta)$. The expected value of $\hat{g}(z)$ is thus equal to $g(\theta)$, i.e. from Lecture 1,

$$\int_{-\infty}^{\infty} \hat{g}(z).\ f(z)dz = g(\theta) \tag{9}$$

Since z is drawn from the process which generates x, it has a p.d.f. of the same form as x:

$$f(z) = f(X;\theta)\big|_{X=Z} \tag{10}$$

This quantity is called the Likelihood Function of the observations z, and is denoted as $L(z,\theta)$. Thus

$$\int_{-\infty}^{\infty} \hat{g}(z).\ L(z,\theta)dz = g(\theta) \tag{11}$$

Differentiating with respect to θ, we get

$$\int_{-\infty}^{\infty} \hat{g}(z)\ \frac{\partial L}{\partial \theta}\ dz = \frac{dg}{d\theta} \tag{12}$$

Since $L(z,\theta)$ is a p.d.f., $\int_{-\infty}^{\infty} L(z,\theta)dz = 1$, so

$$\int_{-\infty}^{\infty} \frac{\partial L}{\partial \theta}\ dz = 0 = \int_{-\infty}^{\infty} g(\theta)\ \frac{\partial L}{\partial \theta}\ dz.$$

Hence $$\int_{-\infty}^{\infty} [\hat{g}(z) - g(\theta)]\ \frac{\partial L}{\partial \theta}\ dz = \frac{dg}{d\theta} \tag{13}$$

Consider the natural logarithm of L, which we will denote as $\mathcal{L}(z,\theta)$ - the *log likelihood function*

Since $$\frac{\partial \mathcal{L}}{\partial \theta} = \frac{1}{L}\ \frac{\partial L}{\partial \theta} \tag{14}$$

$$\int_{-\infty}^{\infty} [\hat{g}(z) - g(\theta)]\ L.\frac{\partial \mathcal{L}}{\partial \theta}\ dz = \frac{dg}{d\theta} \tag{15}$$

so that from the Schwartz inequality,

$$\left\{\int_{-\infty}^{\infty} [\hat{g} - g]^2\ L\ dz\right\} \left\{\int_{-\infty}^{\infty} \left(\frac{\partial \mathcal{L}}{\partial \theta}\right)^2 L\ dz\right\} \geq \left(\frac{dg}{d\theta}\right)^2 \tag{16}$$

The first term on the LHS is the variance of g(z) while the second term is $E(\frac{\partial \mathcal{L}}{\partial \theta})^2$. Hence

$$\mathrm{Var}\ [\hat{g}(z)] \geq \frac{\left[\frac{dg}{d\theta}\right]^2}{E\left[\left(\frac{\partial \mathcal{L}}{\partial \theta}\right)^2\right]} \tag{17}$$

This result is known generally as the *Cramér-Rao variance bound.* The following comments are pertinent:

(i) If $\hat{g}(z)$ is an unbiased estimate of θ itself, that is, if $g(\theta) = \theta$,

$$\mathrm{Var}[\hat{\theta}] \geq \frac{1}{E\left[\left(\frac{\partial \mathcal{L}}{\partial \theta}\right)^2\right]} \tag{18}$$

(ii) The corresponding result for a biased estimate, $\hat{g}(\theta) = \theta + b(\theta)$, say, is

$$\text{Var}[\hat{\theta}] \geq \frac{(1 + \frac{db}{d\theta})^2}{E\left[(\frac{\partial \mathcal{L}}{\partial \theta})^2\right]} \tag{19}$$

(iii) Since, as we have seen already,

$$\int_{-\infty}^{\infty} \frac{\partial L}{\partial \theta} dz = 0,$$

$$\int_{-\infty}^{\infty} \frac{1}{L} \frac{\partial L}{\partial \theta} L dz = 0,$$

so differentiating with respect to θ,

$$\int_{-\infty}^{\infty} (\frac{1}{L} \frac{\partial L}{\partial \theta}) \frac{\partial L}{\partial \theta} dz + \int_{-\infty}^{\infty} \frac{\partial}{\partial \theta} (\frac{1}{L} \frac{\partial L}{\partial \theta}) L\, dz = 0.$$

But since $\frac{1}{L} \frac{\partial L}{\partial \theta} = \frac{\partial \mathcal{L}}{\partial \theta}$, $E\left[(\frac{\partial \mathcal{L}}{\partial \theta})^2\right] = -E\left[\frac{\partial^2 \mathcal{L}}{\partial \theta^2}\right]$.

Hence the Cramér-Rao inequality can be written

$$\text{Var}[\hat{g}(z)] \geqq - \frac{\left[\frac{dg}{d\theta}\right]^2}{E\left[\frac{\partial^2 \mathcal{L}}{\partial \theta^2}\right]} \tag{20}$$

(iv) If x, z and $\boldsymbol{\theta}$ are vectors, $L(z,\theta)$ and $\mathcal{L}(z,\theta)$ become joint p.d.f.'s. In such a case, the Cramér-Rao bound defines a lower bound of the covariance matrix of $\hat{\theta}$, and the quantitites $\frac{\partial \mathcal{L}}{\partial \theta}$ and $\frac{\partial^2 \mathcal{L}}{\partial \theta^2}$ must be interpreted in terms of vectors and matrices respectively.

(v) It can be shown that if the parameter vector θ is chosen in such a way as to achieve a global maximum of the likelihood function, the resulting estimate, known as a *Maximum Likelihood Estimate* (MLE) possesses certain desirable properties. One of these *(asymptotic efficiency)* is that as N (the number of samples) tends to infinity, the variance of a MLE tends to the value corresponding to the Cramér-Rao Bound. These estimates will be discussed further in later sections of this lecture. For present purposes, we introduce an illustrative example to make the concept of a likelihood function more clear.

Example

Suppose that N independent observations x_1, x_2,...,x_N, are made from a normally distributed population.

(a) Assuming that the variance of x has a known value σ^2, find the maximum likelihood estimate of the mean value μ.

(b) Assuming the mean value μ to be known, find the MLE of the variance σ^2.

(c) Using the fact that the variance of a MLE approaches the Cramér-Rao Bound for 'large' samples ($N \to \infty$), deduce the limiting value of the variance of the estimate of the mean (in part (a)).

(a) For a single observation x_k,

$$f(x_k;\mu) = \frac{1}{\sigma\sqrt{2\pi}} \exp(-(x_k - \mu)^2/2\sigma^2)$$

Thus, for N independent observations, the likelihood function is

$$L(\mu) = \frac{1}{(2\pi\ \sigma^2)^{N/2}} \exp(-\frac{1}{2\sigma^2} \sum_{k=1}^{N} (x_k - \mu)^2)$$

Taking natural logarithms, we obtain the log-likelihood function:

$$\mathcal{L}(\mu) = -\frac{N}{2} \log(2\pi\ \sigma^2) - \frac{1}{2\sigma^2} \sum_{k=1}^{N} (x_k - \mu)^2$$

Noting that σ^2 is assumed known, we choose $\hat{\mu}$ such that $\mathcal{L}(\hat{\mu})$ is a maximum, i.e.

$$\frac{\partial\mathcal{L}(\hat{\mu})}{\partial\hat{\mu}} = \frac{1}{\sigma^2} \sum_{k=1}^{N} (x_k - \hat{\mu}) = 0.$$

Thus the MLE of μ turns out to be simply the arithmetic mean of the observations:

$$\hat{\mu} = \frac{1}{N} \sum_{k=1}^{N} x_k.$$

(b) Assuming μ is known, the MLE $(\hat{\sigma}^2)$ of σ^2 is that value which maximises $\mathcal{L}$, so that

$$\frac{\partial\mathcal{L}(\hat{\sigma}^2)}{\partial\hat{\sigma}^2} = -\frac{N}{2\hat{\sigma}^2} + \frac{1}{2\hat{\sigma}^4} \sum_{k=1}^{N} (x_k - \mu)^2 = 0.$$

Thus, for a known mean value μ, the MLE of σ^2 is given by

$$\hat{\sigma}^2 = \frac{1}{N} \sum_{k=1}^{N} (x_k - \mu)^2$$

(c) We have already seen (in Section 4.3) of these notes) that the estimate $\hat{\mu}$ in part (a) is unbiased, and also that the bias of the estimate $\hat{\sigma}^2$ in part (b) tends to zero for large samples ($N \to \infty$). Thus, we could use Eq. (18) to obtain the limiting values of the required variances:

$$\mathrm{Var}[\hat{\theta}] \geq \frac{1}{E\left[\left(\frac{\partial \mathcal{L}}{\partial \theta}\right)^2\right]}$$

For the estimate of the mean, this leads to

$$\mathrm{Var}[\hat{\mu}] \geq E\left[\frac{1}{\sigma^4} \sum_{i=1}^{N} \sum_{j=1}^{N} (x_i - \mu)(x_j - \mu)\right]^{-1}$$

$$= \frac{1}{\sigma^4} \sum_{i=1}^{N} \sum_{j=1}^{N} E[(x_i - \mu)(x_j - \mu)]^{-1}$$

$$E[(x_i - \mu)(x_j - \mu)] = \sigma^2 \quad \text{for } i = j$$

$$= 0 \quad \text{for } i \neq j$$

$$\therefore \quad \mathrm{Var}[\hat{\mu}] \geq \left(\frac{N\sigma^2}{\sigma^4}\right)^{-1} = \frac{\sigma^2}{N} .$$

We note here that the variance of the arithmetic mean deduced in Section 3 actually coincides with this bound for any value of N. Thus, we conclude that there could be no better estimator of the mean value, provided that the observations are independent and normally distributed.

8. OPTIMUM ESTIMATION TECHNIQUES

In this and subsequent sections we concern ourselves with the problem of defining an 'optimum' estimate in terms which are both realistic and convenient in applications. Two distinct approaches are discussed which are considered particularly worthwhile, and which underlie extensive areas of application to control systems.

APPROACH 'A' : CONDITIONAL EXPECTATION[4]

A set of observations $\{z_i\}$ and a set of related but unknown quantitites $\{x_i\}$ are regarded as the elements of suitably dimensioned random vectors:

$$z = [z_1, z_2, \ldots, z_m]^T \tag{21}$$

$$x = [x_1, x_2, \ldots, x_n]^T \tag{22}$$

These vectors are supposed to be characterised by a conditional p.d.f. of *known form*,

i.e. $f(X|z)$ is assumed known (23)

Unlikely as it may seem at this stage, such an assumption may be justified for a significant class of problems which are of interest in control systems analysis.

To simplify the following discussion, but without serious loss of generality, we assume that x and z are simply scalar quantitites; the results we obtain can be shown to hold in the vector case also.

Consider the variance of an unbiased estimate $\hat{x}$ of x, determined as a function of the observations z, i.e. $\hat{x} = \hat{x}(z)$. For given z,

$$\mathrm{Var}[\hat{x}(z)] = \int_{-\infty}^{\infty} [\hat{x}(z) - X]^2 \, f(X|z)dX \tag{24}$$

The quantity $\hat{x}(z)$ in Eq. 24 does not depend on the variable of integration, and so may be regarded as a parameter. Thus, we can write

$$\mathrm{Var}[\hat{x}(z)] = \hat{x}^2(z) \int_{-\infty}^{\infty} f(X|z)dX - 2\hat{x}(z) \int_{-\infty}^{\infty} Xf(X|z)dX + \int_{-\infty}^{\infty} X^2 f(X|z)dX.$$

From an examination of the terms in this equation, and noting that $f(X|z)$ is a p.d.f., we find:

$$\int_{-\infty}^{\infty} f(X|z)dX = 1, \tag{25}$$

$$\int_{-\infty}^{\infty} Xf(X|z)dX = E[x|z], \tag{26}$$

$$\int_{-\infty}^{\infty} X^2 f(X|z)dX = E[x^2|z]. \tag{27}$$

Here, as noted in earlier lectures, the quantity $E[x|z]$ represents the *conditional expectation* of the random variable x, given the observation z.

Thus, Eq. (24) can be re-written as

$$\mathrm{Var}[\hat{x}(z)] = \hat{x}^2(z) - 2\hat{x}(z).E[x|z] + E[x^2|z],$$

and by 'completing the square', we obtain

$$\mathrm{Var}[\hat{x}(z)] = \{\hat{x}(z) - E[x|z]\}^2 + E[x^2|z] - \{E[x|z]\}^2 \tag{28}$$

The last two terms in this expression are unaffected by the choice of $\hat{x}(z)$, thus it is clear that the variance of $\hat{x}$ will be a minimum if and only if we set

$$\hat{x}(z) = E[x|z] \tag{29}$$

That is, *the conditional expectation of the unknown random variable* x *given the observation* z, *is a minimum-variance estimate of* x. *The same result can be shown to hold also for the vector case.*

It must be remarked here that the indicated conditional expectation cannot be evaluated in all cases, but attention may be drawn to two special cases in which it may be evaluated with ease:

(i) x and z are jointly Gaussian vectors

We have already seen (Lecture L1) that the conditional expectation of a Gaussian random vector x is given by

$$\hat{x}(z) = E[x|z] = \mu_x + P_{xz} P_{zz}^{-1} (z - \mu_z) \tag{30}$$

and the corresponding covariance matrix by

$$P = \mathrm{Cov}[\hat{x}] = P_{xx} - P_{xz} P_{zz}^{-1} P_{zx} \tag{31}$$

where μ_x, μ_z, P_{xx}, P_{zz}, P_{xz}, P_{zx} are respectively the vector means and covariance matrices associated with the vectors x and z.

Two important points are worth noting in this case: first, that the conditional expectation of x given z is actually a linear function of z. Second, it can be shown that the selection of $E[x|z]$ as an estimate of x in the Gaussian case actually minimises a much more general class of loss functions than the variance/covariance matrix considered here[4].

(ii) x and z are independent

In this case,

$$E[x|z] = E[x] \tag{32}$$

If a system model can be devised so that it satisfies this condition, it can lead to considerable analytical simplicity.

Example 1

Consider the problem of aircraft take-off and landing on the deck of a ship in rough sea. The pilot is able to guess at certain limited characteristics of the deck motion (say the displacement and its rate of change) at current time; and (assuming the take-off or landing manoeuvre to take a given fixed time τ) he may be concerned with the displacement and velocity of the deck at the projected time $t + \tau$.

A simplified but illuminating model of this situation may be constructed as follows:

Let $d_1 = d(t)$ be the deck displacement time t
Let $v_1 = v(t)$ be the deck velocity at time t
Let $d_2 = d(t + \tau)$ be the deck displacement at time $t + \tau$
Let $v_2 = v(t + \tau)$ be the deck velocity at time $t + \tau$.

Assume that all of these quantities are stationary and jointly normal, with zero mean values, and the following variances and covariances:

$$E[d_1^2] = E[d_2^2] = \sigma_d^2 \ ; \quad E[v_1^2] = E[v_2^2] = \sigma_v^2 \ ;$$

$$E[d_1 d_2] = E[d_2 d_1] = R_{dd}(\tau); \ E[v_1 v_2] = E[v_2 v_1] = R_{vv}(\tau)$$

$$E[d_1 v_2] = R_{dv}(\tau); \quad E[v_1 d_2] = R_{vd}(\tau);$$

$$E[d_1 v_1] = E[d_2 v_2] = R_{dv}(o).$$

One might expect to be able to prescribe these quantities on the basis of experiment and/or analysis.

Now define a vector of observations

$$z = [d_1 v_1]^T$$

and a vector of quantities to be predicted from the observations

$$x = [d_2 v_2]^T$$

If our formulation of the model reflects the true situation with acceptable accuracy, we know from Section 8.4 of Lecture L1 that the vector x will be normally distributed, with mean value, and covariance matrix (as conditioned by the observations z) given by Eqs. (30) and (31) above.

Examining the various matrices appearing in these expressions, we find that

$$\mu_x = \mu_z = 0$$

$$P_{xx} = P_{zz} = \begin{bmatrix} \sigma_d^2 & R_{dv}(o) \\ R_{vd}(o) & \sigma_v^2 \end{bmatrix}$$

$$P_{xz} = \begin{bmatrix} R_{dd}(\tau) & R_{dv}(\tau) \\ R_{vd}(\tau) & R_{vv}(\tau) \end{bmatrix} = P_{zx}^T$$

Thus,

$$P_{zz}^{-1} = \frac{1}{\sigma_d^2 \sigma_v^2 - R_{dv}^2(o)} \begin{bmatrix} \sigma_v^2 & -R_{dv}(o) \\ -R_{vd}(o) & \sigma_d^2 \end{bmatrix}$$

The conditional mean of x is given by

$$\begin{bmatrix} E[d_2|z] \\ \\ E[v_2|z] \end{bmatrix} \begin{bmatrix} R_{dd}(\tau) & R_{dv}(\tau) \\ \\ R_{vd}(\tau) & R_{vv}(\tau) \end{bmatrix} \begin{bmatrix} \frac{\sigma_v^2}{\Delta} & \frac{-R_{dv}(0)}{\Delta} \\ \\ \frac{-R_{vd}(0)}{\Delta} & \frac{\sigma_d^2}{\Delta} \end{bmatrix} \begin{bmatrix} d_1 \\ \\ v_1 \end{bmatrix}$$

and the corresponding covariance matrix is

$$P = \begin{bmatrix} \sigma_d^2 & R_{dv}(o) \\ \\ R_{vd}(o) & \sigma_v^2 \end{bmatrix} - \begin{bmatrix} R_{dd}(\tau) & R_{dv}(\tau) \\ \\ R_{vd}(\tau) & R_{vv}(\tau) \end{bmatrix} \begin{bmatrix} \frac{\sigma_v^2}{\Delta} & \frac{-R_{dv}(o)}{\Delta} \\ \\ \frac{-R_{vd}(o)}{\Delta} & \frac{\sigma_d^2}{\Delta} \end{bmatrix} \begin{bmatrix} R_{dd}(\tau) & R_{vd}(\tau) \\ \\ R_{dv}(\tau) & R_{vv}(\tau) \end{bmatrix}$$

where $\Delta = \sigma_d^2 \sigma_v^2 - R_{dv}^2(o)$.

An important point to note in this example is that the conditional mean is a linear function of the observations. Under the assumption of jointly Gaussian motions, the p.d.f. of the 'future' motions for given values of current displacement and velocity was seen to be Gaussian, with the mean and variance taking values in accordance with Eqs. (30) and (31) of the present notes.

We may now note that the given conditional mean value of the future motions actually constitutes an *optimum estimate* under the stated assumptions, and that the corresponding covariance matrix is the 'smallest' that could be achieved by *any* method of estimation. That is, *the conditional expectation is an efficient estimate of a random variable for a given set of observations.* We see also that the quantities required to determine the optimum estimates in this case are simply the means and correlations of the random variables involved.

A very useful alternative approach to such problems can be developed if the random signal under consideration can be assumed to have been generated by passing white noise w(t) through a known linear filter. That is, we employ a model (in vector-matrix notation for generality) of the form:

$$\dot{x} = Ax(t) + Bw(t). \tag{33}$$

Here, the quantity of interest, say $x_1(t)$, has been embedded in the random vector $x(t)$, the quantities A and B are supposed to be known matrices, and $w(t)$ is a conveniently dimensioned vector of independent "white noise". For simplicity, each component of $w(t)$ is supposed to have zero mean:

$$(E\,[w] = 0), \tag{34}$$

and each component is assumed to be independent of the others and of any time-shifted version of itself:

$$E[w(t_1).w^T(t_2)] = D\,\delta(t_1 - t_2) \tag{35}$$

In this expression, D is a diagonal matrix, and $\delta(\ldots)$ is a Dirac delta function.

Suppose we wish to estimate $x(t+\tau)$, for some prediction interval $\tau \geqq 0$. Using the convolution integral, if $\Phi(t,t_o)$ is the state transition matrix defined by

$$\left.\begin{aligned} \dot{\Phi}(t,t_o) &= A\,\Phi(t,t_o) \\ \Phi(t_o,t_o) &= I, \end{aligned}\right\} \tag{36}$$

we have

$$x(t+\tau) = \Phi(t+\tau,t).x(t) + \int_t^{t+\tau} \Phi(t+\tau,u)B.w(u)\,du \tag{37}$$

Examining the terms on the right side of this equation, we may note that the first one is completely deterministic and involves information which is known at time t. However, the integral contains a term $w(u)$ which is not known at time t, since $t < u \leqq (t+\tau)$.

The minimum-variance estimate of x is given by the conditional expectation:

$$\hat{x}(t+\tau) = E[x(t+\tau)|t] = \Phi(t+\tau,t)x(t) + \int_t^{t+\tau} \Phi(t+\tau,u)B.E[w(u)|t]du \tag{38}$$

where $E[\ldots|t]$ means "conditional expectation based on observations up to time t". But since $w(u)$ is 'white noise'.

$$E[w(u)|t] = E[w(u)] = 0 \tag{39}$$

for $t < u \leqq t+\tau$.

Thus, the optimum estimate of $x(t+\tau)$ is given by

$$\hat{x}(t+\tau) = \Phi(t+\tau,t).x(t) \tag{40}$$

In the case of a constant-coefficient linear system (matrices A, B constant), the state transition matrix in Eq. (40) depends only on the prediction interval τ, so that for this case,

$$\hat{x}(t+\tau) = \Phi(\tau).x(t) \qquad (41)$$

Thus, for a given prediction interval, the optimum predictor is determined as a simple linear function of the observed state vector. It should be noted, however, that the result of Eq. (41) is based on the assumption that the state vector can be measured at all times without significant error. This assumption is not justified in many cases of practical interest, so that a much more difficult problem arises in such cases. This problem is covered in Lecture L.9 on the Kalman filter.

APPROACH 'B' : PARAMETRIC ESTIMATION

It has already been shown (in Section 7 of these notes) that the lower limit to the variance of any estimate is given by the "Cramér-Rao Bound".

If $\hat{\theta}$ is an unbiased estimate of the parameter θ, we have

$$\mathrm{Var}[\hat{\theta}] \geq \frac{1}{E\left[\left(\frac{\partial \mathcal{L}}{\partial \theta}\right)^2\right]} = \frac{-1}{E\left[\frac{\partial^2 \mathcal{L}}{\partial \theta^2}\right]} \qquad (42)$$

where $\mathcal{L}(\theta)$ is the *log-likelihood function:*

$$\left.\begin{aligned} \mathcal{L}(\theta) &= \log_e L(\theta), \\ L(\theta) &= f(X;\theta)\big|_{X=z} \end{aligned}\right. \qquad (43)$$

$L(\theta)$ is the *likelihood function* which is proportional to the probability of occurrence of the observation z.

The results of Eq. (42) may be readily extended to the case where X, z, θ are vectors, as follows.

The Cramer-Rao Bound now becomes[6]

$$\left.\begin{aligned} &\mathrm{Cov}(\hat{\theta}) \geq \left[E\left[\frac{\partial \mathcal{L}}{\partial \theta_i} \cdot \frac{\partial \mathcal{L}}{\partial \theta_j}\right]\right]^{-1} \\ \text{or}\quad & \\ &\mathrm{Cov}(\hat{\theta}) \geq -\left[E\left[\frac{\partial^2 \mathcal{L}}{\partial \theta_i \partial \theta_j}\right]\right]^{-1} \end{aligned}\right\} \qquad (44)$$

In this expression, the log-likelihood function is derived from the joint p.d.f. of the N-vector x, with the substitution X = z. The quantity $\mathrm{Cov}(\hat{\theta})$ now of course represents the covariance matrix of the (assumed unbiased) estimates:

$$\mathrm{Cov}(\hat{\theta}) = E[(\hat{\theta} - \theta)(\hat{\theta} - \theta)^T] \qquad (45)$$

and the relation 'is greater than or equal to' must be taken to mean that the difference (LHS - RHS) is non-negative definite.

Maximum-Likelihood Estimation

It can be shown[7] that if the estimate $\hat{\theta}$ can be chosen in such a way as to yield a unique maximum of the likelihood function $L(\theta)$, then the resulting parameter estimate $\hat{\theta}$ (known as a *maximum-likelihood estimate*) will possess some remarkable and useful properties:

(i) *Asymptotic Unbiasedness*

$$\lim_{N\to\infty} E[\hat{\theta}] = \theta, \tag{46}$$

where N is the sample size.

(ii) *Asymptotic Efficiency*

$$\begin{aligned}\lim_{N\to\infty} \mathrm{Cov}[\hat{\theta}] &= \left[E\left[\frac{\partial \mathcal{L}}{\partial \theta_i} \cdot \frac{\partial \mathcal{L}}{\partial \theta_j}\right]\right]^{-1} \\ &= -\left[E\left[\frac{\partial^2 \mathcal{L}}{\partial \theta_i \partial \theta_j}\right]\right]^{-1}\end{aligned} \tag{47}$$

That is, the covariance matrix of $\hat{\theta}$ actually approaches the Cramér-Rao lower bound defined by Eq. (44) as the sample size increases.

(iii) *Consistency*

$$\lim_{\substack{N\to\infty \\ |\varepsilon|\to 0}} \mathrm{Prob}\,[\|\hat{\theta} = \theta\| < |\varepsilon|] = 1 \tag{48}$$

That is, $\hat{\theta}$ is said to 'converge in probability' towards θ.

(iv) *Invariance*

If $\hat{\theta}$ is a maximum-likelihood estimate (MLE) of θ, and $g(\theta)$ is some given function of θ, then $g(\hat{\theta})$ will be a MLE of $g(\theta)$.

(v) *Asymptotic Normality*

It can be shown that the p.d.f. of $\hat{\theta}$ tends, as $N \to \infty$, to a Gaussian distribution, with mean value θ and covariance matrix given by Eq. (47). This fact facilitates the testing of MLE by means of confidence interval estimation.

Further consideration will be given to the concepts discussed here in later lectures, where the techniques are employed in applications to state and parameter estimation and stochastic control.

Further reading:

1. Spiegel, M.R.: "Probability and Statistics", Schaum Outline Series, McGraw-Hill, 1975.
2. Bendat, J.S. and Piersol, A.G.: "Measurement and Analysis of Random Data", Wiley, 1966.
3. Eykhoff, P.: "System Identification - Parameter and State Estimation", Wiley, 1974.
4. Meditch, J.S.: "Stochastic, Optimal Linear Estimation and Control", McGraw-Hill, 1969.
5. Åström, K.J.: "Introduction to Stochastic Control Theory", Academic Press, 1970.
6. Deutsch, R.: "Estimation Theory", Chapters 11 and 12, Prentice-Hall, 1965.
7. Cramér, H.: "Mathematical Methods of Statistics", Princeton, 1946.
8. Murdoch, J. and Barnes, J.A.: "Statistical Tables for Science, Engineering, Management and Business Studies", Macmillan, 1974.

TUTORIAL EXAMPLES - LECTURE L.2

1. Thirty-one independent observations are collected from a normally-distributed random variable x(k) with the following results:

60	61	47	56	61	63
65	69	54	59	43	61
55	61	56	48	67	65
60	58	57	62	57	58
53	59	58	61	67	62
54					

The sample mean, $\bar{x}(=\hat{\mu}) = \frac{1}{N}\Sigma\, x_i = 58.6$

sample variance, $s^2 = \frac{1}{N}\Sigma(x_i - \bar{x})^2 = 32.4$

Determine 90% confidence intervals for the true mean value and variance of x(k). (ANS: $56.8 \leq \mu \leq 60.4$; $22.9 \leq \sigma^2 \leq 54.2$).

2. Twenty-five independent observations are made of a normally-distributed random variable x(k). The mean of the observations is 10 and estimate of population variance of 4. Ten independent observations are made of a second normally-distributed random variable y(k) with observation mean 100 and estimate of population variance of 8. Determine an interval which will include the ratio of variances of the two populations with a probability of 98%.

(ANS: $0.106 \leq \frac{\sigma_x^2}{\sigma_y^2} \leq 1.63$).

3. A certain random process generates events which are randomly spaced in time. The time T between successive events is distributed with a probability density function of the form

$$f(T) = k\, e^{-kT}$$

where k is a constant parameter.
Show that a maximum-likelihood estimate of k, based on a set of N independent observations of T, i.e. $\{T_1, \ldots, T_N\}$ is

$$\hat{k} = \frac{N}{\sum_{i=1}^{N} T_i}$$

and that for large samples, $(N \rightarrow \infty)$, the variance of $\hat{k}$ tends to a value which satisfies $\frac{\mathrm{var}(\hat{k})}{k^2} = \frac{1}{N}$.

4. A certain random signal x(t) can be modelled in terms of a filtered "white noise" source, w(t), as follows:

$$\ddot{x} + 3\dot{x} + 2x = w(t)$$

The "white noise" process w(t) may be assumed to be stationary, with zero mean value, and with an autocorrelation function of the form $R_{ww}(\tau) = q\delta(\tau)$, where q is a constant and $\delta(\tau)$ a unit delta function.
Assuming that error-free measurements of x(t) and of its derivative can be made up to time t, show how these measurements can be used to obtain an optimum prediction of the value of $x(t + \alpha)$.

TUTORIAL EXAMPLES - SOLUTIONS

1. For $\nu = 30$, $t_{0.05} = 1.697$

so $\text{Prob}\left[-1.697 \leq \frac{(\hat{\mu} - \mu)\sqrt{30}}{s} \leq 1.697\right] = 0.90$

from which the 90% confidence interval is

$$\hat{\mu} - \frac{1.697s}{\sqrt{30}} \leq \mu \leq \hat{\mu} + \frac{1.697s}{\sqrt{30}}$$

i.e. $58.6 - 1.697\sqrt{\frac{32.4}{30}} \leq \mu \leq 58.6 + 1.697\sqrt{\frac{32.4}{30}}$

$$\underline{56.8 \leq \mu \leq 60.4.}$$

For $\nu = 30$, $\chi^2_{0.05} = 43.8$ and $\chi^2_{0.95} = 18.5$

so $\text{Prob}\left[18.5 \leq \frac{31s^2}{\sigma^2} \leq 43.8\right] = 0.90$

from which the 90% confidence interval is

$$\frac{31s^2}{43.8} \leq \sigma^2 \leq \frac{31s^2}{18.5}$$

$$\underline{22.9 \leq \sigma^2 \leq 54.2.}$$

2. $\hat{\sigma}_x^2 = 4, \quad \nu_x = 24$

$\hat{\sigma}_y^2 = 8, \quad \nu_y = 9.$

Directly from Tables, $F_{0.01;24,9} = 4.73$.

$$F_{0.99;24,9} = F_{0.01;9,24}^{-1} = \frac{1}{3.26} \text{ (by interpolation)}$$

so that 98% confidence limits are:

$$\frac{1}{4.73}(0.5) \leq \frac{\sigma_x^2}{\sigma_y^2} \leq 3.26\ (0.5)$$

i.e.

$$0.106 \leq \frac{\sigma_x^2}{\sigma_y^2} \leq 1.63.$$

3. Likelihood function for N independent samples is:

$$L(\hat{k}) = \prod_{i=1}^{N} \hat{k}\, e^{-\hat{k}T_i}$$

so log-likelihood function is

$$\mathcal{L}(\hat{k}) = N \log \hat{k} - \sum_{i=1}^{N} \hat{k}\, T_i$$

Differentiating with respect to $\hat{k}$,

$$\frac{\partial \mathcal{L}}{\partial \hat{k}} = \frac{N}{\hat{k}} - \sum_{i=1}^{N} T_i = 0 \text{ for maximum likelihood}$$

so maximum likelihood estimate $\underline{\hat{k} = \frac{N}{\Sigma\, T_i}}$

Minimum variance bound is achieved by maximum likelihood estimation. As $N \to \infty$,

$$\mathrm{Var}[\hat{k}] \longrightarrow -\frac{1}{E\left[\frac{\partial^2 \mathcal{L}}{\partial k^2}\right]} = \frac{k^2}{N}$$

$$\therefore \quad \underline{\frac{\mathrm{Var}[\hat{k}]}{k^2} = \frac{1}{N} \text{ as } N \to \infty.}$$

4. Write equations of process in matrix form, with $x(t) = x_1$ and $\dot{x}_1 = x_2$.

Then
$$\begin{bmatrix} \dot{x}_1 \\ \dot{x}_2 \end{bmatrix} = \begin{bmatrix} 0 & 1 \\ -2 & -3 \end{bmatrix} \begin{bmatrix} x_1 \\ x_2 \end{bmatrix} + \begin{bmatrix} 0 \\ w(t) \end{bmatrix}.$$

$$\dot{x} = Ax + Bu \text{ in general}$$

If $\Phi(t) = \mathcal{L}^{-1}\,[(sI - A)^{-1}]$

then

$$x(t + a) = \Phi(a)\, x(t) + \int_t^{t+a} \Phi(t+a+\tau)\, B\, u\, (\tau)\, d\tau$$

Take conditional expectation based on information up to time t to get minimum variance estimate of x at t + a.

Since u is zero-mean white noise vector, $E[u(\tau)|t] = 0$, for $\tau > t$, so that best predictor is

$$\hat{x}(t + a) = \Phi(a)\ \ x(t). \quad \text{Now find } \Phi(a).$$

$$(sI - A) = \begin{bmatrix} s & -1 \\ 2 & s+3 \end{bmatrix} \qquad \therefore \quad (sI - A)^{-1} = \begin{bmatrix} s+3 & 1 \\ -2 & s \end{bmatrix} \times \frac{1}{(s+1)(s+2)}$$

$\therefore$

$$\Phi(t) = \left[\begin{array}{c|c} 2e^{-t} - e^{-2t} & e^{-t} - e^{-2t} \\ \hline -2e^{-t} + 2e^{-2t} & 2e^{-2t} - e^{-t} \end{array}\right]$$

$\therefore$

$$\hat{x}(t + a) = \left[\begin{array}{c|c} 2e^{-a} - e^{-2a} & e^{-a} - e^{-2a} \\ \hline 2e^{-2a} - 2e^{-a} & 2e^{-2a} - e^{-a} \end{array}\right] \begin{bmatrix} x_1(t) \\ x_2(t) \end{bmatrix}$$

Therefore, if we can measure $x(=x_1)$ and $v(=x_2)$ we have best estimate of x at time $(t+a) = (2e^{-a} - e^{-2a})\ x_1(t) + (e^{-a} - e^{-2a})x_2(t)$. Note that the coefficients here die away to zero as a increases. We should expect the variances and covariances to increase towards Var[x] as a increases.

SYSTEMS ANALYSIS II

Prof. J.L. Douce

This lecture is mainly concerned with methods for determining the response of dynamic systems to random excitation. It is first necessary, however, to review some important concepts in random process analysis.

1. INTRODUCTION

The concept of a *stochastic process* was introduced in Lecture No. L.1 as a basis for the study of sets of random functions or random signals. Two points of particular significance here are:

(i) Practically all of the usefully available knowledge about the probability structure of a random function may be determined from knowledge of the mean and autocovariance function (or the autocorrelation function). For a Gaussian process, these quantities provide a complete specification of the process characteristics.

(ii) For a *stationary random process*, whether it be 'strictly' or 'weakly' so, the process will have a constant mean value, and the covariance function (or autocorrelation function) will depend on the value of a single time-shift variable τ. That is, for a stationary random signal $x(t)$,

$$\mu = E[x(t)] = \text{Constant} \tag{1}$$

$$R_{xx}(\tau) = E[x(t).x(t+\tau)] \tag{2}$$

$$\begin{aligned} C_{xx}(\tau) &= E[(x(t)-\mu)(x(t+\tau)-\mu)] \\ &= R_{xx}(\tau)-\mu^2 \end{aligned} \tag{3}$$

From Eq. (2) above, the autocorrelation function $R_{xx}(\tau)$ is seen to represent the mean product of the signal $x(t)$ with the time-shifted replica $x(t+\tau)$. The autocovariance function $C_{xx}(\tau)$ differs from $R_{xx}(\tau)$ only through the removal of the mean from $x(t)$. Thus, the two quantities differ only by the constant level μ^2.

In dynamic system analysis, the quantities given by Eqs. (1) to (3) are much more convenient to work with than probability distributions, and this lecture will be principally concerned with their uses in the analysis of linear control systems which are assumed to be subjected to stationary random inputs.

2. THE ERGODIC HYPOTHESIS

For virtually all *stationary* random functions of practical interest, it can be shown that the mathematical expectation operator defined previously is equivalent, under fairly general conditions, to an average performed on any particular realization of the random process, over an infinite time interval, i.e.

$$E[x(t)] = \;= \lim_{T\to\infty} \frac{1}{2T} \int_{-T}^{T} x(t)dt \tag{4}$$

$$E[x(t).x(t+\tau)] = R_{xx}(\tau) = \lim_{T\to\infty} \frac{1}{2T} \int_{-T}^{T} x(t).x(t+\tau)dt \tag{5}$$

3. PROPERTIES OF THE AUTOCORRELATION FUNCTION

(i) $R_{xx}(0) = E\{x^2(t)\}$

This follows immediately from the definition.
Similarly,

$$\text{Cov}[x(t_1),\, x(t_2)] = E[x^2(t_1)] = \sigma_x^2 \text{ if } t_1 = t_2$$

(ii) $R_{xx}(\tau) = R_{xx}(-\tau)$ if the process is stationary.

(iii) $R_{xx}(0) \geqq |R_{xx}(\tau)|$ for all τ.

This follows since

$$E[[x(t)-x(t+\tau)]^2] \geqq 0$$

$$\therefore \quad E[x^2(t)] + E[x^2(t+\tau)] \geqq 2E[x(t).x(t+\tau)]$$

$$\therefore \quad 2\, R_{xx}(0) \geqq 2\, R_{xx}(\tau)$$

(By examining $E[x(t) + x(t+\tau)]^2$, similarly, we can show that

$$R_{xx}(0) \geqq |R_{xx}(\tau)|).$$

(iv) A pure sinusoid has a periodic autocorrelation function of the same period in τ as the original time function.

Illustrative Example 1

(a) Suppose the process is

$$x_i(t) = X \sin(\omega t+\theta_i) = x_1(\theta_i)$$

where X and ω are constants, and θ_i is a random variable, uniformly distributed over the range 0 to 2π

$$x_i(t+\tau) = X \sin (\omega[t+\tau] + \theta_i) = x_2(\theta_i)$$

$$R_{xx}(\tau) = E[x_i(t).x_i(t+\tau)]$$

$$= E[x_1(\theta_i).x_2(\theta_i)]$$

$$= \int_0^{2\pi} x_1(\theta_i).x_2(\theta_i).f(\theta_i)d\theta_i .$$

Since $f(\theta_i) = \frac{1}{2\pi}$,

$$R_{xx}(\tau) = \frac{X^2}{2\pi} \int_0^{2\pi} \sin (\omega t+\theta_i).\sin (\omega[t+\tau] + \theta_i)d\theta_i$$

$$= \frac{X^2}{2} \cos \omega\tau .$$

(b) An Alternative Approach, Invoking the Ergodic Hypothesis

Let $x(t) = X \sin (\omega t+\theta)$

then

$$R_{xx}(\tau) = \lim_{T\to\infty} \frac{1}{2T} \int_{-T}^{+T} X^2 \sin (\omega t+\theta).\sin (\omega[t+\tau] + \theta)dt$$

$$= \lim_{T\to\infty} \frac{1}{2T} \int_{-T}^{+T} \frac{X^2}{2} \{\cos \omega\tau - \cos[\omega(2t+\tau) + 2\theta]\}dt$$

$$= \frac{X^2}{2} \cos \omega\tau .$$

Method (a) evaluates the ensemble average, (b) the time average. Equality of the two results is expected if the process is ergodic.

Example 2

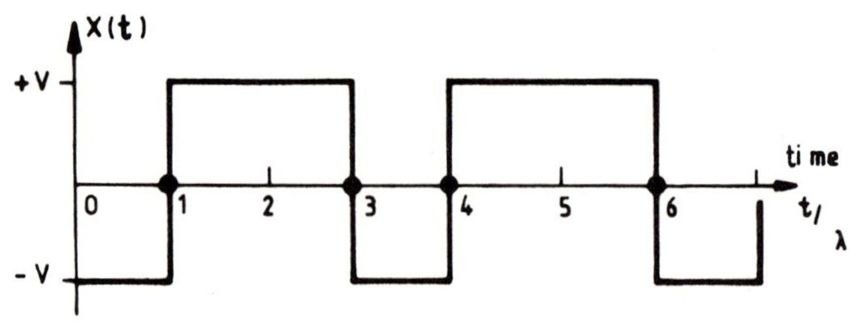

Figure 1

Consider the derivation of the autocorrelation function of the signal x(t) shown in Fig. 1. This signal assumes values ±V for the time intervals of duration λ, changing value with probability 0.5 at regularly spaced 'events points' 0, λ, 2λ etc. Consider the expected value of x(t).x(t+τ). If $|\tau| > \lambda$, an event point occurs with probability one in the time interval t to (t+τ). Thus x(t) and x(t+τ) are independent, and $E[x(t).x(t+\tau)] = E[x(t)].E[x(t+\tau)] = 0$. For $|\tau| < \lambda$ the probability of an event point in the time interval t to (t+τ) is $|\tau|/\lambda$. Therefore the probability of no event point (implying $x(t+\tau) = x(t)$, and $x(t).x(t+\tau) = V^2$) is $1 - {|\tau|}/{\lambda}$.

Thus for $|\tau| < \lambda$

$$E[x(t).x(t+\tau)] = (1 - {|\tau|}/{\lambda}).V^2 + \frac{|\tau|}{\lambda}.0$$

This function is as shown in Fig. 2.

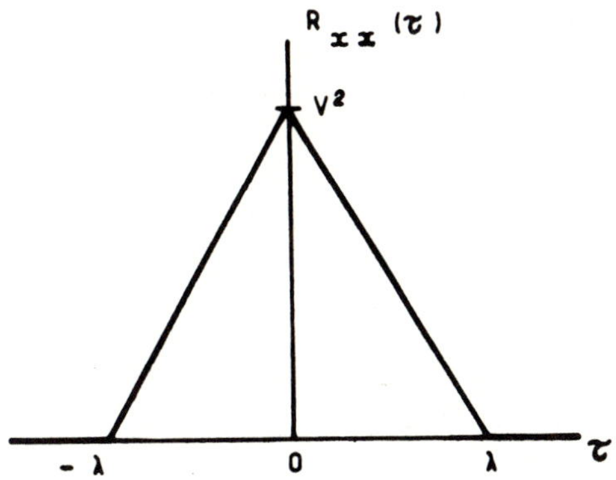

Figure 2

4. THE CROSS-CORRELATION FUNCTION

The cross-correlation function for two stationary random processes x(t) and y(t) is defined as

$$R_{xy}(\tau) = E[x(t).y(t+\tau)] \tag{6}$$

which is equal to

$$\lim_{T\to\infty} \frac{1}{2T} \int_{-T}^{+T} x(t).y(t+\tau)dt \tag{7}$$

Basic properties are:

(i) $R_{xy}(\tau) = R_{yx}(-\tau)$

Unlike the autocorrelation function:

(ii) $R_{xy}(\tau)$ is not, in general, an even function of τ

(iii) $R_{xy}(\tau)$ is not, in general, a maximum at $\tau = 0$.

5. FREQUENCY DOMAIN ANALYSIS

Revision: A repetitive signal x(t) with repetition period 2T can be expressed in terms of the Fourier series

$$x(t) = \frac{a_0}{2} + \sum_{r=1}^{\infty} a_r \cos r\omega_0 t + \sum_{r=1}^{\infty} b_r \sin r\omega_0 t$$

where ω_0 is the fundamental (angular) frequency π/T and the Fourier coefficients a_r, b_r are given by

$$a_r = \frac{1}{T} \int_{-T}^{+T} x(t) \cos r\omega_0 t \, dt$$

$$b_r = \frac{1}{T} \int_{-T}^{+T} x(t) \sin r\omega_0 t \, dt \, .$$

The first equation expresses the repetitive signal as the sum of sinusoidal components of frequency ω_0, $2\omega_0$ etc.

A non-repetitive signal can similarly be expressed by the Fourier transform

$$x(t) = \frac{1}{2\pi} \int_{-\infty}^{+\infty} X(j\omega) \, e^{j\omega t} d\omega$$

and the function $X(j\omega)$ is given by

$$X(j\omega) = \int_{-\infty}^{+\infty} x(t) e^{-j\omega t} dt$$

$X(j\omega)$ is termed the Fourier spectrum of the signal x(t).

Power Spectral Density of a Stationary Random Signal

It has already been observed that the Fourier transform of a random signal having a finite mean square value may not be defined in a strict sense. Even in cases where this difficulty is overcome through the use of limiting processes, the resulting Fourier Transform turns out to be a random quantity which is rather badly behaved statistically.

For a stationary random signal, on the other hand, the Fourier transform of the autocorrelation function can be defined without difficulty:

$$S_{xx}(\omega) = \int_{-\infty}^{\infty} R_{xx}(\tau) . e^{-j\omega\tau} d\tau \qquad (8)$$

The quantity $S_{xx}(\omega)$ is known as the 'Power Spectral Density' (PSD) of the random signal x(t).

Using the inverse Fourier transform, we obtain

$$R_{xx}(\tau) = \frac{1}{2\pi}\int_{-\infty}^{\infty} S_{xx}(\omega)e^{j\omega t}d\omega \qquad (9)$$

or, noting that $R_{xx}(0)$ is the mean square value $\overline{x^2}$, and working in terms of cyclical frequency $f = \omega/2\pi$ instead of radian frequency ω, it follows from Eq. (9) that

$$\overline{x^2} = \int_{-\infty}^{\infty} S_{xx}(2\pi f).df. \qquad (10)$$

It is this property that justifies the physical interpretation of the quantity S_{xx} as an energy density function. If the units of x(t) are volts, and those of time are seconds, then the units of S_{xx} are $(\text{volts})^2$ per cycle per second. The physical significance of the PSD may be further appreciated by considering the idealised situation shown in Fig. 3.

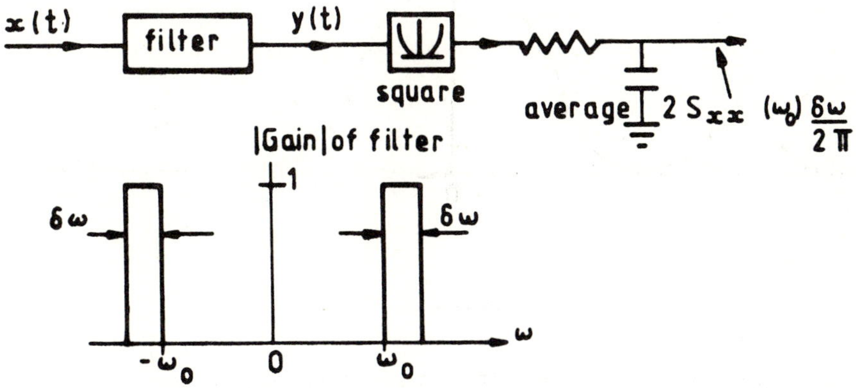

Figure 3

The filter passes the frequency components of x(t) lying in the range $\omega_0 < \omega \leqq \omega_0 + \delta\omega$ and $-(\omega_0 + \delta\omega) < \omega < -\omega_0$ ($\delta\omega$ is small), to give the narrow-band signal y(t). The mean square value of y(t) is then found, and this is the power contained in the original signal within the given frequency band. This is equal to twice the power density at frequency $f_0 (= \omega_0/2\pi)$ times $\delta f\ (= \delta\omega/2\pi)$.

This argument remains valid mathematically when, as here, the power of the signal is considered equally distributed over positive and negative frequencies, for it can be shown that

$$S_{xx}(\omega) = S_{xx}(-\omega) \text{ for all } \omega,$$

and the power gain of any real network can be shown to be an even function of frequency.

Illustrative Examples

(i) Determine the power spectral density of the random square-wave shown in Fig. 1. The autocorrelation function is as given in Fig. 2, and the corresponding PSD is

$$S(\omega) = 2V^2 \int_0^{\lambda} (1 - \tau/\lambda) \cos \omega\tau \, d\tau$$

$$= V^2\lambda \left[\frac{\sin (\frac{\omega\lambda}{2})}{(\frac{\omega\lambda}{2})} \right]^2 \tag{11}$$

This function is shown in Fig. 4.

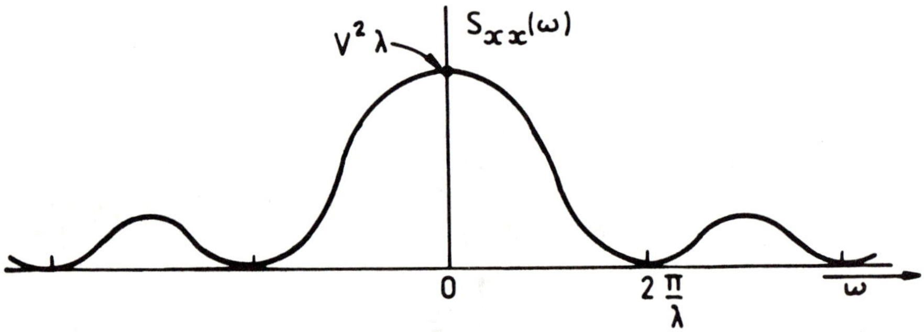

Figure 4

Note the perhaps unexpected result that the power spectral density is zero at the event frequency $f = 1/\lambda$.

(ii) Evaluate the mean square of a signal with power spectrum

$$S_{xx}(\omega) = \frac{S_0}{1 + (\omega/\omega_0)^2} \tag{12}$$

where S_0 is a constant, equal to the power density at zero frequency.

$$\overline{x^2} = \frac{1}{2\pi} \int_{-\infty}^{\infty} \frac{S_0}{1 + (\omega/\omega_0)^2} \, d\omega$$

$$= \frac{S_0\omega_0}{\pi} \int_0^{\pi/2} d\theta \quad \text{where } \tan\theta = \omega/\omega_0$$

$$= S_0\omega_0/2 \, . \tag{13}$$

It is left as an exercise to show that one half of the total power is contained in the frequency range $-\omega_o < \omega < \omega_o$.

(iii) <u>White noise</u>

By analogy with white light, which contains equal intensities of spectral components throughout the visible spectrum, white noise has a constant spectral density at all frequencies. Evidently the power of such a quantity is infinite. Nevertheless, white noise is a useful theoretical and practical concept. In practice a random signal which has a constant spectral density over the entire frequency range of interest may be considered as white noise.

(iv) <u>Band-Limited White Noise</u>

Determine the autocorrelation function and mean square value of a signal for which

$$S_{xx}(\omega) = S_o, \text{ a constant, for } |f| < B$$

$$= 0 \quad \text{for } |f| > B.$$

$$R_{xx}(\tau) = \frac{S_o}{2\pi} \int_{-2\pi B}^{2\pi B} e^{j\omega\tau} d\omega = \frac{S_o}{2\pi} \left[\frac{e^{j\omega\tau}}{j\tau} \right]_{-2\pi B}^{2\pi B}$$

$$= 2B S_o \left(\frac{\sin 2\pi B\tau}{2\pi B\tau}\right)$$

$$\overline{x^2} = 2B S_o \tag{14}$$

The results for this problem are illustrated in Fig. 5.

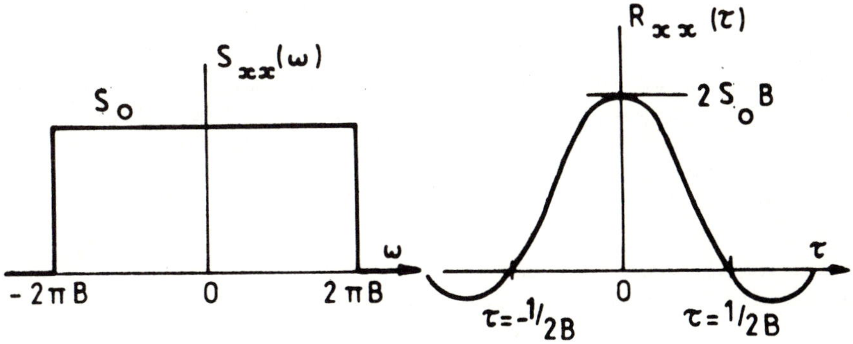

Figure 5

For Discussion

Why is the autocorrelation function

$$R(\tau) = \text{constant for } |\tau| < T$$

$$= 0 \text{ otherwise} \qquad \text{physically unrealisable?}$$

The cross-spectral density function

The cross-spectral density function for two signals x(t) and y(t) written $S_{xy}(j\omega)$ is defined as the Fourier transform of the cross-correlation function, so that

$$S_{xy}(j\omega) = \int_{-\infty}^{+\infty} R_{xy}(\tau)e^{-j\omega\tau}d\tau \qquad (15)$$

$$R_{xy}(\tau) = \frac{1}{2\pi}\int_{-\infty}^{+\infty} S_{xy}(j\omega)\, e^{j\omega\tau}d\omega \qquad (16)$$

The physical significance of this function may be appreciated by noting the manner in which the cross spectrum is determined in practice (using digital processing).

Consider two signals x(t) and y(t). Let each signal be passed through identical narrow band filters, of gain unity for $\omega_0 < \omega < \omega_0 + \delta\omega$, and zero at other frequencies.

The outputs of the two filters are multiplied together. The average value of the product, divided by the (small) bandwidth of the filter is the real part of the cross-spectral density at the frequency ω_0.

The imaginary part of the cross-spectral density at frequency ω_0 is obtained as above, except that the signal y(t) is phase-shifted by 90° before multiplication.

The figure shows how Im $[S_{xy}(\omega_0)]$ can be obtained

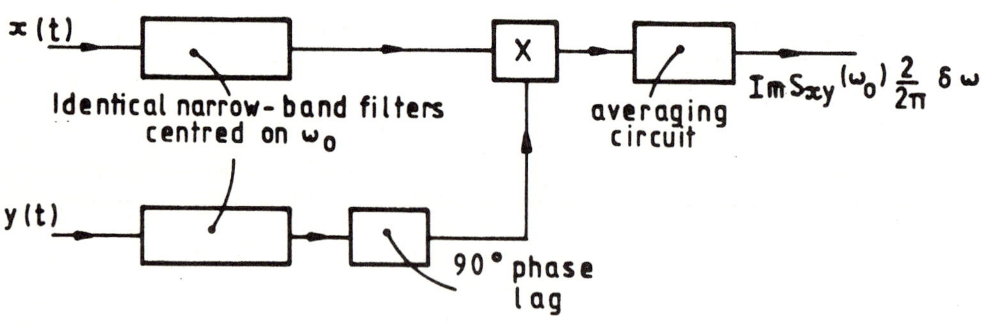

Figure 6

Basic Properties of Cross-Spectral Density Functions

1. If y(t) = x(t) then the cross-spectral density function becomes the power spectral

density function $S_{xx}(\omega) = S_{yy}(\omega)$, which is real-valued, non-negative, and an even function of frequency.

2. If $y(t) \neq x(t)$, then the cross-spectral density function will be generally complex and unsymmetrical about the frequency origin. The notation $S_{xy}(j\omega)$ is normally employed to highlight the distinction from the real quantities $S_{xx}(\omega)$ and $S_{yy}(\omega)$.

Coherence function

A useful measure of the statistical interdependence of the two signals x(t) and y(t) is given by the coherence function $\gamma^2_{xy}(\omega)$, defined by

$$\gamma^2_{xy}(\omega) = \frac{|S_{xy}(j\omega)|^2}{S_{xx}(\omega)\, S_{yy}(\omega)} \quad . \tag{17}$$

It can be shown that the value of this quantity must always lie between zero and one.

When $\gamma^2_{xy}(\omega) = 1$ for all ω, x(t) and y(t) are said to be fully coherent. When $\gamma^2_{xy}(\omega) = 0$ at a particular frequency, x(t) and y(t) are said to be incoherent at that frequency.

The concepts of cross-spectral densities and coherency are found to be of considerable interest in the spectral analysis of input/output data for dynamic systems. A full discussion of this topic will be presented in Lecture L8.

6. RESPONSE OF LINEAR SYSTEMS TO STATIONARY RANDOM EXCITATION

We now consider the problem of evaluating the response of a dynamic system to random excitation. For simplicity, we assume that stationary random excitation is applied to a *linear time-invariant* system. The analysis can be undertaken primarily in the time domain or the frequency domain, each approach aiding the understanding of system behaviour in terms of the functions discussed earlier.

Relationships in the Time Domain

The fundamental feature of the response of a linear system in the time domain is the response to a unit impulse, written h(t). The response y(t) of the same system to an arbitrary function of time, x(t) is given, (neglecting transients due to initial conditions), by the superposition integral:

$$y(t) = \int_{-\infty}^{\infty} h(u).x(t-u)du \tag{18}$$

It is noted, in passing, that h(u) is zero for all negative values of u for all physically realisable systems.

Similarly,

$$y(t+\tau) = \int_{-\infty}^{\infty} h(v)x(t+\tau-v)dv. \tag{19}$$

$$\therefore \quad R_{yy}(\tau) = \lim_{T\to\infty} \frac{1}{2T} \int_{-T}^{T} dt\left\{\int_{-\infty}^{\infty} h(u)x(t-u)du \int_{-\infty}^{\infty} h(v)x(t+\tau-v)dv\right\}$$

Performing first the integration with respect to time,

$$R_{yy}(\tau) = \int_{-\infty}^{\infty}\int_{-\infty}^{\infty} h(u)\; h(v)R_{xx}(\tau+u-v)du\; dv \tag{20}$$

Example

Determine the autocorrelation function of the signal at the output of a system of transfer function $\frac{1}{1+sT}$ when the input is white noise.

The autocorrelation function of white noise is an impulse at the origin, written $A.\delta(t)$, where A is one half of the measurable "power" per unit bandwidth of the signal. The impulse response of the system is

$$h(t) = 0,\; t < 0$$

$$h(t) = \frac{1}{T}\, e^{-t/T},\; t > 0$$

$$\therefore \quad R_{yy}(\tau) = \frac{A}{T^2}\int_{-\infty}^{\infty}\int_{-\infty}^{\infty} e^{-u/T}.e^{-v/T}.\;\delta(\tau+u-v)du\; dv.$$

The delta function is zero except for $u = v - \tau$

$$\therefore \quad R_{yy}(\tau) = \frac{A}{T^2}\int_{a}^{\infty} e^{(\tau-v)/T}.e^{-v/T}\; dv$$

where the lower limit of integration is

$a = \tau$ for $\tau > 0$

and $a = 0$ for $\tau < 0$

$$\therefore \quad R_{yy}(\tau) = \frac{A}{T^2}\, e^{\tau/T}\left[\frac{e^{-2v/T}}{-2/T}\right]_a^{\infty}$$

For $\tau < 0$, $R_{yy}(\tau) = \frac{A}{2T}\, e^{\tau/T}$

For $\tau > 0$, $R_{yy}(\tau) = \frac{A}{2T}\, e^{-\tau/T}$

i.e. $$R_{yy}(\tau) = \frac{A}{2T}\, e^{-|\tau|/T} \text{ for all } \tau\,. \tag{21}$$

A physical interpretation of the qualitative features of the autocorrelation function of a signal can be obtained by comparing (a) the autocorrelation of the signal obtained by passing white noise through a linear network with (b) the result obtained by recording the impulse response of this network and applying the signal in reverse time to the system.

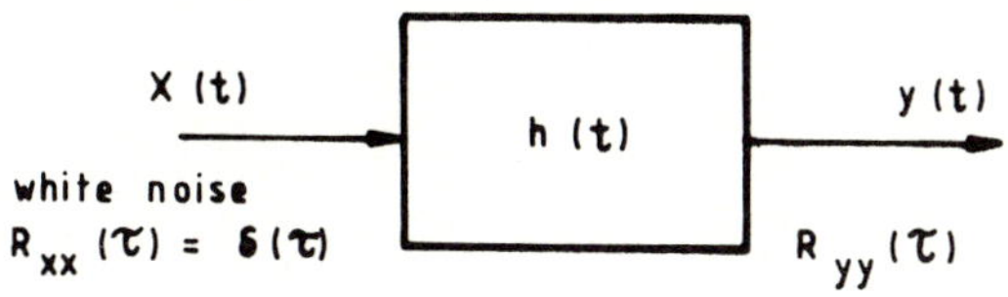

Figure 7

(a) From Eq. 20, illustrated in Figure 7,

$$R_{yy}(\tau) = \int_{-\infty}^{\infty}\int_{-\infty}^{\infty} h(u)\ h(v)\ R_{xx}(\tau+u-v)du\ dv$$

for $\quad R_{xx}(\tau) = \delta(\tau),$

$$R_{yy}(\tau) = \int_{-\infty}^{\infty}\int_{-\infty}^{\infty} h(u)\ h(v)\ \delta(\tau+u-v)\ du\ dv$$

$$= \int_{-\infty}^{\infty} h(v)\ h(v-\tau)\ dv$$

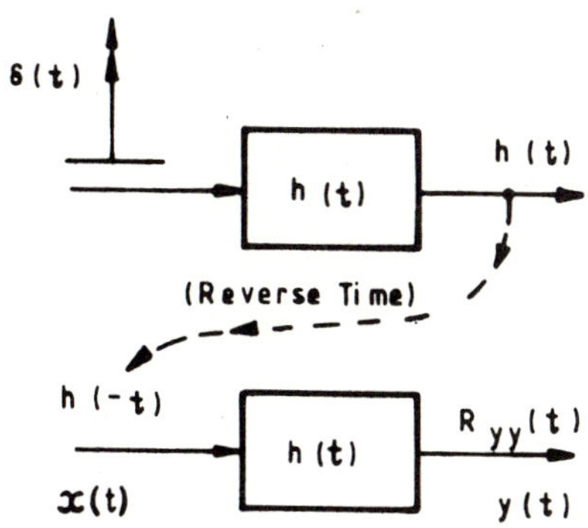

Figure 8

(b) Using the convolution integral to obtain y(t) in terms of x(t), with x(t) = h(-t),

$$y(t) = \int_{-\infty}^{\infty} h(v).\ x(t-v)\ dv$$

$$= \int_{-\infty}^{\infty} h(v).\ h(v-t)\ dv.$$

Correlation coefficient

For later use, we define the correlation coefficient $\rho_{xx}(\tau)$ as the ratio

$$\rho_{xx}(\tau) = \frac{R_{xx}(\tau)}{R_{xx}(0)} \tag{22}$$

From previous results,

$$|\rho_{xx}(\tau)| \leq 1.$$

Cross-correlation between input and output of a time-invariant linear system

The cross-correlation is obtained schematically as shown in Fig. 9.

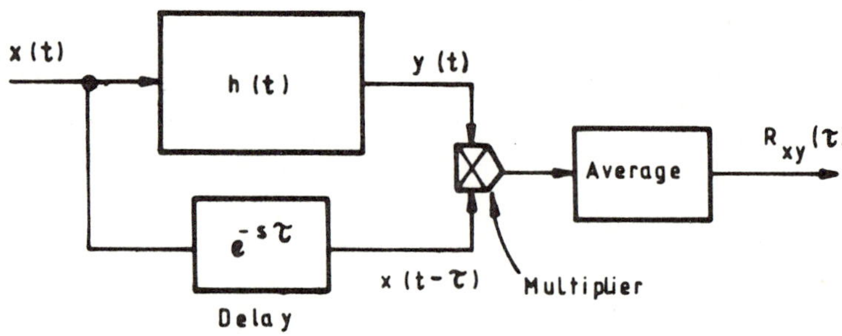

Figure 9

Again using the convolution integral to obtain an expression for y(t) and substituting this into the definition of $R_{xy}(\tau)$,

$$y(t) = \int_{-\infty}^{\infty} h(u).\ x(t-u)\ du$$

$$R_{xy}(\tau) = \lim_{T\to\infty} \frac{1}{2T} \int_{-T}^{T} x(t) \left\{ \int_{-\infty}^{\infty} h(u).x(t+\tau-u)du \right\} dt$$

Reversing the order of integration gives

$$R_{xy}(\tau) = \int_{-\infty}^{\infty} h(u)\ R_{xx}(\tau-u)\ du \tag{23}$$

which can be visualised by the schematic diagram of Fig. 10.

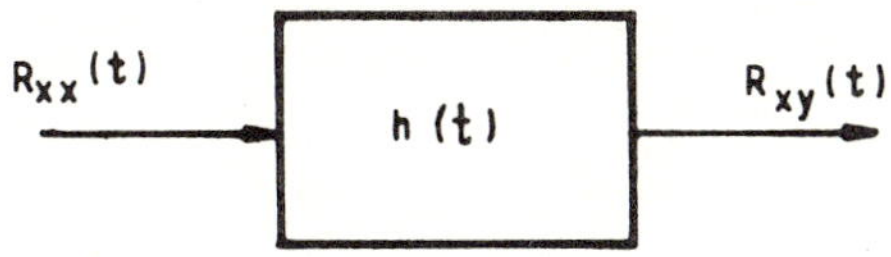

Figure 10

In particular, if the input signal approximates to white noise, so that $R_{xx}(\tau) = \delta(\tau)$, then the cross-correlation function is proportional to the impulse response of the system.

This result provides a useful practical method for determining the impulse response of a system subjected to a variety of disturbances. A relatively small white-noise signal is injected and cross-correlated with the system output. The user-generated test signal is uncorrelated with the other system disturbances so that a good estimate of the impulse response can be obtained.

Relationships in the frequency domain

In the frequency domain, the response of a linear system is characterised by the frequency response function $H(j\omega)$. This function is the Fourier transform of the impulse response $h(t)$. For deterministic signals, the Fourier transforms of input and output, $X(j\omega)$ and $Y(j\omega)$ respectively, are related by

$$Y(j\omega) = H(j\omega).X(j\omega)$$

The amplitude gain at any frequency ω, defined as the ratio (output amplitude)/(input amplitude), is $|H(j\omega)|$. At this same frequency, since power is proportional to $(\text{amplitude})^2$, the power gain, defined as the ratio (output power) / (input power) is $|H(j\omega)|^2$. For systems with real parameters, $H(-j\omega)$ is the complex conjugate of $H(j\omega)$. Hence $|H(-j\omega)|$ is identical to $|H(j\omega)|$, and the power gain is thus an even function of frequency.

If the input to this system has a power spectrum $S_{xx}(\omega)$ then the power spectrum $S_{yy}(\omega)$ of the output signal $y(t)$ is given by

$$S_{yy}(\omega) = S_{xx}(\omega).\ |H(j\omega)|^2 \qquad (24)$$

A more rigorous derivation of Eq. 24 is obtained by taking the Fourier transform of the autocorrelation function of the output signal. This is expressed in terms of the input spectral density and the frequency response function of the system using Eq. 20.

Example

If white noise of constant power per unit bandwidth S_o is applied to a first order system of transfer function

$$\frac{Y}{X} = H_s(s) = \frac{1}{1 + Ts}$$

then the power spectrum of the output signal is

$$S_{yy}(\omega) = S_o \left|\frac{1}{1 + j\omega T}\right|^2 = \frac{S_o}{1 + \omega^2 T^2} .$$

When the input signal to a network has a specified spectrum $S_{xx}(\omega)$, it is convenient to regard this signal as being produced by applying white noise, of unity power density, to a network of transfer function $H_i(s)$, such that

$$S_{xx}(\omega) = |H_i(j\omega)|^2.$$

If the signal x(t) is applied to a linear network of transfer function $H_s(s)$ then the power spectrum of the output y(t) is given by

$$S_{yy}(\omega) = S_{xx}(\omega).\ |H_s(j\omega)|^2$$

$$= |H_i(j\omega)|^2.|H_s(j\omega)|^2$$

$$= |H_i(j\omega).H_s(j\omega)|^2$$

The mean square value of the response is

$$\overline{y^2} = \frac{1}{2\pi}\int_{-\infty}^{\infty} |H_i(j\omega).H_s(j\omega)|^2\ d\omega$$

or, writing s for $j\omega$, ds = j $d\omega$

$$\overline{y^2} = \frac{1}{2\pi j}\int_{-j\infty}^{j\infty} |H_i(s).H_s(s)|^2\ ds.$$

Usually, $H_i(s)$ and $H_s(s)$ are ratios of polynomials in s. (Pure delay terms of the form $e^{-s\tau}$ in the numerator may be ignored). The integral may be evaluated in several ways, for example using pole-zero methods, or by using standard tables. In the latter method, the expression for $\overline{y^2}$ is written in the form

$$\overline{y^2} = \frac{1}{2\pi j}\int_{-j\infty}^{j\infty} \left|\frac{c_{n-1}\ s^{n-1} + \ldots + c_1 s + c_o}{d_n\ s^n + \ldots + d_1 s + d_o}\right|^2 ds \qquad (25)$$

$$= I_n.$$

This integral has been tabulated in terms of c_0 to c_{n-1} and d_0 to d_n for n = 1 to 7 by James, Nichols and Phillips[3], (errors have been reported in I_7), and for n = 1 to 10 in (a) Newton, Gould and Kaiser "Analytical Design of Linear Feedback Controls"[2], and (b) Siefert and Steig "Control Systems Engineering"[4] (I_{10} requires 4 pages of tabulation). A Fortran program for evaluation of the integral is available in reference 4.

The notes for this lecture conclude with a tabulation for I_1 through I_4 (Table I). It should be noted that such tables can only be used if:

(i) The denominator polynomial is of higher degree than the numerator (otherwise the value of the integral is infinite).

(ii) The transfer function $H_i(s).H_s(s)$ describes a stable system. If it is not known that $H_i(s).H_s(s)$ describes a stable system, this should be checked, e.g. by the Routh-Hurwitz criterion. If the system is unstable, the evaluation may or may not give a negative answer for $\overline{y^2}$.

Examples of the use of the Integral Tables

(i) Consider the response of a linear system to a random signal having a power spectrum

$$S_{xx}(\omega) = \frac{S_0}{1 + (\omega/\omega_0)^2}$$

where S_0 is a constant, and the linear system has a transfer function

$$\frac{Y}{X} = \frac{1}{1 + sT} \;.$$

Determine the ratio $(\overline{y^2}) \div (\overline{x^2})$ as a function of $\omega_0 T$.

The input power is

$$\overline{x^2} = \frac{S_0}{2\pi j} \int_{-j\infty}^{j\infty} \left|\frac{1}{1 + s/\omega_0}\right|^2 ds$$

$$= \frac{S_0\omega_0}{2} \quad \text{(using } I_1 \text{ in Table I)}$$

The output power is

$$\overline{y^2} = \frac{S_0}{2\pi j} \int_{-j\infty}^{j\infty} \left|\frac{1}{1 + s/\omega_0} \cdot \frac{1}{1 + sT}\right| ds$$

$$= \frac{S_0}{2\pi j} \int_{-j\infty}^{j\infty} \left|\frac{1}{1 + s(T + 1/\omega_0) + s^2 T/\omega_0}\right|^2 ds$$

Since the denominator is of second order, use I_2 with $d_2 = T/\omega_o$, $d_1 = (T + 1/\omega_o)$, $d_o = 1$, $c_1 = 0$, $c_o = 1$.

$$\overline{y^2} = \frac{S_o T/\omega_o}{(2T/\omega_o)(T + 1/\omega_o)}$$

$$= \frac{S_o\omega_o}{2(1 + T\omega_o)}$$

$$= \frac{\overline{x^2}}{1 + T\omega_o} = \frac{1}{1 + T\omega_o} \cdot \overline{x^2}$$

(ii) A signal having a power spectrum

$$S_{xx}(\omega) = \frac{S_o}{1 + (\omega/\omega_o)^2}$$

is applied to a linear feedback system having an open-loop transfer function

$$\frac{C}{E} = \frac{K}{1 + s/\omega_o}$$

as shown in Fig. 11.

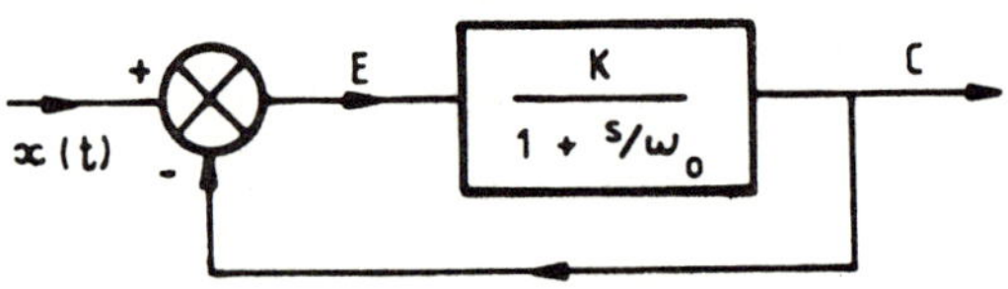

Figure 11

Find the value of K such that the mean square value of the error signal is equal to $\frac{1}{10}$ of the mean square value of the input signal.

As before,

$$\overline{x^2} = \frac{S_o\omega_o}{2}.$$

Since $\quad \dfrac{C}{E} = \dfrac{K}{1 + s/\omega_o}, \quad E = X - C$

it follows that

$$\frac{E(s)}{X(s)} = \frac{1 + s/\omega_o}{(1 + K) + s/\omega_o}$$

$$\overline{e^2} = \frac{S_o}{2\pi j}\int_{-j\infty}^{j\infty} \left| \frac{1 + s/\omega_o}{(1 + K) + s/\omega_o} \cdot \frac{1}{1 + s/\omega_o} \right|^2 ds$$

$$= \frac{S_o}{2\pi j}\int_{-j\infty}^{j\infty} \left| \frac{1 + s/\omega_o}{(1 + K) + s(\frac{2+K}{\omega_o}) + s^2/\omega_o^2} \cdot \right|^2 ds$$

Using I_2, with $c_1 = 1/\omega_o$, $c_o = 1$, $d_2 = 1/\omega_o^2$, $d_1 = (2+K)/\omega_o$, $d_o = (1+K)$,

$$\overline{e^2} = \frac{S_o\omega_o}{2(1 + K)}$$

$\therefore$ The required ratio $\frac{1}{10} = \frac{1}{1+K}$

$\therefore$ $K = 9$.

To complete this section, we note the most significant application of the cross-spectral density function $S_{xy}(j\omega)$, in the case (considered previously) where x(t) and y(t) represent respectively the input and output of a linear dynamic system.

From the definitions in Eqs. (15) and (23),

$$S_{xy}(j\omega) = \int_{-\infty}^{\infty}\int_{-\infty}^{\infty} h(u) \left[R_{xx}(\tau-u)e^{-j\omega(\tau-u)}\right] e^{-j\omega u}\, d\tau du$$

where the inner bracket has been multiplied by $e^{j\omega u}$. Changing one variable of integration from τ to $(\tau-u)$ enables us to write

$$S_{xy}(j\omega) = \int_{-\infty}^{\infty} h(u).e^{-j\omega u}du \,.\, \int_{-\infty}^{\infty} R_{xx}(\tau-u)e^{-j\omega(\tau-u)}d(\tau-u).$$

The first term is the frequency response function $H(j\omega)$ and the second is the power spectral density $S_{xx}(\omega)$. Hence

$$S_{xy}(j\omega) = H(j\omega).\, S_{xx}(\omega) \tag{26}$$

It follows that the frequency response function of a linear system of the type considered (open-loop) can be estimated by taking the ratio $S_{xy}(j\omega)/S_{xx}(\omega)$ at the frequencies of interest. It may be rather simpler to do this operation than to work in the time domain and perform the so-called 'deconvolution' of

$$R_{xy}(\tau) = \int_{-\infty}^{\infty} h(u)\ R_{xx}(\tau-u)du$$

to determine the impulse response when x(t) cannot be approximated to white noise.

We conclude by stating without proof that the frequency response of elements of a closed loop system with uncorrelated additive noise disturbances can be obtained simply. In Figure 12, if $n_1(t)$ and $n_2(t)$ are disturbances uncorrelated with x(t) then

$$H_1(j\omega) = \frac{S_{xy}(j\omega)}{S_{xe}(j\omega)}$$

and (27)

$$H_2(j\omega) = \frac{S_{xz}(j\omega)}{S_{xy}(j\omega)}$$

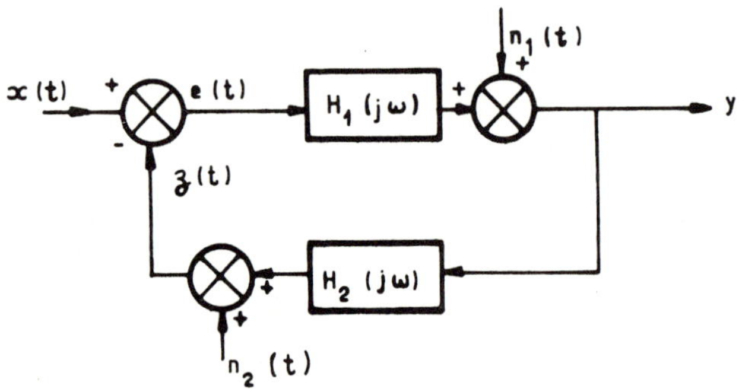

Figure 12

Note that in this system, it is a serious error to assume that $H_1(j\omega)$ is estimated by the ratio $Y(j\omega)/E(j\omega)$. (Consider $x(t) = n_2(t) = H_1(j\omega) = 0$; $n_1(t) = \sin \omega t$.)

Parameter Optimisation

An important classical optimisation technique is based on the theory developed above. We shall not consider the powerful techniques which allow us to answer the important question - 'What is the best impulse response such that some given cost function is minimised?' since this is best considered within a different theoretical framework. Rather, we consider the case in which the structure of a dynamic system is given, and we wish to choose the values of free parameters of the system to obtain the 'best' possible value of some suitable performance measure.

A common measure of performance of a control system subjected to a random input is the mean square error. Given a system with parameter values $K_1 \ldots K_M$ to be chosen by the designer, the mean square error can be evaluated in terms of the input power spectrum and these parameters.

The resulting expression may then be minimised with respect to the parameters by setting the partial derivatives to zero and solving the resulting algebraic expressions.

Example

A unity gain control system has a closed-loop transfer function

$$\frac{C}{R} = \frac{1}{s^2/\omega_o^2 + 2\zeta s/\omega_o + 1}$$

and is subjected to a random signal with power spectrum

$$S_{rr}(\omega) = \frac{S_o}{1 + (\omega/\omega_i)^2}$$

Determine the value of damping factor ζ which minimises the mean square value of the error signal $e = c - r$.

Since $$\frac{E}{R} = \frac{s^2/\omega_o^2 + 2\zeta s/\omega_o}{s^2/\omega_o^2 + 2\zeta s/\omega_o + 1}$$

$$S_{ee}(\omega) = S_o \left| \frac{s^2/\omega_o^2 + 2\zeta s/\omega_o}{s^3/\omega_i\omega_o^2 + s^2\left(\frac{1}{\omega_o^2} + \frac{2\zeta}{\omega_o\omega_i}\right) + s\left(\frac{1}{\omega_i} + \frac{2\zeta}{\omega_o}\right) + 1} \right|^2_{s = j\omega}$$

Using I_3 from the table of integrals gives

$$\overline{e^2} = S_o \frac{\omega_i}{4\zeta} \frac{(\mu + 2\zeta + 4\mu\zeta^2)}{(1 + \mu^2 + 2\mu\zeta)}$$

where $\mu = \omega_o/\omega_i$ is a normalised measure of the bandwidth of the input signal.

Differentiating the above expression with respect to ζ and setting the result to zero gives

$$\zeta = \frac{1 + \sqrt{2 + \mu^2}}{2\mu} .$$

Inspection of this result shows that for small input bandwidths ($\mu \to \infty$) the optimum damping factor is one half critical, and the damping increases monotonically with increasing input bandwidth.

This technique has been extended in a practically useful way to handle constraints on the mean square values of system variables, using Lagrange multipliers (see, for instance, Ref. (2)).

7. DISCRETE-TIME SYSTEMS

The preceding sections have considered continuous-time signals and systems. Discrete-time systems, involving one or more sampling devices, are of practical importance, and we now review the most important properties of sampled signals and the relationships between the input and output of sampled data systems with random inputs.

We may consider a stochastic process to generate an ensemble of random functions $x_i(t)$ with t assuming integer values (... -1, 0, 1, ...).

For this process, (assumed to be stationary and ergodic),

$$\mu_x = E[x(t)] = \lim_{n\to\infty} \frac{1}{2n+1} \sum_{t=-n}^{n} x(t)$$

$$R_{xx}(r) = \lim_{n\to\infty} \frac{1}{2n+1} \sum_{t=-n}^{n} x(t).x(t+r).$$

Discrete-time white noise has the property that $R_{xx}(r) = 0$ for $r \neq 0$.

The spectral density of this discrete process may be defined by

$$S_{xx}(\omega) = \sum_{n=-\infty}^{\infty} R_{xx}(n)e^{-jn\omega} .$$

Note that this is a periodic function of ω, with period $2\pi/\omega = 1$, since $S(\omega+2\pi k) = S(\omega)$ for all integer k.

The inverse relationship is

$$R_{xx}(r) = \frac{1}{2\pi} \int_{-\pi}^{\pi} S_{xx}(\omega)e^{jn\omega}\, d\omega.$$

This expression differs from the corresponding relationship for the continuous case particularly as regards the limits of integration. The values of the limits here are associated with the fact that a sampled signal, sampled at a frequency f (unity in our case) can be represented completely by frequency components in the range $-\frac{f}{2} < \frac{\omega}{2\pi} < \frac{f}{2}$.

The above relationships may be expressed in terms of the variable z defined by

$z = e^{j\omega}$ to give

$$S_{xx}(z) = \sum_{n=-\infty}^{\infty} R_{xx}(n)\, z^{-n}$$

$$R_{xx}(n) = \frac{1}{2\pi J} \oint S_{xx}(z).z^{n} \frac{dz}{z} .$$

The integral is taken around the unit circle in an anticlockwise direction.

Response of discrete-time systems

The response of a discrete-time system is specified by its impulse response at sampling instants, h(n) for n = 0 to ∞, or by its pulse transfer function

$$H(z) = \sum_{n=0}^{\infty} z^{-n}\, h(n) \text{ where } z = e^{s}.$$

For such a system, the cross-correlation function between input and output is

$$R_{xy}(r) = \lim_{n\to\infty} \frac{1}{2n+1} \sum_{t=-n}^{n} x(t)y(t+r)$$

Substituting

$$y(t) = \sum_{m=0}^{\infty} h(m)\, x(t-m) \text{ gives}$$

$$R_{xy}(r) = \sum_{m=0}^{\infty} h(m)\, R_{xx}(r-m).$$

Similarly, $R_{yy}(r) = \sum_{m=0}^{\infty} \sum_{n=0}^{\infty} h(m)\, h(n)\, R_{xx}(r-n+m).$

The power spectral density of the output is, from the definition,

$$S_{yy}(z) = \sum_{r=-\infty}^{\infty} R_{yy}(r) z^{-r}$$

$$= H(z).H(z^*)\, S_{xx}(z)$$

$$= |H(z)|^2.S_{xx}(z)$$

and $S_{xy}(z) = H(z).S_{xx}(z).$

Evaluation of the mean square error

To find the mean square value of the output of a stable time-invariant sampled system in response to a statistically stationary random signal, we integrate the output power spectral density with respect to frequency. This is equivalent to determination of the output autocorrelation function at zero time shift. Thus

$$\overline{y^2} = R_{yy}(0) = \frac{1}{2\pi j} \oint S_{yy}(z) \frac{dz}{z}.$$

Now $S_{yy}(z) = |H(z)|^2\, S_{xx}(z).$

For $z = e^{j\omega}$

$$|H(z)|^2 = H(z).H(z^{-1}).$$

The spectral density function $S_{xx}(z)$ can, for a wide range of functions, be represented by the output of a linear system subjected to discrete time white noise (see, for example, Åström, Ref. 5, p. 101). Hence the above integral reduces to

$$\sigma_y^2 = \frac{1}{2\pi j} \oint \frac{C(z).C(z^{-1})}{D(z).D(z^{-1})} \frac{dz}{z} .$$

C(z) and D(z) are polynomials in z, such that the ratio C(z)/D(z) is the overall pulse transfer function of two linear systems in cascade, representing first the modelling of the given input signal spectrum and the response of the dynamic system under consideration.

The above integral is available in tabulated form (Jacobs, Ref. 6, p. 110). The evaluation is described in detail, with a Fortran program listing in Ref. 5.

8. CONCLUDING COMMENTS

These notes have outlined the main concepts and techniques which arise in the analysis of response of single-input, single-output linear systems to stationary random excitation.

To cover the case of system nonlinearities and the use of vector-matrix models, additional developments are necessary. These are to be covered where necessary in subsequent lectures.

9. REFERENCES

1. Papoulis, A.: "Probability, Random Variables, and Stochastic Processes", McGraw-Hill, 1965.
2. Newton, G.N., Gould, L.A. and Kaiser, J.F.: "Analytical Design of Linear Feedback Controls". Wiley, 1957.
3. James, H.M., Nichols, N.B. and Philips, R.S.: "Theory of Servo-mechanisms". MIT Radiation Laboratory Series, Vol. 25, McGraw-Hill, 1947.
4. Siefert, W.W. and Steeg, C.W.: "Control Systems Engineering". McGraw-Hill, 1960.
5. Åström, K.J.: "Introduction to Stochastic Control Theory". Academic Press, 1970.
6. Jacobs, O.L.R.: "Introduction to Control Theory". Oxford U.P., 1974.
7. Jones, N.B. (ed.).: "Digital Signal Processing". Peter Perigrinus 1982.
8. Åström, K.J. and Wittenmark, B.: "Computer Control Systems-Theory and Design". Prentice Hall, 1984.

TABLE I

The integral

$$I_n = \frac{1}{2\pi j} \int_{-j\infty}^{j\infty} \left| \frac{c_{n-1} s^{n-1} + \ldots + c_1 s + c_o}{d_n s^n + \ldots + d_1 s + d_o} \right| \quad ds$$

is given by

$$I_1 = \frac{c_o^2}{2d_o d_1}$$

$$I_2 = \frac{c_1^2 d_o + c_o^2 d_2}{2d_o d_1 d_2}$$

$$I_3 = \frac{c_2^2 d_o d_1 + (c_1^2 - 2c_o c_2) d_o d_3 + c_o^2 d_2 d_3}{2d_o d_3 (d_1 d_2 - d_o d_3)}$$

$$I_4 = \frac{\left\{ c_3^2(-d_o^2 d_3 + d_o d_1 d_2) + (c_2^2 - 2c_1 c_3) d_o d_1 d_4 + (c_1^2 - 2c_o c_2) d_o d_3 d_4 + c_o^2(-d_1 d_4^2 + d_2 d_3 d_4) \right\}}{2d_o d_4 (-d_o d_3^2 - d_1^2 d_4 + d_1 d_2 d_3)}$$

TUTORIAL PROBLEMS

1.

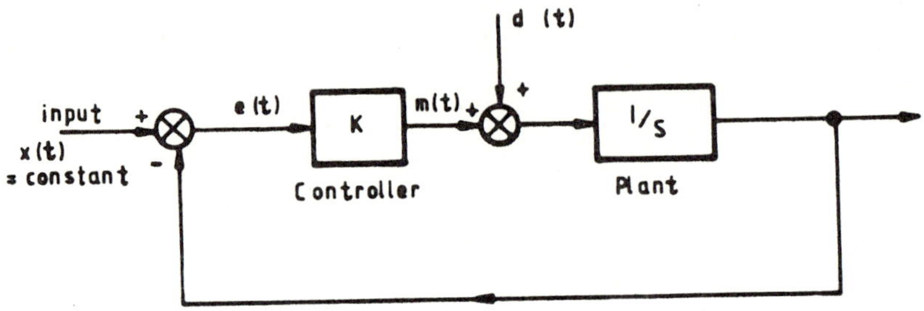

Fig. Problem 1

In the feedback control system shown, the disturbance may be assumed white noise, with power spectral density N. The gain K of the proportional controller is to be chosen such that $(\overline{e^2} + \overline{m^2})$ is a minimum.

Show that the appropriate value of K is unity, and that with this value of K the mean square error is N/2.

2. A control system has an open-loop transfer function

$$\frac{C}{E} = \frac{K}{sT(1 + sT)}$$

Show that when the input signal to the closed-loop system has a power spectrum

$$S_{rr}(\omega) = \frac{S_0}{1 + \omega^2 T^2}$$

then the mean square value of the error signal is equal to that of the input, for all positive values of K.

Why does this result not hold for negative values of K?

3. A signal with autocorrelation function

$$R_{xx}(\tau) = A(1 - |\tau|/\Delta); \quad |\tau| < \Delta$$

$$= 0; \quad |\tau| \geq \Delta$$

is applied to a first order system with impulse response

$\frac{1}{T} e^{-t/T}$ where $T >> \Delta$.

Sketch carefully the cross-correlation function between input and output for

$$- \Delta \leqq \tau \leqq \Delta \quad .$$

Consider the relevance of this result to procedures for the determination of the impulse response by correlation of the system output with a 'white noise' input.

TUTORIAL PROBLEMS - SOLUTIONS

1. $\frac{E}{D} = \frac{-1}{k+s}$ $\qquad\qquad$ $\frac{M}{D} = \frac{-k}{k+s}$

$$\overline{e^2} = \frac{N}{2\pi j} \int \left| \frac{1}{k+s} \right|^2 ds \qquad\qquad \overline{m^2} = \frac{N}{2\pi j} \int \left| \frac{k}{k+s} \right|^2 ds$$

$$= \frac{N}{2k} \text{ (using } I_1) \qquad\qquad = \frac{Nk^2}{2k} \text{ (similarly).}$$

$\overline{e^2} + \overline{m^2} = \frac{N}{2} (\frac{1}{k} + k)$. $\qquad$ This is min. ω.r.t. k when k = 1

with $\quad k = 1, \overline{e^2} = N/2$.

2. Input power $= \frac{S_o}{2\pi j} \int_{-j\infty}^{j\infty} \left|\frac{1}{1+sT}\right|^2 ds = \frac{S_o}{2T}$ (using I_1).

$$\frac{E}{R} = \frac{sT(1+sT)}{s^2T^2+sT+k}$$

$$\therefore \quad \overline{e^2} = \frac{S_o}{2\pi j} \int \left| \frac{sT(1+sT)}{s^2T^2+sT+k} \cdot \frac{1}{1+sT} \right|^2 ds = \frac{S_o}{2\pi j} \int \left| \frac{sT}{s^2T^2+sT+k} \right|^2 ds.$$

$$= S_o \left| \frac{T^2k}{2T^3k} \right| = \frac{S_o}{2T} \text{ , using } I_2.$$

Table inapplicable if $k < 0$, since closed-loop system unstable.

3. $$R_{xy}(\tau) = \int_0^\infty h(u)\ R_{xx}(\tau-u)du$$

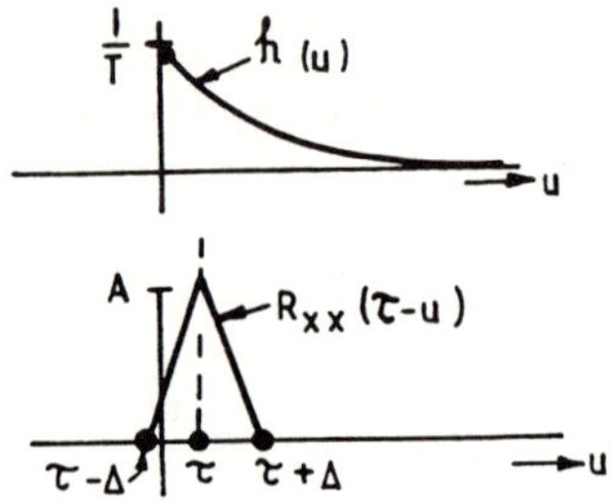

Convolution is area under product of two. Since T > > Δ, impulse response

$\approx \frac{1}{T}$ for $0 < u < 2\Delta$, so we convolve R_{xx} with

constant $\frac{1}{T}$ for $u > 0$

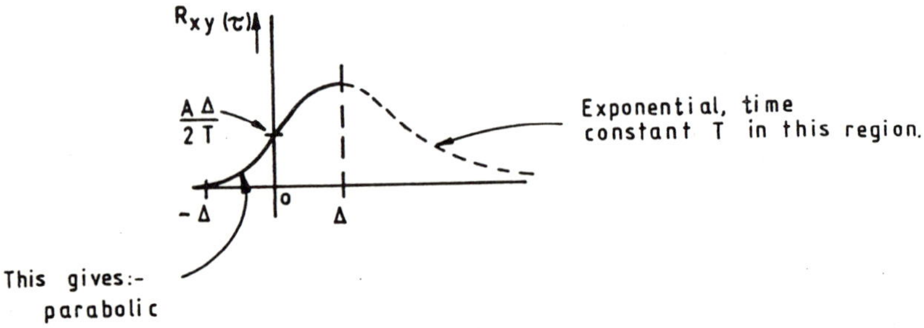

Important features are that measured R_{xy} (and estimate of h(t):

(a) non-zero for negative τ

(b) is half initial value (at $t = 0^+$) at $\tau = 0$.

Implication: bandwidth of test signal must be adequate to give required resolution in the time domain.

SIGNAL ANALYSIS II

Dr. R.P. Jones

1. INTRODUCTION

Lecture R1 outlined a number of basic signal analysis procedures that are of interest in many applications. In that lecture we saw that the quantities of interest included the signal mean value, mean square value and variance, correlation functions, power spectra, and probability density functions. When any of these quantities are estimated by experiment, only a finite amount of data is available. This has the effect of introducing errors into the estimation procedure, which are conveniently distinguished as *bias* (systematic errors) and *variance* (random scatter).

An important problem which commonly arises in the planning of experiments is that of deciding, in advance, how much data must be collected to achieve a given accuracy. The objective of an experiment should not be thwarted by large errors due to insufficient data, nor should time and effort be wasted by collecting and analysing more data than is necessary to achieve the required confidence in the result. In order to determine the quantity of data that has to be analysed in order to resolve any given signal characteristic to a given degree of accuracy, it is necessary to know at, least approximately, the relationships existing between the record characteristics and the probable errors, for particular data analysis procedures.

Such relationships have been discussed extensively by Bendat and Piersol [1,3]. In this lecture we outline the considerations that affect the question and show how a quantitative analysis leads to certain useful principles and guidelines for the design of experimental procedures involving random data.

2. BASIC ESTIMATION PROCEDURES

It is useful at this stage to note that data may be analysed using either analogue or digital techniques. The principal differences between the two methods may be summarised as follows:

(i) Analogue techniques assume continuous records, whereas digital techniques assume the data to be sampled at regular time intervals, each sample being converted into a digital representation for subsequent analysis. Decreasing the sample interval and hence increasing the number of samples tends to decrease the loss of information due to the sampling process, but it also increases the time and/or machine capacity required to perform the analysis. It can also introduce errors due to 'roundoff' effects in numerical processing. Thus, the choice of a suitable sampling interval for digital processing often requires careful consideration.

(ii) Analogue methods require special-purpose equipment for each procedure (e.g., square-law elements for measuring mean square value, time delays and multipliers for correlation analysis, tuned filters for frequency-domain analysis). Digital techniques use, in the main, general-purpose logical elements or computers, augmented by analogue-to-digital converters, with either software or hard-wired programs for analysis. It is thus more convenient to change parameters of the analytical process and to implement novel procedures digitally.

(iii) Digital apparatus is less prone to inaccurate operation due to drift etc., than most analogue equipment.

Both analogue and digital estimation techniques involve the use of finite data sets corresponding to either a finite number of discrete samples or a continuous record measured over a finite time interval.

The estimation of any quantity from a finite data sample will be inherently subject to errors which may be conveniently represented in terms of a statistical variance, reflecting limited sample duration, and a bias error, reflecting limited resolution.

3. ILLUSTRATIVE EXAMPLES

3.1 Discrete Process

Suppose we have N *independent* observed values, $x_1,\ldots,x_N$, of a random variable x and we estimate the mean and variance of the process by using

$$\hat{\mu}_x = \frac{1}{N}\sum_{i=1}^{N} x_i$$

and

$$\hat{\sigma}_x^2 = \frac{1}{N}\sum_{i=1}^{N} (x_i - \hat{\mu}_x)^2$$

We now consider each of these estimations in turn.

(a) Mean value

The expected value of the estimate is

$$E[\hat{\mu}_x] = E[\frac{1}{N}\sum_{i=1}^{N} x_i] = \frac{1}{N}\sum_{i=1}^{N} E[x_i]$$

$$= \frac{1}{N} \cdot N\,\mu_x = \mu_x.$$

and therefore the estimator is unbiased.

The variance of this estimate is

$$\mathrm{Var}[\hat{\mu}_x] = E[(\hat{\mu}_x - \mu_x)^2]$$

$$= E\left[\frac{1}{N^2} \sum_{i=1}^{N} (x_i - \mu_x)^2\right]$$

$$= \frac{\sigma_x^2}{N} \to 0 \text{ as } N \to \infty ,$$

which shows the estimator is consistent.

(b) Variance

The expected value of the estimate is

$$E[\hat{\sigma}_x^2] = \frac{1}{N} E\left[\sum_{i=1}^{N} (x_i - \hat{\mu}_x)^2\right]$$

Consider, first,

$$\sum_{i=1}^{N} (x_i - \hat{\mu}_x)^2 = \sum_{i=1}^{N} (x_i - \mu_x + \mu_x - \hat{\mu}_x)^2$$

$$= \sum_{i=1}^{N} (x_i - \mu_x)^2 - 2(\hat{\mu}_x - \mu_x)\sum_{i=1}^{N}(x_i - \mu_x) + \sum_{i=1}^{N} (\hat{\mu}_x - \mu_x)^2$$

$$= \sum_{i=1}^{N} (x_i - \mu_x)^2 - 2(\hat{\mu}_x - \mu_x).N(\hat{\mu}_x - \mu_x) + N(\hat{\mu}_x - \mu_x)^2$$

$$= \sum_{i=1}^{N} (x_i - \mu_x)^2 - N(\hat{\mu}_x - \mu_x)^2$$

Then using $E[(x_i - \mu_x)^2] = \sigma_x^2$

and, as shown above, $E(\hat{\mu}_x - \mu_x)^2 = \dfrac{\sigma_x^2}{N}$ it follows that

$$E[\hat{\sigma}_x^2] = \frac{1}{N} [N\sigma_x^2 - \sigma_x^2]$$

$$= \sigma_x^2 [1 - \frac{1}{N}].$$

Hence the estimator is biased.

In this case, an unbiased estimator may be obtained using

$$\hat{\sigma}_x^2 = \frac{1}{N-1} \sum_{i=1}^{N} (x_i - \hat{\mu}_x)^2.$$

3.2 Continuous Process

Consider the random process defined by

$$x(t) = X \cos (\omega t + \theta)$$

where X and ω are unknown constants, and θ is uniformly distributed over the range 0 to 2π. We estimate the mean square value by analysing a single time record of length T. Clearly,

$$x^2(t) = X^2 \cos^2 (\omega t + \theta)$$

$$= \frac{X^2}{2} [1 + \cos 2(\omega t + \theta)]$$

Our estimator $\hat{\sigma}_x^2$ is given by

$$\hat{\sigma}_x^2 = \frac{1}{T} \int_0^T x^2(t)dt$$

$$= \frac{X^2}{2T} \int_0^T [1 + \cos 2(\omega t + \theta)]dt$$

$$= \frac{X^2}{2} [1 + \frac{\sin 2(\omega T + \theta) - \sin 2\theta}{2\omega T}]$$

The estimate is unbiased, since $E[\hat{\sigma}_x^2] = \int_0^{2\pi} \hat{\sigma}_x^2 \cdot \frac{1}{2\pi} d\theta$

$$= \int_0^{2\pi} \frac{X^2}{2} [1 + \frac{\sin 2(\omega T + \theta) - \sin 2\theta}{2\omega T}] \cdot \frac{1}{2\pi} d\theta$$

$$= \frac{X^2}{2}$$

$$= \int_0^{2\pi} \frac{X^2}{2} [1 + \cos 2(\omega t + \theta)] \cdot \frac{1}{2\pi} d\theta$$

$$= \sigma_x^2.$$

The estimate is also consistent since the maximum error decreases to zero as $T \to \infty$.

The variance of this estimate is $\text{Var}[\hat{\sigma}_x^2] = E[(\hat{\sigma}_x^2 - \sigma_x^2)^2]$

$$= (\frac{X^2}{2})^2 \cdot \frac{1}{2\pi} \int_0^{2\pi} [\frac{\sin 2(\omega T + \theta) - \sin 2\theta}{2\omega T}]^2 d\theta$$

$$= \frac{1}{2} (\frac{X^2}{2})^2 (\frac{\sin \omega T}{\omega T})^2$$

As expected, the variance is zero when the sample length is an integral number of half-periods of the signal, i.e. $T = \frac{N\pi}{\omega}$.

4. ESTIMATION OF MEAN SQUARE VALUE

In this section, we shall derive the necessary fundamental relationships for a normal process with zero mean for the cases where the sample consists of

(i) a single spot value
(ii) N samples taken at equal time intervals
(iii) a continuous record.

The latter result will then be modified to take into account the situation where the mean value of the process is non-zero.

4.1 Single spot value

For the case of a single observation x_1, if the mean value is taken to be zero, then the estimated mean square value is

$$\hat{\sigma}_x^2 = x_1^2$$

and the true mean square value is

$$\sigma_x^2 = E[x_1^2].$$

Thus, the estimate, $\hat{\sigma}_x^2$ is unbiased and the variance is given by

$$\begin{aligned} \text{Var}\,[\hat{\sigma}_x^2] &= E[(x_1^2 - \sigma_x^2)^2] \\ &= E[x_1^4 - 2\sigma_x^2 x_1^2 + \sigma_x^4] \\ &= E[x_1^4] - \sigma_x^4 \end{aligned}$$

For a normal process with zero mean,

$$f(X) = \frac{1}{\sigma_x\sqrt{2\pi}} \exp(-X^2/2\sigma_x^2),$$

and evaluation of the integral

$$E[x_1^4] = \int_{-\infty}^{\infty} X^4 f(X)\,dX = 3\sigma_x^4$$

yields

$$\text{Var}[\hat{\sigma}_x^2] = 2\sigma_x^4$$

4.2 N Samples taken at equal time intervals

Consider N samples, x_1 to x_N, taken at equal time intervals λ. Note that it is not now assumed that the samples are independent.

The estimated mean square value is again the same as the estimated variance, since the mean is taken to be zero. Thus,

$$\hat{\sigma}_x^2 = \frac{1}{N}\sum_{i=1}^{N} x_i^2$$

The variance is $\mathrm{Var}[\hat{\sigma}_x^2] = E[(\hat{\sigma}_x^2 - \sigma_x^2)^2]$, and substituting for $\hat{\sigma}_x^2$ and collecting terms gives

$$\mathrm{Var}[\hat{\sigma}_x^2] = E[\frac{1}{N^2}(x_1^4 + x_2^4 + \ldots + x_N^4)] + E[\frac{2}{N^2}(x_1^2.x_2^2 + \ldots + x_{N-1}^2.x_N^2)]$$

$$+ \ldots + E[\frac{2}{N^2}(x_1^2.x_N^2)] - 2\sigma_x^2\, E[\frac{1}{N}(x_1^2 + \ldots + x_N^2)] + (\sigma_x^2)^2.$$

The last two terms combine to give $-\sigma_x^4$. Each term of the form $E[\Sigma\, x_i^2\, x_j^2]$ is a fourth order moment of the process, and it has been stated previously that

$$E[x_i^q.x_j^r] = \int_{-\infty}^{\infty}\int_{-\infty}^{\infty} x_i^q.x_j^r f(x_i,x_j)dx_i\, dx_j$$

For a normal process with zero mean, such that $x_i = x(t)$; $x_j = x(t+\tau)$, $f(x_i,x_j)$ can be expressed in terms of the correlation coefficient

$$\rho(\tau) = \frac{R_{xx}(\tau)}{R_{xx}(0)}$$

as $$f(x_i,x_j) = \frac{1}{2\pi\sigma_x^2(1-\rho^2)^{\frac{1}{2}}}\exp\left(-\left[\frac{x_1^2 + x_2^2 - 2\,x_1\,x_2}{2\sigma_x^2(1-\rho^2)}\right]\right)$$

and evaluation gives

$$E[x_k^2.x_{k+r}{}^2] = \sigma_x^4\,[1 + 2\rho^2(r.\lambda)].$$

Substituting

$$\mathrm{Var}[\hat{\sigma}_x^2] = \sigma_x^4\,\{\tfrac{1}{N}[1 + 2\rho^2(0)] + \frac{2(N-1)}{N^2}[1+2\rho^2(\lambda)] + \frac{2(N-2)}{N^2}[1+2\rho^2(2\lambda)]$$

$$+ \ldots \frac{2}{N^2}[1 + 2\rho^2[(N-1)\lambda]] \quad -1\}.$$

$$\mathrm{Var}\,[\hat{\sigma}_x^2] = 2\sigma_x^4\,\{\frac{1}{N} + 2\sum_{i=1}^{N-1}\frac{(N-i)}{N^2}\,\rho^2\,(i\lambda)\}.$$

4.3 Example

Show how the variance of an estimate of the mean square value varies with sample size and time interval when the process is obtained by passing white noise through a transfer function $\frac{1}{1 + sT_s}$. The mean value of the process is zero, and the mean square value is the same as the variance σ_x^2. For this process, $\rho(\tau) = e^{-|\tau|/T_s}$ and hence

$$\mathrm{Var}[\hat{\sigma}_x^2] = 2\sigma_x^4 \left\{\frac{1}{N} + \frac{2}{N^2} \sum_{i=1}^{N-1} (N - i)e^{-2i\frac{\lambda}{T_s}}\right\}.$$

$N\lambda$ is the length of the record, and $L = 2(\frac{N.\lambda}{T_s})$ is a covenient normalised length of record.

In terms of L,

$$\mathrm{Var}[\hat{\sigma}_x^2] = 2\sigma_x^4 \left\{\frac{N(1 - e^{-2L/N}) - 2(1 - e^{-L}).e^{-L/N}}{N^2(1 - e^{-L/N})^2}\right\}.$$

This expression is sketched in Figure 1.

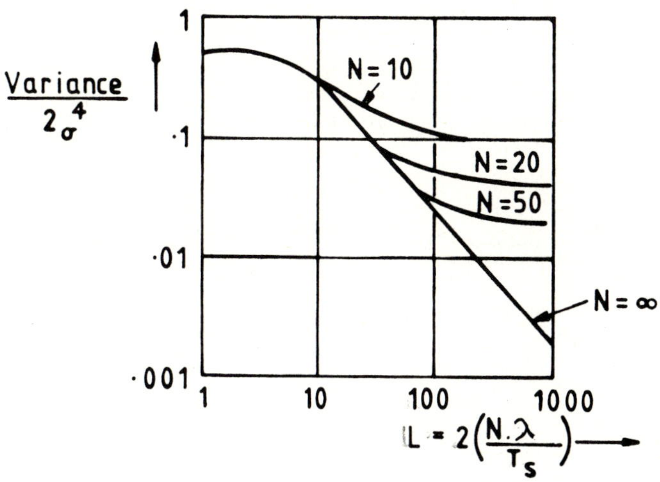

Figure 1. Variance of Estimates of Mean Square Value

Note that:

(i) For $N \to \infty$ with L fixed,

$$\mathrm{Var}[\hat{\sigma}_x^2] \to 2\sigma_x^4 \left\{\frac{N(\frac{2L}{N}) - 2(1 - e^{-L})}{N^2(\frac{L}{N})^2}\right\}$$

$$= \frac{4\sigma_x^4(L - 1 + e^{-L})}{L^2} \quad .$$

This implies that there is a lower useful limit to the sampling interval λ. From the figure it can be seen that little can be gained by decreasing λ below T_s, for a given length of record.

(ii) For long records ($L >> 1$) with $\lambda = T_s$,

$$\mathrm{Var}[\hat{\sigma}_x^2] \approx \frac{2}{N}\sigma_x^4 .$$

4.4 Continuous Record

The original expression for the variance of a sample of size N can be used to obtain results for a continuous record by letting $N \to \infty$, $\lambda \to 0$ such that $N\lambda = T$ (the observation time), and $i\lambda = \tau$.

The original expression

$$\mathrm{Var}[\hat{\sigma}_x^2] = 2\sigma_x^4 \left\{\frac{1}{N} + 2\sum_{i=1}^{N-1} \frac{(N - i)\,\rho^2(i\lambda)}{N^2}\right\}$$

becomes

$$\mathrm{Var}[\hat{\sigma}_x^2] = 4\sigma_x^4 \int_0^T \left(\frac{T - \tau}{T^2}\right)\rho^2(\tau)d\tau. = \frac{4\sigma_x^4}{T}\int_0^T \left(1 - \frac{\tau}{T}\right)\rho^2(\tau)d\tau.$$

If $\rho(\tau) \to 0$ for $\tau << T$, and $\int_0^\infty \rho^2(\tau)d\tau$ is finite, then the expression may be approximated by

$$\mathrm{Var}[\hat{\sigma}_x^2] = \frac{4\sigma_x^4}{T}\int_0^\infty \rho^2(\tau)d\tau.$$

When the process has a mean value μ_x, the modified exact and approximate expressions are

$$\mathrm{Var}[\hat{\sigma}_x^2] = \frac{4}{T}\int_0^T \left(1 - \frac{\tau}{T}\right)(C_{xx}{}^2(\tau) + 2\mu_x{}^2\, C_{xx}(\tau))d\tau$$

$$\approx \frac{4}{T}\int_0^\infty (C_{xx}{}^2(\tau) + 2\mu_x{}^2\, C_{xx}(\tau))d\tau.$$

where $C_{xx}(\tau)$ is the auto-covariance function of x.

In the following two examples we assume that the process is normally distributed with zero mean, and estimate the variance of an estimate of mean square value in terms of the power density spectrum of the process and the observation time, using the approximate expression derived above.

4.5 Example

Consider a process with power spectrum given by

$$S_{xx}(\omega) = \frac{A}{1 + (\omega/\omega_0)^2}, \text{ for constant A.}$$

Then

$$R_{xx}(\tau) = \frac{A\omega_0}{2} e^{-\omega_0|\tau|}$$

and

$$\rho(\tau) = e^{-\omega_0|\tau|}.$$

Then

$$\mathrm{Var}[\hat{\sigma}_x^2] = \frac{4(\sigma_x^2)^2}{T} \int_0^\infty e^{-2\omega_0\tau} d\tau.$$

$$= \frac{2}{\omega_0 T} (\sigma_x^2)^2$$

Thus to achieve a standard deviation not greater than 10% of the true value we require

$$\mathrm{Var}[\hat{\sigma}_x^2] < 0.01\,(\sigma_x^2)^2$$

i.e.

$$T > \frac{200}{\omega_0}$$

(For a standard deviation of 1%, $T = \frac{2 \times 10^4}{\omega_0}$).

4.6 Example: Band-limited white noise

In this case

$$S_{xx}(\omega) = A, \qquad -\omega_0 < \omega < \omega_0, \qquad A \text{ constant}$$

$$= 0 \text{ elsewhere}$$

giving

$$R_{xx}(\tau) = \frac{A}{\pi} \frac{\sin \omega_o\tau}{\tau}$$

and

$$\rho(\tau) = \frac{\sin \omega_o\tau}{\omega_o\tau} .$$

Then

$$\mathrm{Var}[\hat{\sigma}_x^2] = \frac{4(\sigma_x^2)^2}{T\omega_o} \int_0^\infty \frac{\sin^2 \omega_o\tau}{(\omega_o\tau)^2} d(\omega_o\tau)$$

and from standard integral tables,

$$\mathrm{Var}[\hat{\sigma}_x^2] = \frac{2\pi}{T\omega_o} (\sigma_x^2)^2$$

This result is often quoted in terms of the bandwidth B in cycles per second,

$$B = \frac{\omega_o}{2\pi} .$$

In this case, we have the particularly simple result

$$\frac{\mathrm{Var}[\hat{\sigma}_x^2]}{(\sigma_x^2)^2} = \frac{1}{B.T}$$

Qualitatively, the greater variance when compared with Example 4.5 is due to the absence of high-frequency components in this case.

5. SPECTRAL ANALYSIS

We now examine the variance of an estimate of the power density spectrum for the particular example of a random signal passed through a filter which has unity power gain for $\omega_o < |\omega| < \omega_o + \delta\omega$ and zero gain at other frequencies. If $\delta\omega$ is sufficiently small so that the input power spectrum may be assumed constant for $\omega_o < |\omega| < \omega_o + \delta\omega$, then the output of the filter has a power spectral density:

$$S_{xx}(\omega) = A, \quad \omega_o < |\omega| < \omega_o + \delta\omega$$

$$0 \text{ elsewhere.}$$

The autocorrelation function is

$$R_{xx}(\tau) = \frac{\sigma_x^2}{\delta\omega\cdot\tau} [\sin (\omega_o + \delta\omega)\tau - \sin \omega_o\tau]$$

$$\rho(\tau) = \frac{1}{\delta\omega\cdot\tau} [\sin (\omega_o + \delta\omega)\tau - \sin \omega_o\tau]$$

Substituting in

$$\text{Var}\,[\hat{\sigma}_x^2] = \frac{4\sigma_x^4}{T} \int_0^{\infty} \rho^2(\tau)d\tau$$

gives

$$\text{Var}[\hat{\sigma}_x^2] = \frac{2\pi}{T.\delta\omega}\,(\sigma_x^2)^2$$

or

$$\frac{\text{Var}\,[\hat{\sigma}_x^2]}{(\sigma_x^2)^2} = \frac{1}{BT}$$

as in Example 4.6.

For example, if it is required to analyse a signal with a resolution of 0.1 Hz, with a standard deviation of 10%, corresponding to a variance of 0.01 $(\sigma_x^2)^2$, then B.T = 100 and T = 1,000 seconds.

6. VARIANCE OF CORRELATION AND CROSS-SPECTRAL DENSITY FUNCTIONS

For a Gaussian signal with zero mean value it can be shown that the variance of an estimate of the cross-correlation function $R_{xy}(\tau)$ obtained by measurements over a time interval of duration T is given by

$$\text{Var}\,[\hat{R}_{xy}(\tau)] = \frac{1}{T} \int_{-\infty}^{+\infty} [R_{xx}(u).R_{yy}(u) + R_{xy}(u+\tau).R_{xy}(\tau-u)]du$$

Letting y = x gives, the variance of an estimate of the autocorrelation function as

$$\text{Var}\,[\hat{R}_{xx}(\tau)] = \frac{1}{T} \int_{-\infty}^{+\infty} [R_{xx}{}^2(u) + R_{xx}(u+\tau).R_{xx}(u-\tau)]du$$

As before, when $\tau = 0$,

$$\text{Var}\,\hat{R}_{xx}(0) = \text{Var}\,[\hat{\sigma}_x^2] = \frac{4}{T} \int_0^{\infty} R_{xx}{}^2(u)du$$

The cross-spectral density may be expressed in terms of its real and imaginary parts as

$$S_{xy}(j\omega) = \text{Re}\,[S_{xy}(j\omega)] + j\ \text{Im}[S_{xy}(j\omega)]$$

and the variance of each component is bounded by

$$\frac{S_{xx}(\omega)\,.\,S_{yy}(\omega)}{B.T}$$

where T is the observation time and B is the filter bandwidth in Hz.

Finally, the variance of an estimate of the cross-correlation function of a discrete Gaussian process with zero mean can be approximated as

$$\mathrm{Var}[\hat{R}_{xy}(r)] = \frac{1}{N-r}\,\sigma_x^2\,\sigma_y^2$$

BIBLIOGRAPHY

1. J.S. Bendat and A.G. Piersol, "Measurements and Analysis of Random Data", Wiley, 1966.
2. R.B. Blackman and J.W. Tukey, "The Measurement of Power Spectra", Dover, 1958.
3. J.S. Bendat and A.G. Piersol, "Engineering Applications of Correlation and Spectral Analysis", Wiley, 1980.

DESIGN AND IMPLEMENTATION OF DIGITAL FILTERS

Dr. J.P. Norton

1. INTRODUCTION

Analogue controllers, still the mainstay of many industrial control systems, are rapidly being superseded by digital controllers. Digital controllers may be designed either by transforming a continuous-time controller specification into digital form or by a completely discrete-time design procedure, based on a discrete-time process model and performance index. In either case the broader field of digital filtering has useful techniques to offer. Continuous-time design is appealing because it allows us to start off with familiar classical methods, but it leaves the non-trivial problem of meeting a continuous-time specification with a digital implementation. All-digital design avoids that problem, but leaves the equally important one of ensuring accurate enough implementation in the face of quantisation and restricted-precision arithmetic. Both problems have received close attention in digital filtering, but the results are less well known among control engineers than their practical importance merits.

These notes will consider digital filtering under the headings of filter structure, design methods and the implications of quantisation and rounding. We shall assume linear, constant dynamics and uniform-in-time sampling. Implementation by computer rather than special-purpose digital or discrete-time analogue hardware will be considered. References 1 to 4 deal with the material in increasing order of detail and completeness. Reference 5, from a previous SERC Vacation School, covers much the same ground but says more about hardware realisation. The terminology of digital filtering is defined in [10], with useful explanations in most cases.

2. FILTER STRUCTURE

Design methods to produce a practical digital filter from a specification such as an impulse response or frequency response differ according to whether the filter is purely moving-average (MA) or autoregressive-moving-average (ARMA).

An MA filter has a z-transform transfer function of the form

$$H(z) = h_0 + h_1 z^{-1} + h_2 z^{-2} + \ldots + h_N z^{-N} \tag{2.1}$$

corresponding to a unit-pulse response (u.p.r.)

$$\{h\} = h_0 \text{ at time } 0,\ h_1 \text{ at time } T,\ h_2 \text{ at time } 2T, \ldots h_N \text{ at time } NT \tag{2.2}$$

where T is the sampling period. Such filters are often called *finite-impulse-response* (FIR) filters as {h} has finite duration, N+1 samples. Other names are transversal filters and non-recursive filters. The input sequence {u} and output sequence {y} of an FIR filter are related by

$$y(k) = h_0u(k) + h_1u(k-1) + \dots + h_Nu(k-N) \tag{2.3}$$

where y(k) indicates the output at sample instant kT, and similarly for the input.

An ARMA filter has a rational polynomial transfer function of the form

$$H(z) \equiv \frac{B(z)}{1 + A(z)} \equiv \frac{b_0 + b_1z^{-1} + \dots + b_mz^{-m}}{1 + a_1z^{-1} + a_2z^{-2} + \dots + a_nz^{-n}}$$

$$= \frac{b_0(1-\beta_1z^{-1})(1-\beta_2z^{-1}) \dots (1-\beta_mz^{-1})}{(1-\alpha_1z^{-1})(1-\alpha_2z^{-1}) \dots (1-\alpha_nz^{-1})}$$

$$= \frac{C_1}{1-\alpha_1z^{-1}} + \frac{C_2}{1-\alpha_2z^{-1}} + \dots + \frac{C_n}{1-\alpha_nz^{-1}}$$

$$= h_0 + h_1z^{-1} + h_2z^{-2} + \dots \infty \tag{2.4}$$

As its u.p.r. is of infinite duration, it is an *infinite-impulse-response* (IIR) filter. Its input-output relation can be written as a recursion for {y}:

$$\begin{aligned} y(k) = &-a_1y(k-1) - a_2y(k-2) - \dots - a_ny(k-n) \\ &+ b_0u(k) + \dots + b_mu(k-m) \end{aligned} \tag{2.5}$$

where the terms in u on the right-hand side make up the moving-average part, and the terms in y the autoregressive part. With an eye on hardware realisation, an IIR filter can be arranged as Direct Form 1, Fig. 1(a), or Direct Form 2, Fig. 1(b).

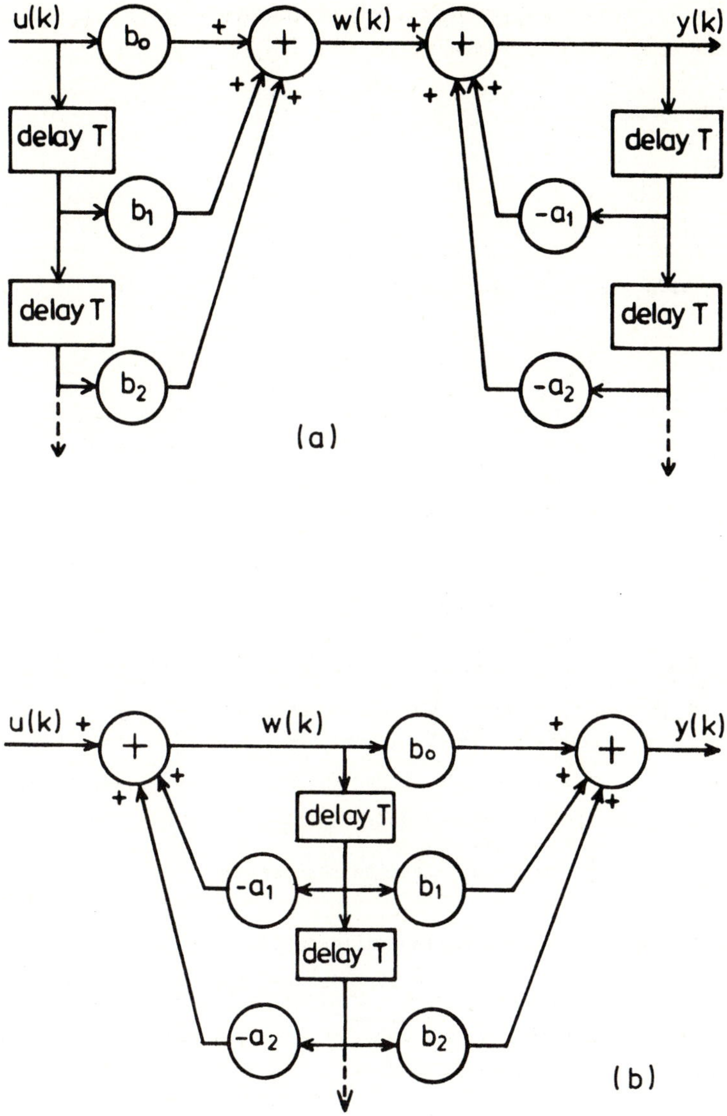

Figure 1(a) Direct Form 1 realisation of IIR filter
(b) Direct Form 2 realisation.

If the filter is realised by software, the difference is small, Direct Form 2 requiring the same operations to compute

$$w(k) = -a_1 w(k-1) - \dots - a_n w(k-n) + u(k);$$

$$y(k) = b_o w(k) + \dots + b_n w(k-m) \tag{2.6}$$

as taken by Direct Form 1 for

$$w(k) = b_o u(k) + \dots + b_m u(k-m);$$

$$y(k) = -a_1 y(k-1) - \dots -a_n y(k-n) + w(k) \tag{2.7}$$

but requiring storage of fewer values, max(m,n) per time step against m + n + 1. Direct use of (2.5) looks simpler anyway, but turns out to have high sensitivity to rounding error, as we shall see later. A better arrangement is to split H(z) into sections in cascade, Fig. 2. Here the realisation for m = n and all poles and zeros

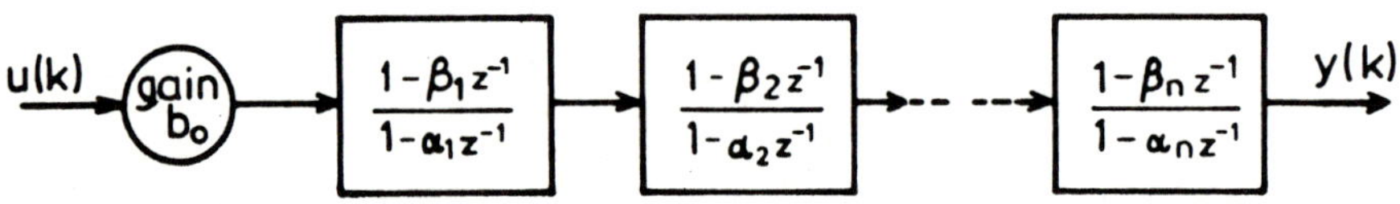

Figure 2. Cascade realisation

real is shown for simplicity. A further alternative is to put H(z) together from its partial fractions as in (2.4), in the parallel realisation form shown in Fig. 3.

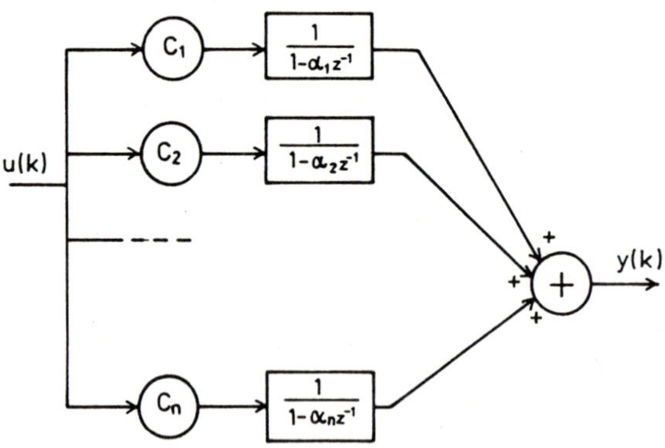

Figure 3. Parallel realisation

A Direct Form 2 realisation of a first-order section with transfer function

$$\frac{Y_i(z)}{U_i(z)} = \frac{1-\beta_i z^{-1}}{1-\alpha_i z^{-1}} \tag{2.8}$$

is shown in Fig. 4. It is the building block for the cascade and parallel realisations (with β_i zero where appropriate).

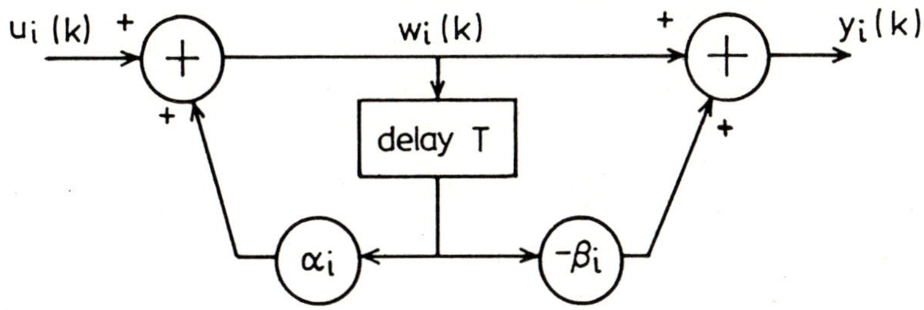

Figure 4. Direct Form 2 realisation of 1st-order section

As complex poles or zeros can only occur in complex-conjugate pairs, the only other building block we need is the second-order section

$$\frac{Y_i(z)}{U_i(z)} = \frac{(1-\beta_i z^{-1})(1-\beta_i^* z^{-1})}{(1-\alpha_i z^{-1})(1-\alpha_i^* z^{-1})} \tag{2.9}$$

(or a simpler version with α_i^* or β_i^* zero), realised without complex arithmetic as

$$w_i(k) = u_i(k) + (\alpha_i + \alpha_i^*)w_i(k-1) - \alpha_i\alpha_i^* \, w_i(k-2);$$

$$y_i(k) = w_i(k) - (\beta_i + \beta_i^*)w_i(k-1) + \beta_i\beta_i^* \, w_i(k-2) \tag{2.10}$$

An obvious advantage of splitting an IIR filter into first and second-order sections is that the effects of finite precision in any one coefficient are confined to the gain and poles or zeros of that section. If B(z) and 1 + A(z) are realised all in one piece, one coefficient affects all zeros or poles.

3. DESIGN METHODS

For FIR filters, the filter weights (u.p.r. ordinates) are calculated so as to match a frequency-response specification. The weights may be determined either by a computer algorithm to minimise, for instance, the maximum deviation from the ideal frequency response, or by analytical design as described in Section 3.2.

By contrast IIR filter design starts from an analogue prototype in the form of an impulse response or Laplace transfer function, and transforms it directly but approximately into a digital design. As the attractions of transforming a classical continuous-time controller design straight into discrete time are considerable and the procedure is quite straightforward, we examine IIR filter design first.

3.1 IIR filter design

3.1.1 Impulse-invariant design We can make the u.p.r. of the digital filter coincide exactly with the sample-instant values of the impulse response of a continuous-time design by breaking the impulse response into components with known z transforms. That is, we find the u.p.r. $\{h\}$ of the digital filter as

$$h_0 + h_1 z^{-1} + h_2 z^{-2} + \dots = Z[h(t)] = Z[\mathcal{L}^{-1}(\sum_{i=1}^{n} \frac{A_i e^{-st_i}}{s + \gamma_i})]$$

$$= \sum_{i=1}^{n} \frac{A_i z^{-d_i}}{1-\alpha_i z^{-1}} \tag{3.1}$$

where α_i is $\exp(-\gamma_i T)$ and dead time t_i is rounded up to the next multiple $d_i T$ of the sample period T. This is known as *impulse (-response)-invariant design.* Let us see how it works in an example.

Example 3.1

If the prototype transfer function is

$$H(s) = \frac{1}{(10s+1)(50s+1)}$$

and the sampling period is T, the continuous-time impulse response is

$$h(t) = \mathcal{L}^{-1}[\frac{1}{40}(\frac{1}{s+0.02} - \frac{1}{s+0.1})]$$

$$= \frac{1}{40}(\exp(-0.02t) - \exp(-0.1t))$$

so the sampled version has z transform

$$H(z) = \frac{1}{40}(\frac{1}{1-\alpha_1 z^{-1}} - \frac{1}{1-\alpha_2 z^{-1}}) = \frac{0.025(\alpha_1-\alpha_2)z^{-1}}{1-(\alpha_1+\alpha_2)z^{-1} + \alpha_1\alpha_2 z^{-2}}$$

where

$$\alpha_1 = \exp(-0.02T), \quad \alpha_2 = \exp(-0.1T).$$

The input-output relation of the filter is, from H(z),

$$y(k) = (\alpha_1+\alpha_2)y(k-1) - \alpha_1\alpha_2 y(k-2) + 0.025(\alpha_1-\alpha_2)u(k-1)$$

and the u.p.r. is easily seen to be

$h_0 = 0$

$h_1 = 0.025\ (\alpha_1 - \alpha_2)$

$h_2 = 0.025\ (\alpha_1^2 - \alpha_2^2)$

$h_3 = 0.025\ (\alpha_1^2 + \alpha_1\alpha_2 + \alpha_2^2)(\alpha_1 - \alpha_2) = 0.025\ (\alpha_1^3 - \alpha_2^3)$

$h_4 = 0.025\ (\alpha_1^2 + \alpha_1\alpha_2 + \alpha_2^2 - \alpha_1\alpha_2)(\alpha_1^2 - \alpha_2^2) = 0.025\ (\alpha_1^4 - \alpha_2^4)$

These are indeed the sample-instant values of $h(t)$. Nevertheless, the filter does not behave entirely as we would like. For instance, its d.c. gain is

$$H(z)\Big|_{z=1} = \frac{0.025(\alpha_1 - \alpha_2)}{1 - \alpha_1 - \alpha_2 + \alpha_1\alpha_2}$$

which for various T gives

T	0.1	1	2	5	10	20	33.3	50
d.c. gain	10.00	0.9998	0.4997	0.1992	0.09837	0.04692	0.02545	0.01438
T×d.c. gain	1.000	0.9998	0.9993	0.9959	0.9837	0.9384	0.8485	0.7190

The continuous-time prototype has a d.c. gain 1.

The example shows that the digital filter gain is too low by a factor of T, and there is a further discrepancy increasing with T. The reasons are not hard to find. Recall that sampling a signal gives rise to replicas of its spectrum, repeated at intervals of 1/T along the frequency axis and each scaled by 1/T. When, as in the example, we obtain a u.p.r. by sampling a unit-impulse response, the frequency response is correspondingly scaled by 1/T from the original, and replicated at frequency intervals of 1/T. The scaling is easily put right by inserting an extra gain T. The replication causes the remaining discrepancy and is less readily dealt with. At a frequency f, say, within the passband of the prototype (unsampled) filter, the sampled filter response will have superimposed on it replicas of the responses originally integer multiples of 1/T away from f. If the original passband extends beyond ± 1/2T, the response of the digital filter will be affected by this superimposition or *aliasing*, as in Fig. 5.

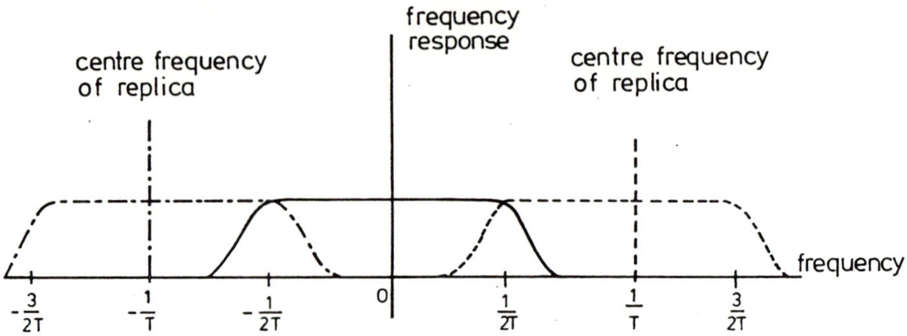

Figure 5. Aliasing affecting frequency response of impulse-invariant digital filter

---

EXAMPLE 3.2

We see in Example 3.1 that a sampling rate of 0.03 (i.e. T = 33.3), ten times the original 3dB bandwidth, still gives 15% error in d.c. gain due to aliasing. For the error to be reduced below 1%, we have to sample at 45 times the 3dB bandwidth.

The inaccuracy of the frequency response of impulse-invariant filters at reasonable sampling rates encourages us to look for an alternative.

3.1.2 Bilinear transformation One way to avoid inaccuracy due to aliasing is to apply a non-linear frequency transformation to the prototype filter specification before generating the digital filter from it, so as to squash the whole passband of the prototype into the range from -1/2T to 1/2T. There will then be no overlap of successive replicas. Whatever the original bandwidth, we are safe if we transform the range $[-\infty,\infty]$ into [-1/2T, 1/2T]. Positive frequencies should transform to positive, negative to negative and zero to zero. A simple transformation with the right properties is from angular frequency ω to

$$\omega' \triangleq \frac{2}{T}\tan^{-1}\left(\frac{\omega}{C}\right) . \tag{3.2}$$

We could make ω' equal ω at any desired frequency by choice of C, but the simplest choice is to make ω' close to ω at low frequencies. In other words, we choose C so that $d\omega'/d\omega$ tends to 1 as ω tends to zero:

$$\left.\frac{d\omega'}{d\omega}\right|_{\omega=0} = \frac{2}{T}\cdot\left.\frac{\frac{1}{C}}{1+\left(\frac{\omega}{C}\right)^2}\right|_{\omega=0} = \frac{2}{TC} = 1 \tag{3.3}$$

giving 2/T for C. The transformation is then

$$\frac{\omega' T}{2} = \tan^{-1}\left(\frac{\omega T}{2}\right) \tag{3.4}$$

which is

$$\frac{\omega T}{2} = \tan\frac{\omega' T}{2} = \frac{1}{j}\;\frac{\exp(\frac{j\omega' T}{2}) - \exp(-\frac{j\omega' T}{2})}{\exp(\frac{j\omega' T}{2}) + \exp(-\frac{j\omega' T}{2})}\;. \tag{3.5}$$

Putting s for $j\omega$ and s' for $j\omega'$ we obtain the transformation to apply to a prototype Laplace transfer function:

$$\frac{sT}{2} = \frac{\exp(\frac{s'T}{2}) - \exp(-\frac{s'T}{2})}{\exp(\frac{s'T}{2}) + \exp(-\frac{s'T}{2})} \tag{3.6}$$

so we have the *bilinear transformation*

$$s = \frac{2}{T}\cdot\frac{1-z^{-1}}{1+z^{-1}} \tag{3.7}$$

where z is exp(s'T), the z appropriate to the transformed specification.

To summarise, by putting $\frac{2}{T}\cdot\frac{1-z^{-1}}{1+z^{-1}}$ for s everywhere in the original transfer-function specification, we obtain the z-transform transfer function H'(z) of a digital filter which approximates the original H(s) well at low frequencies and suffers no aliasing, but diverges from the prototype at frequencies approaching 1/2T.

Example 3.3

The transfer function H(s) of Example 3.1 transforms to

$$H'(z) = \frac{1}{(\frac{20}{T}\cdot\frac{1-z^{-1}}{1+z^{-1}} + 1)(\frac{100}{T}\cdot\frac{1-z^{-1}}{1+z^{-1}} + 1)}$$

$$= \frac{(1 + z^{-1})^2 T^2}{(20+T + (T-20)z^{-1})(100+T + (T-100)z^{-1})}\;.$$

Setting z^{-1} to 1 we find the d.c. gain to be 1 for any T. The poles of H'(z), $z = (20-T)/(20+T)$ and $z = (100-T)/(100+T)$, are close to $1-T/10$ and $1-T/50$, and hence close to the values $\alpha_1 = \exp(-T/10)$ and $\alpha_2 = \exp(-T/50)$ obtained by impulse-invariant design in Example 3.1, so long as $T \ll 20$.

We arrived at the bilinear transformation by frequency-response considerations, but we can interpret it more broadly. First, notice that

$$s = \frac{2}{T}\cdot\frac{1-z^{-1}}{1+z^{-1}} = \frac{2}{T}\cdot\frac{z-1}{z+1} \quad (z \neq 0) \tag{3.8}$$

implying that

$$z = \frac{1 + \frac{sT}{2}}{1 - \frac{sT}{2}} \quad . \tag{3.9}$$

Hence the interior of the unit circle in the z plane, $|z| < 1$, transforms to

$$|1 + \frac{sT}{2}| < |1 - \frac{sT}{2}| \tag{3.10}$$

which is the left-hand half s plane, since (3.10) says that $sT/2$ is closer to $(-1,0)$ than to $(1,0)$. A stable original $H(s)$ therefore guarantees a stable digital $H'(z)$.

A time-domain interpretation of the bilinear transformation is also possible. We can loosely regard $1/s$ as an integration operator, which the transformation replaces by $\frac{T}{2} \cdot \frac{1+z^{-1}}{1-z^{-1}}$, an approximate integration operator. To see this, note that the trapezoidal integration formula

$$w(k) = w(k-1) + T(\frac{v(k) + v(k-1)}{2}) \tag{3.11}$$

corresponds to

$$\frac{W(z)}{V(z)} = \frac{T}{2} \cdot \frac{1+z^{-1}}{1-z^{-1}} \quad . \tag{3.12}$$

We can thus view the bilinear transformation as merely replacing each integrator in an analogue realisation of $H(s)$ by a discrete-time trapezoidal integrator.

3.2 FIR filter design

We now look at two approaches to FIR filter design which aim to meet a frequency-response specification.

3.2.1 Design by Fourier series and windowing

The idea is to find what u.p.r. weights $\{h\}$ would be needed to produce the required frequency response perfectly, then modify them to obtain a practicable design, i.e. one which is causal (non-anticipatory, with h_i zero for all negative i) and has a small enough number of filter weights.

We start by examining the frequency response given by an arbitrary u.p.r. without initially worrying whether the u.p.r. is practically realisable. The digital filter with transfer function

$$H(z) \equiv \sum_{n=-\infty}^{\infty} h_n z^{-n} \tag{3.13}$$

has the frequency response

$$H(e^{j\omega T}) = \sum_{n=-\infty}^{\infty} h_n\, e^{-jn\omega T} = \sum_{n=-\infty}^{\infty} h_n\,(\cos n\omega T - j\sin n\omega T)$$

$$= H_r + jH_i \text{ say.} \tag{3.14}$$

The real part H_r, composed wholly of cos terms with real weights h_n, is an even function of ω, and the imaginary part H_i is entirely sine terms and therefore odd. If we could choose $\{h\}$ to make H_i zero, we should obtain a frequency response with zero phase change at all frequencies, and if at the same time we could fit H_r to the desired amplitude frequency response, the design method would be perfect for many filter applications. If conversely H_r were made zero and H_i tailored, we could meet a specification requiring quadrature phase shift at all frequencies, as in an integrator or differentiator (over a finite bandwidth in practice, of course). All that is necessary for H_i to be zero is to make h_n an even function of n, causing $h_{-n}\sin(-n\omega T)$ to cancel $h_n \sin n\omega T$ at each n. Each h_n is then set to half the Fourier coefficient in the cosine series for the desired (real) $H(e^{j\omega T})$:

$$H(e^{j\omega T}) = a_o + \sum_{n=1}^{\infty} a_n \cos n\omega T = a_o + \sum_{n=1}^{\infty} \frac{a_n}{2}(z^n + z^{-n})$$

$$= \sum_{n=-\infty}^{\infty} h_n z^{-n} \tag{3.15}$$

with

$$h_o = a_o;\ h_n = \frac{a_n}{2} = h_{-n},\ n > 0. \tag{3.16}$$

Two remaining difficulties are that an infinity of u.p.r. terms h_n are required, and the u.p.r. is non-causal, starting at time $-\infty$. The solution is to truncate the series in (3.15) at some acceptable n, say N, leaving 2N+1 u.p.r. terms, then delay the u.p.r. by NT to make it start at time zero (or a little over NT to allow for computation delay). The delay multiplies $H(e^{j\omega T})$ by $e^{-j\omega NT}$ and so produces a phase lag increasing linearly with frequency. This may not be a significant drawback in some filtering applications, but might be troublesome in closed-loop control systems. Truncation of the u.p.r. is equivalent to multiplying the u.p.r. by a time-window

$$\left.\begin{aligned} w(t) &= 1, \quad -\tau \leqq t \leqq \tau \\ &= 0, \quad t < -\tau \text{ and } t > \tau \end{aligned}\right\} \quad NT < \tau < (N+1)T \tag{3.17}$$

The effect in the frequency domain is to convolve the spectrum $H(e^{j\omega T})$ with $W(j\omega)$, the Fourier transform of $w(t)$:

$$W(j\omega) = \int_{-\infty}^{\infty} w(t)e^{-j\omega t}\,dt = \left[-\frac{1}{j\omega}e^{-j\omega t}\right]_{-\tau}^{\tau} = \frac{2}{\omega}\sin\omega\tau = 2\tau\,\mathrm{sinc}\,\frac{\omega\tau}{\pi} \tag{3.18}$$

The rectangular-window spectrum is shown in Fig. 6.

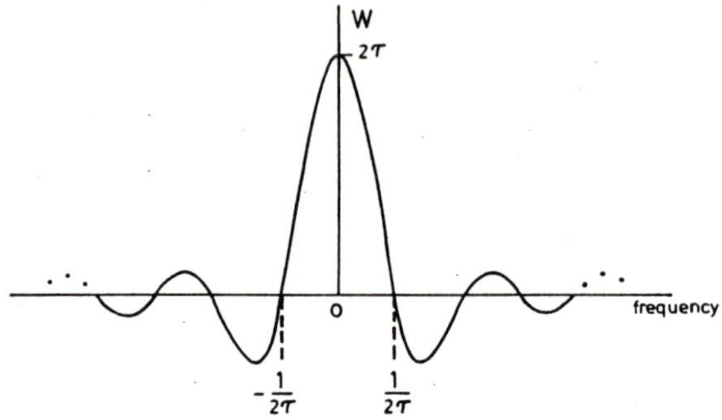

Figure 6. Frequency response of rectangular window

We see that truncation of $\{h\}$ distorts the filter frequency response, since the response at any frequency is influenced by leakage through the side lobes of $W(j\omega)$ from other frequencies, and blurred by adjacent frequencies through the finite-width main lobe. The cure is to employ not a rectangular window but a window designed to have small side lobes. The window must trade smallness of the side lobes against narrowness of the main lobe; a wider main lobe implies less sharp transition between pass and stop bands, and greater blurring of any sharp peak or notch in $H(e^{j\omega T})$.

Popular windows include

(i) generalised Hamming window with sample-instant values

$$\left.\begin{aligned} w_n &= \alpha + (1-\alpha)\cos\frac{2\pi n}{2N+1} \text{ for } -N \le n \le N \\ &= 0 \text{ for } n < -N \text{ and } n > N \end{aligned}\right\} \qquad (3.19)$$

This includes as special cases the *Hamming window* with $\alpha = 0.54$ and the *Hanning Window* with $\alpha = 0.5$. The side-lobe peak of the Hamming window is about 40dB below the main-lobe peak, against 14dB for the rectangular window;

(ii) the Kaiser window

$$w_n = \frac{I_0(\beta)\sqrt{1 - (\frac{n}{N})^2}\,)}{I_0(\beta)}, \quad -N \le n \le N. \qquad (3.20)$$

Here β determines the compromise between side-lobe height and main-lobe width and I_0 is the modified zero-order Bessel function of the first kind;

(iii) the Blackman window

$$w_n = 0.42 + 0.5 \cos \frac{2\pi n}{2N+1} + 0.08 \cos \frac{4\pi n}{2N+1}, \ -N \leq n \leq N \qquad (3.21)$$

giving side lobes more than 80dB down from the main-lobe peak.

Details are given by Ziemer, Tranter and Fannin [2] and Gold [4].

3.2.2 Computer-optimised FIR filters The design methods discussed so far are essentially pencil-and-paper techniques giving adequate, but not in any defined sense optimal, filters. A variety of computational algorithms to optimise the filter weights $\{h\}$ has been developed for FIR filters [3,4]. One possibility is to minimise the peak deviation of the actual from the ideal frequency response, i.e. the Chebyshev or L_∞ norm, over the bandwidth up to 1/2T, given the filter order and the frequencies at the edges of the pass and stop bands. The most widely recommended algorithm is that of McClellan and Parks [4]. Optimisation algorithms are applied more to FIR than to IIR filters because of the computational simplification resulting from linearity of the frequency response in the filter weights.

3.3 Choice between IIR and FIR filters

Briefly, an IIR filter can realise a long-duration u.p.r. more economically, since each pole of its rational transfer function gives rise to a u.p.r. component with an indefinitely long tail. IIR filters also have the attraction of being designed by direct transformation of continuous-time designs arrived at by familiar classical methods. On the other hand, control of the phase response is more straightforward for FIR filters and, as we have seen, linear phase-versus-frequency characteristics are readily achieved. FIR filtering can also make use of the extensive discrete-Fourier-transform technology [3, Chapter 10] embracing window selection and economical computer algorithms.

4. QUANTISATION AND ROUNDING

We have not yet taken into account the fact that a digital filter will be implemented with finite and often quite limited precision, with all signals and filter weights quantised. Before we can accept a filter design, we must be sure that quantisation and rounding will not alter its performance too much from that of the nominal design or introduce excessive "noise" or other uncertainty.

4.1 Input quantisation

If each sample of the input is quantised to the nearest integer multiple of the quantisation interval q, and if on average the input varies from one sample to the next by several times q, we can represent the error introduced by quantisation as a zero-mean random variable

$$\tilde{u}(k) \triangleq u(k) - u_q(k) \tag{4.1}$$

with the sequence $\{\tilde{u}\}$ not autocorrelated, i.e. with

$$E[\tilde{u}(k)\tilde{u}(j)] = 0, \quad k \neq j. \tag{4.2}$$

If we also take $\{\tilde{u}\}$ as uniformly distributed between $-q/2$ and $q/2$, its probability density is $1/q$ over that range and its mean-square value, equal to its variance, is

$$\sigma_{\tilde{u}}^2 = E[(\tilde{u}(k) - E\tilde{u}(k))^2] = \int_{-q/2}^{q/2} \frac{1}{q}\, \tilde{u}^2(k)\, d\tilde{u}(k)$$

$$= \frac{1}{q}\left[\frac{\tilde{u}^3(k)}{3}\right]_{-q/2}^{q/2} = \frac{q^2}{12}, \text{ all } k. \tag{4.3}$$

Let us see the effect of this input quantisation noise on the output of a digital filter with transfer function

$$H(z) \equiv h_0 + h_1 z^{-1} + h_2 z^{-2} + \ldots \infty \tag{4.4}$$

The filter is linear, so we can consider the output component $\{v\}$ due to quantisation noise separately from the rest. We have

$$v(k) = h_0\tilde{u}(k) + h_1\tilde{u}(k-1) + \ldots \infty \tag{4.5}$$

so the mean-square output noise due to $\{\tilde{u}\}$ is

$$\sigma_v^2 \triangleq E[(v(k) - Ev(k))^2] = E[v^2(k)]$$

$$= E[(h_0\tilde{u}(k) + h_1\tilde{u}(k-1) + \ldots \infty)^2]$$

$$= E[h_0^2\, \tilde{u}^2(k) + h_1^2\tilde{u}^2(k-1) + \ldots \infty]$$

$$= (h_0^2 + h_1^2 + h_2^2 + \ldots \infty)\sigma_{\tilde{u}}^2. \tag{4.6}$$

since the expected values of all products of error samples at different instants are zero, as $\{\tilde{u}\}$ is not autocorrelated. Thus we have only to square and add the u.p.r. weights to obtain the mean-square noise gain of a filter. For a FIR filter the sum is certainly finite and will be large only if the u.p.r. includes large weights and/or has a long duration. An IIR filter may easily have a u.p.r. small at all lags but a large noise-power gain. For example, a first-order IIR filter section with transfer function

$$H(z) = \frac{1}{1-\alpha z^{-1}} \equiv 1 + \alpha z^{-1} + \alpha^2 z^{-2} + \ldots \infty \tag{4.7}$$

gives

$$\sigma_v^2 = (1 + \alpha^2 + \alpha^4 + \ldots \infty)\ \sigma_{\tilde{u}}^2 = \frac{\sigma_{\tilde{u}}^2}{1-\alpha^2} \qquad (4.8)$$

which is large if α is close to the real axis and also close to the unit circle. Poles close to (1,0) are common, and we shall consider the problems they cause again later. Similarly, a second-order IIR filter section with complex-conjugate poles close to the unit circle will also have a large noise-power gain. We forego the details, which involve some complex-variable theory.

4.2 Product roundoff error

When a signal sample in the form of an M_1-bit word is multiplied by an M_2-bit filter coefficient, the $M_1 + M_2$-bit product is rounded to the subsequent word length, probably M_1 bits. Much as for input quantisation error, the roundoff error can be represented by an uncorrelated ("white") uniformly distributed noise, added to the result of each multiplication. The mean-square value of the noise is $2^{-2M}/12$ if the product is rounded to M bits (not including the sign bit). Fig. 7 shows a second-order section with these noise sources.

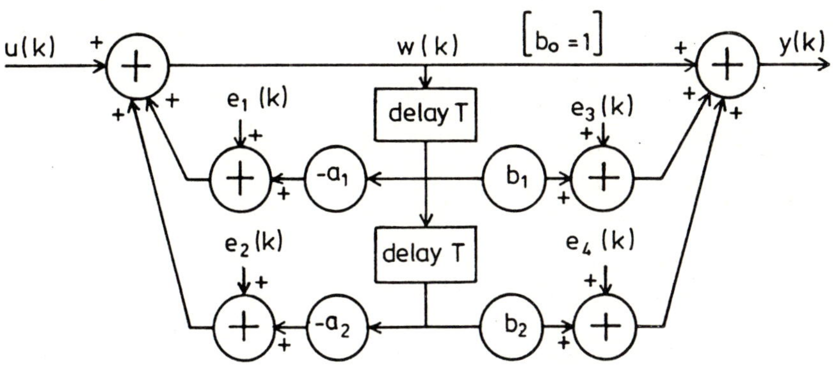

Figure 7. Product roundoff error in second-order section

To find the mean-square noise at the output of any section we must calculate the u.p.r. at the output for each noise source. If the noises from separate sources are assumed not to be cross-correlated, the total output noise due to sources 1 to m + n + 1 of an ARMA section with numerator order m and denominator order n has mean-square value

$$\sigma_v^2 = \sum_{i=1}^{m+n+1} \sigma_{ei}^2 \sum_{k=1}^{\infty} h_{ki}^2 \qquad (4.9)$$

in obvious notation, since all cross-products of noises from two sources average to zero. From Fig. 7 we see that $\{e_1\}$ and $\{e_2\}$ have the same effect as if they were additive at the input so, as discussed in Section 4.1, they may produce large contributions to the output noise if either pole is close to the unit circle. Noises $\{e_3\}$ and $\{e_4\}$ are effectively additive at the output and are unaffected by the dynamics of the section. In a FIR filter, all product noise is additive at the output, an advantage.

In cascades, our simple-minded analysis of noise-power gain can be applied confidently only to the first section or to the cascade as a whole, as it assumes that the noise entering a section is not autocorrelated; to calculate the noise-power gain section by section would involve considerable extra work to account for the autocorrelation once through the first section.

4.3 Frequency-response error due to coefficient quantisation

A further effect of finite word length is that the coefficients in the filter are not implemented exactly, with the result that the frequency response differs from nominal. We should bear in mind the possibility of employing a coefficient word length different from that of the signals, to control frequency-response accuracy independently of signal-noise ratio.

A bound on frequency-response error due to coefficient quantisation can readily be established in terms of the u.p.r; if the error between nominal h_k and quantised h_{kq} is

$$\tilde{h}_k \triangleq h_k - h_{kq} \tag{4.10}$$

then

$$|\tilde{h}_k| \leq \frac{q}{2} \tag{4.11}$$

and the error in the entire u.p.r. $\{h\}$ is given by

$$\tilde{H}(z) = \tilde{h}_0 + \tilde{h}_1 z^{-1} + \tilde{h}_2 z^{-2} + \ldots \tag{4.12}$$

which corresponds to a frequency-response error

$$\tilde{H}(e^{j\omega T}) = \tilde{h}_0 + \tilde{h}_1 e^{-j\omega T} + \tilde{h}_2 e^{-2j\omega T} + \ldots \tag{4.13}$$

This is bounded, since all the exponentials have modulus 1, by

$$|\tilde{h}|_{max} = |\tilde{h}_0| + |\tilde{h}_1| + |\tilde{h}_2| + \ldots \tag{4.14}$$

For a FIR filter, the coefficients are the u.p.r. ordinates, so it is easy to compute $|\tilde{h}|_{max}$ and adjust the coefficient word length until it is acceptable. The autoregressive coefficients a_1 to a_n of an IIR filter, on the other hand, enter nonlinearly into the u.p.r., so calculation and adjustment of $|\tilde{h}|_{max}$ is a little less easy, and it is worth examining errors in pole and zero positions instead.

4.4 Error in pole and zero positions due to coefficient quantisation

An important factor in selecting a filter structure is sensitivity of pole and zero positions to rounding error in the filter coefficients. The sensitivity problem is well illustrated by considering what error in a coefficient would cause a pole in a direct-form filter to cross from inside to outside the unit circle in the z-plane, rendering the filter unstable. Commonly the filter is lowpass and has a bandwidth small enough for all poles to be fairly close to $z = 1$. This is another way of saying that the sampling period is short compared with the shortest time constant in the u.p.r. The transfer-function denominator is then

$$1+A(z) \equiv 1 + a_1 z^{-1} + \dots + a_n z^{-n} = \prod_{i=1}^{n} (1-p_i z^{-1}) = \prod_{i=1}^{n} (1-(1-\delta_i)z^{-1}) \quad (4.15)$$

with

$$|\delta_i| << 1 \quad i = 1,2,\dots,n \quad (4.16)$$

If coefficient a_k is rounded up by $\tilde{a}_k$, the denominator is altered to

$$1 + A'(z) = 1 + A(z) + \tilde{a}_k z^{-k} \quad (4.17)$$

The effect is to shift all the poles. Let us assume that as a_k is increased by shortening the word length, one pole crosses the unit circle at $z = 1$ (as is quite likely). At that value of $\tilde{a}_k$,

$$1 + A'(1) = 1 + A(1) + \tilde{a}_k = 0 \quad (4.18)$$

so the error in a_k which will induce instability is

$$\tilde{a}_k = -1-A(1) = -\prod_{i=1}^{n} (1-p_i) = -\prod_{i=1}^{n} \delta_i \quad (4.19)$$

Example 4.1

The bilinear-transformation-designed IIR filter of Example 3.3 has the transfer function

$$H'(z) = \frac{(1+z^{-1})^2 T^2}{(20+T+(T-20)z^{-1})(100+T+(T-100)z^{-1})} .$$

The poles and $\tilde{a}_k$ given by (4.19) are, for various values of T:

T	poles		$\tilde{a}_k$ to cause instability
0.1	0.99005,	0.998002	-1.988×10^{-5}
1	0.9048,	0.9802	-1.886×10^{-3}
2	0.8182,	0.9608	-7.130×10^{-3}
5	0.6,	0.9048	-3.810×10^{-2}
10	0.3333,	0.8182	-0.1212

Even when the sampling rate is no higher than 10 per shortest time constant of the prototype H(s), i.e. T=1, a very small error in a denominator coefficient in H'(z) is enough to cause instability.

We conclude from (4.18) that the more poles are close to z=1, so that the smaller $\prod_{i=1}^{n} \delta_i$ is, the smaller is the coefficient error which will induce instability. This is a cogent reason for preferring a cascade or parallel combination of low-order filter sections to a single high-order direct-form realisation, so that the effect of quantisation in any coefficient is confined to the poles of one low-order section.

Even if stability is ensured, we must, of course, check whether the poles and zeros are realised accurately enough. For a filter with poles all distinct, a very simple analysis [3] gives the sensitivity of a pole $z = p_j$ to error in a_k:

$$\frac{\partial p_j}{\partial a_k} = \frac{\partial A(z)}{\partial a_k} \Big/ \frac{\partial A(z)}{\partial p_j} = \frac{z^{-k}}{-z^{-1} \prod_{\substack{i=1 \\ i \neq j}}^{n} (1-p_i z^{-1})} \tag{4.20}$$

so at the pole $z = p_j$,

$$\left.\frac{\partial p_j}{\partial a_k}\right|_{z=p_j} = -\frac{p_j^{1-k}}{\prod_{\substack{i=1 \\ i \neq j}}^{n} \left(1 - \frac{p_i}{p_j}\right)} \tag{4.21}$$

If one or more of the poles p_i is close to p_j, the product in (4.21) is small and the sensitivity of p_j to a_k is high. The effect is more pronounced the more poles there are close together, and again direct realisation of a high-order filter is inadvisable.

4.5 The delta operator and sensitivity

Recently Goodwin [6] has brought to the attention of the control-engineering community the numerical advantages, pointed out previously in digital filtering [7,8], of working with the *delta operator*

$$\delta \triangleq \frac{z-1}{T} \tag{4.22}$$

rather than z in filters with poles close to z = 1. The filter is described in terms of δ^{-1} rather than z^{-1}. The operation δ^{-1} is easy to implement, since

$$w_t = \delta^{-1} u_t \text{ implies } w_t = w_{t-1} + Tu_{t-1} \tag{4.23}$$

The δ^{-1} operation is seen to be Euler integration. In [6] the worst-case change in pole location due to coefficient rounding in an nth-order filter implemented in direct form is compared with that for the same filter realised in terms of δ^{-1}. Fixed-point arithmetic is assumed, with all coefficients scaled into the range from 0 to 1 with maximum rounding error ε. First derivatives as in (4.21) are taken to determine the changes, and the poles are assumed all to fall within a circle of radius $\frac{1}{2}$ centred at z=1. This last assumption is a bandwidth constraint imposed to avoid aliasing, and says roughly that the sampling frequency must be at least 10 times the highest frequency of interest. In the filter employing z^{-1}, the worst-case change, when all poles are at z=1, gives the bound

$$\tilde{p}_{max} \leq \varepsilon \sup_i \{ \prod_{k \neq i}^{n} |p_i - p_k|^{-1} \sum_{j=0}^{n-1} {}^nC_j \, |p_i|^j \} \tag{4.24}$$

Using δ^{-1} instead, the worst case is with all poles at $\delta = -0.5/T$, i.e. z = 0.5, giving the bound

$$\tilde{p}_{max} \leq \varepsilon \sup_i \{ \prod_{k \neq i}^{n} |p_i - p_k|^{-1} \sum_{j=0}^{n-1} {}^nC_j \, (0.5)^{n-j} \, |p_i - 1|^j \} \tag{4.25}$$

As $|p_i - 1| < |p_i|$ for each pole, the filter using δ^{-1} is clearly less sensitive to coefficient quantisation.

4.6 Dead band and limit-cycle oscillation

Our discussion so far has assumed that the sequence of signal-rounding errors is not autocorrelated. While this is reasonable for a signal changing by several quantisation intervals per sample, it is not true for slowly changing or constant signals, rounding of which can give rise to effects not yet considered. For instance, an IIR filter has feedback (of earlier output samples) and non-linearity in the feedback loop (rounding and perhaps overflow), so the possibility exists of limit-cycle oscillation, i.e. constant-amplitude, constant-frequency unforced oscillation. [No such problem arises in FIR filters, as they have no feedback].

Another possibility is that, for constant input, the output will "stick" at a value not equal to the ideal output but determined by initial conditions. Let us see how. A constant input $\bar{u}$ to a stable filter with nominal transfer function $B(z)/(1+A(z))$ would, after an initial transient, give a constant output $\bar{y}$ satisfying

$$\bar{y} = -a_1\bar{y} - \ldots - a_n\bar{y} + b_o\bar{u} + \ldots + b_m\bar{u}$$

$$= -A(1)\bar{y} + B(1)\bar{u} \tag{4.26}$$

so

$$\bar{y} = \frac{B(1)}{1+A(1)}\bar{u} \; . \tag{4.27}$$

Assume for simplicity that the filter coefficients are already rounded in $A(z)$ and $B(z)$ and that no overflow occurs in computing $y(k)$, and consider the situation when $y(k-n)$ to $y(k-1)$ have all been rounded to y_R. In that case

$$y(k) = \bar{y} - A(1)(y_R-\bar{y}) = y_R + (1+A(1))(\bar{y}-y_R) \tag{4.28}$$

from (2.5) and (4.26) and $y(k)$ will be rounded to y_R if

$$-\frac{q}{2} \leq (1+A(1))(\bar{y}-y_R) \leq \frac{q}{2} \; . \tag{4.29}$$

Now if, as usual, the filter has any poles close to $z=1$, $1+A(1)$ may be small compared with 1, and (4.29) indicates that the output will be rounded to y_R even if y_R is quite a few quantisation steps from the ideal steady-state output $\bar{y}$. Evidently there is a "dead band" about $\bar{y}$ in which substantial error in y can persist with a steady input. The dead-band effect is more pronounced the higher the order of the direct-form filter implementation and the narrower the bandwidth.

More generally the output may "hang up" at the edge of the dead band, cross it rapidly or oscillate in limit cycles, depending on the filter and initial conditions. Such behaviour is not usually easy to analyse, but various bounds on limit-cycle amplitude can be calculated [3]. The alternative is exhaustive simulation.

Finally, we should note one other potential source of oscillation in a naive filter implementation. The commonest fixed-point number representation is 2's-complement, with each M-bit number scaled until between -1 and $1-2^{-M+1}$. In this representation, the most significant bit is read as -1 if it is 1 and the number otherwise interpreted normally. Subtraction then only entails adding a negative number. However, at the top of the positive-number range, successive ordinary binary numbers correspond to the following numbers in 2's-complement form:

bit	M	M-1	M-2	...	2	1	
weight	-1 or 0	$\frac{1}{2}$	$\frac{1}{4}$	...	2^{-M+2}	2^{-M+1}	
so	0	1	1	...	1	1	means $1-2^{-M+1}$
	1	0	0	...	0	0	means -1
	1	0	0	...	0	1	means $-1+2^{-M+1}$

The number represented jumps downwards by the whole range if the result of an addition just exceeds the top of the range. This amounts to step forcing by the adder, and may lead to full-scale oscillations, which turn out to be self-sustaining. They can be avoided, at a price, by either altering the adder so that an out-of-range result is replaced by the top-of-range number, or scaling the operands to obviate overflow. The price paid is non-linear distortion or loss of resolution, respectively [9].

REFERENCES

[1] Gabel, R.A. and Roberts, R.A. *Signals and linear systems*. 2nd ed., 1980, Wiley, New York.

[2] Ziemer, R.E., Tranter, W.H. and Fannin, D.R. *Signals and systems*, 1983, Macmillan, New York & London.

[3] Tretter, S.A. *Introduction to discrete-time signal processing*, 1976, Wiley, New York.

[4] Rabiner, L.R. and Gold, B. *Theory and application of digital signal processing*, 1975, Prentice-Hall, Englewood Cliffs, New Jersey.

[5] Bozic, S.M. Chapters 5 to 7 of *Digital signal processing*, IEE Control Engineering Series, 22, ed. N.B. Jones, 1982, Peter Peregrinus, London.

[6] Goodwin, G.C. Some observations on robust estimation and control, 7th IFAC/IFORS Symp. on Identification & System Parameter Estimation, York, U.K. 3-7 July, 1985, 851-859.

[7] Agarwal, R.C. and Burrows, C.S. New recursive digital filter structures having very low sensitivity and roundoff noise, *IEEE Trans. Circuits & Systems*, CAS-22, 12, 1975.

[8] Orlandi, G. and Martinelli, G. Low sensitivity recursive digital filters obtained via delay replacement, *IEEE Trans. Circuits and Systems*, CAS-31, 7, 1984.

[9] Oppenheim, A.V. and Schafer, R.W. *Digital signal processing*, 1975, Prentice-Hall, Englewood Cliffs, New Jersey.

[10] Rabiner, L.R. and 8 others. Terminology in digital signal processing, *IEEE Trans. Audio Electroacoust.*, AU-20, 1972, 322-337.

PARAMETER ESTIMATION

Dr. M.T.G.Hughes

1. INTRODUCTION

The main aim of this lecture is to outline the statistical basis for available methods by which the characteristics of unknown dynamic systems may be estimated by analysis of records of the input and output signals.

The basic assumptions of the problem are as follows. We have a dynamic system with characteristics that are more or less unknown, and we wish to deduce something about the unknown characteristics by analysing measured values of the accessible signals $\{x_t\}$ and $\{y_t\}$. According to the circumstances of the particular experiment, the observable input signal $\{x_t\}$ might be a specially generated test signal, or it might be one component of the total normal operating input. The observable output signal $\{y_t\}$ is generally assumed to include extraneous components arising from unmeasurable inputs and errors of measurement. We assume that the aggregate of all these effects may be represented by a single "random" disturbance signal $\{\varepsilon_t\}$ acting at a single point in the system, as shown in Fig. 1.

2. GAUSSIAN PROCESS MODELS

The estimation of models for a large class of physical systems can be approached generally as indicated schematically in Fig. 1.

An unknown system is subjected to a set of measurable inputs (which may be represented by a vector, say, x), and to an unmeasurable set of inputs (which may be represented as an unknown random vector ε). Knowledge about the vector ε is generally confined to information related to its probability structure. The combined effect of inputs x and ε is to produce a measurable output, which may be denoted as the vector y in Fig. 1.

The system is assumed to be characterised by a set of equations of known form, so that complete characterisation requires merely evaluation of the unknown parameter vector θ. This is not always true of physical situations of course, but no demonstrably better approach to modelling has yet been found.

On the basis of the foregoing assumptions, it is further assumed (or arranged) that the equations relating vectors x, y, and ε, can be 'inverted' in the model of Fig. 1, so that a set of 'residuals', denoted as vector $\hat{\varepsilon}$, can be calculated using the given quantities x, y, together with an *estimate* $\hat{\theta}$ of the unknown parameter vector θ. According to all these assumptions, in the unlikely event that the parameter

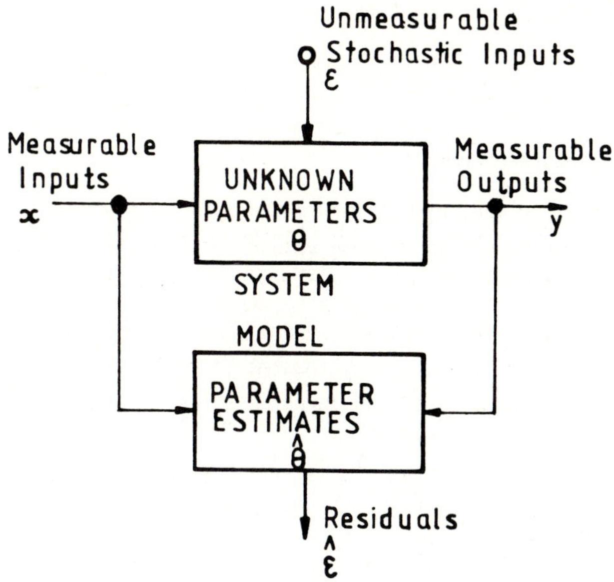

Figure 1. Process Modelling Schematic

estimate $\hat{\theta}$ coincided exactly with the true parameter vector θ, the resulting vector of residuals $\hat{\varepsilon}$ would coincide with the stochastic vector ε.

Subject to further assumptions concerning the probability structure of ε, it is possible to devise powerful procedures for the estimation of the system parameters.

The most commonly used procedures are based on an assumption that ε can be modelled as a set of N independent, normally distributed quantities, each having zero mean value and a single constant variance σ^2. If the possible values of $\{\varepsilon_i\}$ are denoted as $\{X_i\}$, the p.d.f. will have the form

$$f(X) = \left(\frac{1}{\sigma\sqrt{2\pi}}\right)^N \exp\left(\frac{-1}{2\sigma^2} \sum_{i=1}^{N} X_i^2\right) \tag{1}$$

Setting $X = \hat{\varepsilon}$ to obtain the Likelihood function, and then taking natural logarithms, we obtain the log-likelihood function $\mathcal{L}(\theta)$:

$$\mathcal{L}(\theta) = -N\log\sigma - \frac{N}{2}\log(2\pi) - \frac{1}{2\sigma^2} \sum_{i=1}^{N} \hat{\varepsilon}_i^2 \tag{2}$$

In this expression, the unknown parameters θ are supposed to occur in some way in the computation of the residuals $\hat{\varepsilon}$. The MLE of θ are chosen so as to maximise the likelihood function $L(\theta)$. However, since the logarithmic relationship is monotonic, the same value of $\theta = \hat{\theta}$ will be reached if we maximise $\mathcal{L}(\theta)$. This is usually more convenient than working directly with $L(\theta)$.

By examination of Eq. (2), it can be seen that maximisation of $\mathcal{L}(\theta)$ and thus of $L(\theta)$ can be achieved, for given values of N and σ, by choosing $\theta = \hat{\theta}$ so as to *minimise* the quantity:

$$S = \sum_{i=1}^{N} \hat{\varepsilon}_i^2 \tag{3}$$

This highlights an important result:

when the random errors or disturbances are independent and normally distributed, the MLE of system parameters are simply those values which minimise the sum of squares of residuals.

In a case where the disturbances were normal but correlated, the MLE would be obtained by minimising a *weighted* sum of squares of residuals, in which the weighting factors corresponded to elements of the inverted covariance matrix of the disturbances[2,3].

Referring to Eq. (2) we see that if the variance of ε (viz. σ^2) is unknown (as usually would be the case in practice), we could obtain a MLE of it as follows:

$$\left(\frac{\partial \mathcal{L}}{\partial \sigma}\right)_{\sigma=\hat{\sigma}} = -\frac{N}{\hat{\sigma}} + \frac{1}{\hat{\sigma}^3} \sum_{i=1}^{N} \hat{\varepsilon}_i^2 = 0$$

Thus, $$\hat{\sigma}^2 = \frac{1}{N} \sum_{i=1}^{N} \hat{\varepsilon}_i^2 \, . \tag{4}$$

Although this quantity, as a MLE, will be *asymptotically* unbiased, it may contain significant bias in the case where a large number (p) of parameters is estimated from a sample of size (N) which may not be very much larger than p. In such a case, it is more common to use an *unbiased* estimate of σ^2, which is given by

$$\hat{\sigma}^2_{(\text{unbiased})} = \frac{1}{N-p} \sum_{i=1}^{N} \hat{\varepsilon}_i^2 \tag{5}$$

Example 1 : Linear Regression

Consider the problem of estimating the *pulse-response sequence* of the system model shown in Fig. 2.

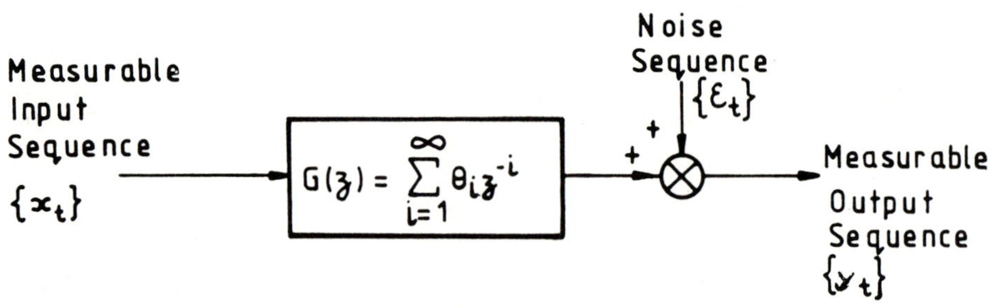

Figure 2. Model for Example 1

Suppose we have available a set of measured input samples $\{x_t\}$, and a set of measured output samples $\{y_t\}$, and we wish to analyse these in order to estimate the set of unknown parameters $\{\theta_i\}$ for i = 1,2,..., etc. From the superposition property of linear systems, it is known that

$$y_t = \sum_{i=1}^{\infty} \theta_i x_{t-i} + \varepsilon_t \tag{6}$$

for t = 1,2,..., etc.

Now it is clearly impossible to estimate the infinite set of unknown parameters in Eq. (6); and in any case, it is known that for a stable system, the magnitues of terms θ_i for $i > p$ will be negligibly small, where p is a number defining the effective 'memory' of the system. Thus, changing the upper limit of the summation to p, and expanding Eq. (6) for t = 1,2,...,N, we get

$$\left.\begin{array}{l} y_1 = x_0\theta_1 + x_{-1}\theta_2 + \ldots + x_{1-p}\theta_p + \varepsilon_1 \\ y_2 = x_1\theta_1 + x_0\theta_2 + \ldots + x_{2-p}\theta_p + \varepsilon_2 \\ \ldots\ldots\ldots\ldots\ldots\ldots \\ \ldots\ldots\ldots\ldots\ldots\ldots \\ y_N = x_{N-1}\theta_1 + x_{N-2}\theta_2 + \ldots + x_{N-p}\theta_p + \varepsilon_N \end{array}\right\} \tag{7}$$

Using vector-matrix notation, this can be written as

$$y = X\theta + \varepsilon \tag{8}$$

which is the general form of model for linear regression problems. For a given set of parameter estimates, the residual errors are given by

$$\hat{\varepsilon} = y - X\hat{\theta} \tag{9}$$

and we have seen that if the elements of ε are independent and normal, the MLE $\hat{\theta}$ will be that value which minimises the sum of squaresof residuals:

$$S = \sum_{i=1}^{N} \hat{\varepsilon}_i^2 = \hat{\varepsilon}^T\hat{\varepsilon} = (y - X\hat{\theta})^T(y - X\hat{\theta}) \tag{10}$$

It was shown in the example at the end of Revision Lecture R3 that the value of $\hat{\theta}$ which minimises S is given by

$$\hat{\theta} = (X^TX)^{-1}X^Ty \tag{11}$$

This is the linear least-squares estimate of θ; and in the case of normal, independent (white) noise, it is also a MLE.

From the structure of the matrices X and y, it is readily verified that the matrix X^TX is composed of terms which are representative of the autocorrelation function of the sequence $\{x_t\}$, while the elements of the vector X^Ty represent statistical cross-correlations between the sequences $\{x_t\}$ and $\{y_t\}$:

$$\left.\begin{aligned} [X^TX]_{i,j} &= [\sum_t x_t \cdot x_{t+j-i}] \\ [X^Ty]_i &= [\sum_t x_t \cdot y_{t-i}] \end{aligned}\right\} \tag{12}$$

It is sometimes possible to structure the input sequence $\{x_t\}$ in such a way that the matrix X^TX is approximately diagonal:

$$\left.\begin{aligned} \sum_t x_t \cdot x_{t+j-i} &= N.\sigma_x^2 \quad \text{for } i = j \\ &= 0 \quad\quad \text{for } i \neq j \end{aligned}\right\} \tag{13}$$

When this can be done, the solution of Eq. (11) is made very straightforward, and we find that

$$\hat{\theta}_r = \frac{\hat{R}_{xy}(r)}{\hat{R}_{xx}(0)} \tag{14}$$

where

$$\hat{R}_{xy}(r) = \frac{1}{N} \sum_{t=1}^{N} y_t \cdot x_{t-r} \tag{15}$$

$$\hat{R}_{xx}(0) = \frac{1}{N} \sum_{t=1}^{N} x_t^2 \tag{16}$$

This is the basis of the well-known method of *correlation analysis* for the determination of impulse response.

In addition to (or as an alternative to) the use of specially tailored input signals, it is often helpful to employ Fourier transform techniques, to avoid the necessity for inversion of large-dimensional matrices in the solution of the normal equations. Such techniques involve the important concepts of *spectral analysis* which are covered in greater detail in Lecture L8.

Example 2

Consider the linear model

$$y_t = \beta_1 x_t + \beta_2 x_{t-1} + \varepsilon_t.$$

(i) Develop explicit expressions for the least-squares estimates $\hat{\beta}_1$ and $\hat{\beta}_2$ based on N observations of x_t and y_t (x_1 to x_N and y_1 to y_N).

(ii) Apply these expressions, in the following cases, both with $\{\varepsilon_t\} = 0$:–

Case A

$\{x_t\} = \{+1, +1, +1, -1, -1, -1, +1, -1, -1, +1, -1\}$

$\{y_t\} = \{+2, +3, +3, -1, -3, -3, +1, -1, -3, +1, -1\}$

Case B

$\{x_t\} = \{+1, -1, +1, -1, +1, -1, +1, -1, +1, -1, +1\}$

$\{y_t\} = \{+2, -1, +1, -1, +1, -1, +1, -1, +1, -1, +1\}$

In both cases, $\{x_t\}$ is a "one-shot" sequence with $x_t = 0$ except for $1 \le t \le 11$.

(iii) Now add a noise sequence

$\{\varepsilon_t\} = \{-1, +1, -1, -1, +1, -1, -1, -1, +1, +1, +1\}$

to the observations and again obtain least squares estimates for β_1 and β_2.

Solution

(i) Writing the model equations in the form

$$y = X\beta + \varepsilon$$

then from the standard least-squares theory above, $\varepsilon^T\varepsilon$ is minimised if $\hat{\beta} = (X^TX)^{-1}X^Ty$.

(ii) In both cases, the sequence $\{y_t\}$ has been generated from $\{x_t\}$ with $\beta_1 = 2$ and $\beta_2 = 1$. With $\{\varepsilon_t\} = 0$, the equations in β_1 and β_2 may be written as

$$\begin{bmatrix} y_1 \\ y_2 \\ \vdots \\ y_{11} \end{bmatrix} = \begin{bmatrix} x_1 & x_0 \\ x_2 & x_1 \\ \vdots & \vdots \\ x_{11} & x_{10} \end{bmatrix} \begin{bmatrix} \beta_1 \\ \beta_2 \end{bmatrix}$$

Thus, for *Case A*,

$$X^TX = \begin{bmatrix} 1 & 1 & 1 & -1 & -1 & -1 & 1 & -1 & -1 & 1 & -1 \\ 0 & 1 & 1 & 1 & -1 & -1 & -1 & 1 & -1 & -1 & 1 \end{bmatrix} \begin{bmatrix} 1 & 0 \\ 1 & 1 \\ 1 & 1 \\ -1 & 1 \\ -1 & -1 \\ -1 & -1 \\ 1 & -1 \\ -1 & 1 \\ -1 & -1 \\ 1 & -1 \\ -1 & 1 \end{bmatrix}$$

$$= \begin{bmatrix} 11 & 0 \\ 0 & 10 \end{bmatrix}$$

X^Ty is similarly $\begin{bmatrix} 22 \\ 10 \end{bmatrix}$

so that $\hat{\beta} = \begin{bmatrix} 11 & 0 \\ 0 & 10 \end{bmatrix}^{-1} \begin{bmatrix} 22 \\ 10 \end{bmatrix} = \begin{bmatrix} \frac{1}{11} & 0 \\ 0 & \frac{1}{10} \end{bmatrix} \begin{bmatrix} 22 \\ 10 \end{bmatrix} = \begin{bmatrix} 2 \\ 1 \end{bmatrix}$

Thus $\hat{\beta} = \beta$ in this noise-free case. The input sequence $\{x_t\}$ is well designed for the estimation of β, with the off-diagonal elements of $(X^TX)^{-1}$ being zero.

For *Case B*,

$$X^TX = \begin{bmatrix} 1 & -1 & 1 & -1 & 1 & -1 & 1 & -1 & 1 & -1 & 1 \\ 0 & 1 & -1 & 1 & -1 & 1 & -1 & 1 & -1 & 1 & -1 \end{bmatrix} \begin{bmatrix} 1 & 0 \\ -1 & 1 \\ 1 & -1 \\ -1 & 1 \\ 1 & -1 \\ -1 & 1 \\ 1 & -1 \\ -1 & 1 \\ 1 & -1 \\ -1 & 1 \\ 1 & -1 \end{bmatrix}$$

$$= \begin{bmatrix} 11 & -10 \\ -10 & 10 \end{bmatrix}$$

$$(X^TX)^{-1} = \begin{bmatrix} 1.0 & 1.0 \\ 1.0 & 1.1 \end{bmatrix}$$

$X^Ty = \begin{bmatrix} 12 \\ -10 \end{bmatrix}$ and $\hat{\beta} = \begin{bmatrix} 1.0 & 1.0 \\ 1.0 & 1.1 \end{bmatrix} \begin{bmatrix} 12 \\ -10 \end{bmatrix} = \begin{bmatrix} 2 \\ 1 \end{bmatrix}$

In this case, we have been able to obtain correct estimates for β_1 and β_2 but X^TX

is poorly conditioned and we could expect considerable inaccuracies in our parameter estimates when noise is present. In other cases, it is quite possible for X^TX to be singular; a simple example where this is so is for a cyclic sequence $\{x_t\}$ of even length of alternating +1's and -1's.

(iii) For *Case A*, with the given noise sequence, the output sequence $\{y_t\}$ is now

$$\{+1, +4, +2, -2, -2, -4, 0, -2, -2, +2, 0\}$$

and $X^Ty = \begin{bmatrix} 21 \\ 8 \end{bmatrix}$

$$\therefore \quad \hat{\beta} = \begin{bmatrix} \frac{1}{11} & 0 \\ 0 & \frac{1}{10} \end{bmatrix} \begin{bmatrix} 21 \\ 8 \end{bmatrix} = \begin{bmatrix} 1.91 \\ 0.80 \end{bmatrix}$$

For *Case B*, the output sequence is now

$$\{+1, 0, 0, -2, +2, -2, 0, -2, +2, 0, +2\}$$

and $X^Ty = \begin{bmatrix} 13 \\ -12 \end{bmatrix}$

$$\therefore \quad \hat{\beta} = \begin{bmatrix} 1.0 & 1.0 \\ 1.0 & 1.1 \end{bmatrix} \begin{bmatrix} 13 \\ -12 \end{bmatrix} = \begin{bmatrix} 1 \\ -0.2 \end{bmatrix}$$

Note the much larger error in $\hat{\beta}$ in Case B, as we had expected from the poor conditioning of X^TX.

With the noise present,

$$\begin{aligned} \hat{\beta} &= (X^TX)^{-1}X^Ty \\ &= (X^TX)^{-1}X^T(X\beta+\varepsilon) \\ &= \beta + (X^TX)^{-1}X^T\varepsilon \end{aligned}$$

so that the errors in our estimates are given by

$$\delta\beta = (X^TX)^{-1}\,X^T\varepsilon.$$

For Case A, $X^T\varepsilon = \begin{bmatrix} -1 \\ -2 \end{bmatrix}$

and $\delta\beta = \begin{bmatrix} \frac{1}{11} & 0 \\ 0 & \frac{1}{10} \end{bmatrix} \begin{bmatrix} -1 \\ -2 \end{bmatrix} = \begin{bmatrix} -\frac{1}{11} \\ -\frac{2}{10} \end{bmatrix}$

as expected. Similarly for Case B,

$$\delta\beta = \begin{bmatrix} -1 \\ -1.2 \end{bmatrix}$$

Statistical Assessment of $\hat{\theta}$.

$$\hat{\theta} = \theta + (X^T X)^{-1} X^T \varepsilon$$

$$\therefore \quad b(\hat{\theta}) = E[\hat{\theta}-\theta] = E[(X^T X)^{-1} X^T \varepsilon]$$

Thus $\hat{\theta}$ is unbiased if there is no correlation between the elements of X and ε.

The covariance matrix, assuming $b(\hat{\theta}) = 0$, is given by

$$\begin{aligned} \text{Cov}(\hat{\theta}) &= E[(\hat{\theta}-\theta)\,(\hat{\theta}-\theta)^T] \\ &= E[(X^T X)^{-1} \varepsilon\varepsilon^T X (X^T X)^{-1}] \\ &= (X^T X)^{-1} X^T V X (X^T X)^{-1} \end{aligned}$$

where $V = E[\varepsilon\varepsilon^T] = \text{cov}(\varepsilon)$.

If the elements of ε are uncorrelated with standard deviation σ,

$$V = \sigma^2 I$$

and then $\text{cov}(\hat{\theta}) = \sigma^2 (X^T X)^{-1}$.

In general, σ^2 is not known exactly and is estimated from

$$\hat{\sigma}^2 = \frac{1}{N-p} \sum_{i=1}^{N} \hat{\varepsilon}_i^2$$

where p is the number of parameters to be estimated in the model. Then $\text{cov}[\hat{\theta}] \simeq \hat{\sigma}^2 (X^T X)^{-1}$ is a useful approximation to the covariance matrix of $\hat{\theta}$.

In the rather artificial example given above, we knew what ε is so we do not have to use the estimated value for σ^2. In both cases $\sigma^2 = 1$ so that for Case A

$$\mathrm{cov}(\hat{\beta}) = \begin{bmatrix} \frac{1}{11} & 0 \\ 0 & \frac{1}{10} \end{bmatrix}$$

Thus the standard deviation of $\hat{\beta}_1$ is 0.3015 while that of $\hat{\beta}_2$ is 0.3162. There is no crosscovariance between $\hat{\beta}_1$ and $\hat{\beta}_2$ so that a known high (low) estimate for β_1 would tell us nothing about the estimate for β_2 and vice-versa.

For Case B,

$$\mathrm{cov}(\hat{\beta}) = \begin{bmatrix} 1.0 & 1.0 \\ 1.0 & 1.1 \end{bmatrix}$$

so that the standard deviation of $\hat{\beta}_1$ is 1 while that of $\hat{\beta}_2$ is 1.049. There is now a substantial positive crosscovariance between $\hat{\beta}_1$ and $\hat{\beta}_2$ so that a known high (low) estimate for β_1 would indicate a similarly high (low) estimate for β_2 and vice-versa.

The values of $\delta\beta$ found in part (iii) of the Example show that the covariance matrix at least predicts the order of magnitude of actual variations.

3. RECURSIVE LEAST SQUARES

In some applications, it is required to update parameter estimates on receipt of each new piece of information. In such cases, the matrix inversion involved in the direct use of Eq. (11) is to be avoided wherever possible.

One well-known method which avoids the direct matrix inversion results when we consider the effect of adding an additional equation to the basic matrix equation (8):

$$\begin{bmatrix} y \\ \hline y_{N+1} \end{bmatrix} = \begin{bmatrix} X \\ \hline x^T \end{bmatrix} \begin{bmatrix} \theta \end{bmatrix} + \begin{bmatrix} \varepsilon \\ \hline \varepsilon_{N+1} \end{bmatrix} \tag{17}$$

Here, the scalar y_{N+1} is the additional observation, x^T is the additional row of known quantities appended to the matrix X, and ε_{N+1} is the additional measurement error.

Denoting as $\hat{\theta}_N$ the 'best' estimate of θ based on the original N equations, the new updated estimate $\hat{\theta}_{N+1}$ is given by

$$\hat{\theta}_{N+1} = \left(\begin{bmatrix} X^T \,\vdots\, x \end{bmatrix} \begin{bmatrix} X \\ \hline x^T \end{bmatrix} \right)^{-1} [X^T \,\vdots\, x] \begin{bmatrix} y \\ \hline y_{N+1} \end{bmatrix} \tag{18}$$

Expanding this expression, we obtain

$$\hat{\theta}_{N+1} = (X^T X + xx^T)^{-1} (X^T y + y_{N+1} x) \tag{19}$$

Now, let us define

$$X^T X = P_N^{-1} \tag{20}$$

(It can be shown that this matrix is closely related to the covariance matrix of the estimate $\hat{\theta}_N$.)

Using Eq. (20), the first bracket on the RHS of Eq. (19) can be written as

$$(X^T X + xx^T)^{-1} = P_N (I + xx^T P_N)^{-1}$$

Expanding the RHS using the Binomial Series, we get

$$P_N (I + xx^T P_N)^{-1} = P_N (I - xx^T P_N + (xx^T P_N)^2 - (xx^T P_N)^3 + \ldots)$$

$$= P_N \{ I - x[1 - x^T P_N x + (x^T P_N x)^2 - (x^T P_N x)^3 + (x^T P_N x)^4 - \ldots] x^T P_N \}$$

The quantity within the square brackets above will be recognised as the series for the *scalar* quantity $(1 + x^T P_N x)^{-1}$, so we can replace Eq. (19) by

$$\hat{\theta}_{N+1} = P_N \left(I - \frac{xx^T P_N}{1 + x^T P_N x} \right) (X^T y + x y_{N+1}) \tag{21}$$

On multiplying out and simplifying, this expression leads to the final *recursive least-squares algorithm:*

$$\hat{\theta}_{N+1} = \hat{\theta}_N + \frac{P_N x}{1 + x^T P_N x} (y_{N+1} - x^T \hat{\theta}_N) \tag{22}$$

with P_N determined by the recursion

$$P_{N+1} = P_N \left(I - \frac{xx^T P_N}{1 + x^T P_N x} \right) \tag{23}$$

As noted earlier, the matrix P_N is related to the covariance of the parameter estimates. The relevant equations are:

$$\hat{\theta} - \theta = (X^TX)^{-1} X^T\varepsilon \tag{24}$$

$$Cov(\hat{\theta}) = E[(\hat{\theta} - \theta)(\hat{\theta} - \theta)^T] = E[(X^TX)^{-1} X^T\varepsilon\varepsilon^TX(X^TX)^{-1}] \tag{25}$$

If ε is a 'white-noise' vector, with variance σ_e^2,

$$E[\varepsilon\varepsilon^T] = \sigma_e^2 I \tag{26}$$

So that Eq. (25) becomes

$$Cov(\hat{\theta}) = \sigma_e^2(X^TX)^{-1} = \sigma_e^2 P_N \tag{27}$$

4. NONLINEAR REGRESSION

The foregoing techniques of linear regression work well in some circumstances, but in others they can lead to seriously biased estimates. This is particularly evident in cases where it is required to estimate the values of coefficients in both the numerator and denominator of a rational pulse transfer function. The main problem in such cases arises from spurious correlations among the elements of the X matrix and the noise vector. These can be avoided only at the expense of introducing nonlinear elements into the regression equations. This is probably best illustrated by means of a simple example.

Example 2 : Nonlinear Regression

We now consider the problem of estimating values of the parameters α and β in the model shown in Fig. 3, given records of the sampled input and output for $t = 1,2,\ldots,N$.

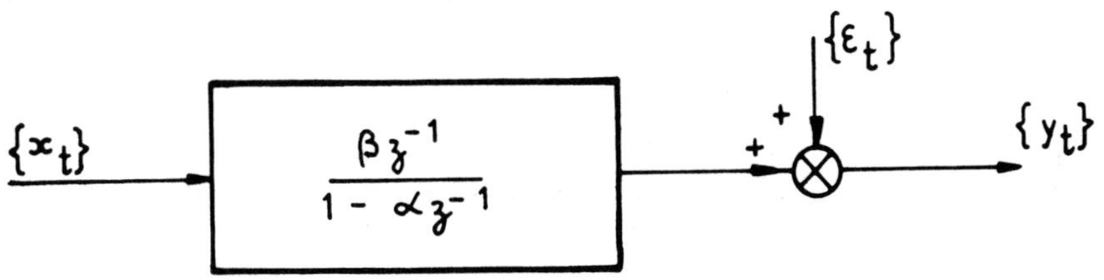

Figure 3. Model for Example 2

For this system we have, operationally,

$$y_t = \frac{\beta z^{-1}}{1 - \alpha z^{-1}} x_t + \varepsilon_t.$$

It is assumed that $\{\varepsilon_t\}$ is a sequence of independent, normal random errors with zero mean and unknown variance σ_ε^2.

Solving the system equations for ε_t, we have operationally

$$(1 - \alpha z^{-1})\ \varepsilon_t = (1 - \alpha z^{-1})y_t - \beta z^{-1}x_t,$$

or

$$\varepsilon_t = \alpha\ \varepsilon_{t-1} + y_t - \alpha y_{t-1} - \beta x_{t-1}$$

for t = 1,2,...,N.

Noting that evaluation of ε_t requires knowledge of the values of α, β, and the initial condition ε_0, we define as our vector of parameters to be estimated:

$$\theta = [\alpha,\beta,\varepsilon_0]^T$$

and in the absence of exact information, employ the estimate

$$\hat{\theta} = [\hat{\alpha},\hat{\beta},\hat{\varepsilon}_0]^T$$

to obtain a sequence of residual errors

$$\hat{\varepsilon}_t = \hat{\alpha}\hat{\varepsilon}_{t-1} + y_t - \hat{\alpha}y_{t-1} - \hat{\beta}x_{t-1}$$

for t = 1,2,...,N.

To find the MLE of θ we must find values of $\hat{\alpha}$, $\hat{\beta}$, $\hat{\varepsilon}_0$ which make the quantity $\hat{\varepsilon}^T\ \hat{\varepsilon}$ a minimum. Since in the example considered here, the relationship between the observations and the estimated parameters is nonlinear, we must either employ some method of trial-and-error search, such as Newton's method, conjugate gradient search, Simplex search, etc. or we must resort to a reformulation of the model or some form of approximation by a *linear* estimation procedure.

Such considerations tend to depend on particular applications, thus the student is referred to the extensive literature on the subject for further information.

5. REFERENCES

1. Kendall, M.G. and Stewart, A.: "The advanced Theory of Statistics", Vol. 2. Griffin, 1973.
2. Cramer, H.: "Mathematical Methods of Statistics", Princeton , 1946.
3. Deutsch, R.: "Estimation Theory", Prentice-Hall, 1965.
4. Eykhoff, P.: "System Identification - Parameter and State Estimation", Wiley 1974.
5. Sage, A.P. and Melsa, J.L.: "Systems Identification", Academic Press, 1971.

RECURSIVE METHODS IN IDENTIFICATION

Dr. K. Warwick

1. Introduction

The objective of System Identification is to find values for the parameters within a model which has a similar structure to the system it is wished to control, such that the model responds as nearly as possible like the real system for a set of input signals. For off-line Identification a batch of data, usually input-output, is collected from the system and this is subsequently employed within a previously specified algorithm in order to arrive at the desired model.

In many practical situations though, the system characteristics can vary with respect to time. If the resultant system parameter variations are small or only vary rather slowly over a long period of time, the controller operating on the system need only be retuned occasionally to ensure adequate performance. Indeed this is a standard procedure carried out on many industrial PID controllers. However, the variations can be quite large in magnitude and occur rapidly and unpredictably. In these cases, in order to continue controlling the system sufficiently well, it is often desirable and/or necessary to modify the system model and controller in line with the changes that have taken place in the system, thus providing an adaptive control scheme. This periodic updating of the system model is called Recursive Identification and is also useful in adaptive filtering and signal processing applications.

Recursive Identification algorithms can however be used on batch data as an alternative to the more conventional off-line techniques, generally though several passes through the collected data are required in order to achieve a suitable model. A big advantage of the recursive methods over off-line procedures is that as an up-to-date model of the system is required only the relatively recently acquired data is deemed to be important, hence only a small amount of memory is necessary such that this data can be temporarily stored.

When Recursive Identification is employed on time invariant systems the accuracy of the model obtained is rarely as good as that from an off-line calculation, although for time variant systems the converse is true. A standard part of off-line Identification is the formulation of the most applicable model order. Unfortunately incorporating a recursive form of model order testing is usually computationally inefficient and hence for Recursive Identification it is standard practice to select one model order and to retain this for all time. Therefore, if the system structure changes with respect to time, a more complicated procedure is generally required in order to cope with this eventuality.

This chapter is set out as follows. In section 2 recursive forms of some off-line identification techniques are discussed, in particular the method of Recursive Least Squares is considered in terms of possible simplifications and approximations. In the following three sections the recursive approaches of Stochastic Approximation, Bayesian Estimation and Model Reference schemes are covered and this is followed in section 6 by an overview of the problems encountered with time varying systems and how recursive methods can be modified to cope with them more precisely. Finally, a brief look at convergence analysis and numerical stability properties is taken in section 7.

Basic Definitions

The system to be identified is, in general, assumed to be of the form

$$A(z^{-1})y(t) = B(z^{-1})u(t) + \bar{e}(t) \tag{1.1}$$

where $\{y(t)\}$ is the sequence of system output signals, $\{u(t)\}$ is the sequence of system input signals and $\bar{e}(t)$ is a disturbance affecting the system. Also the delay operator polynomials are defined as:

$$\begin{aligned} A(z^{-1}) &= 1 + a_1 z^{-1} + \ldots\ldots\ldots + a_n z^{-n} \\ B(z^{-1}) &= b_1 z^{-1} + b_2 z^{-2} + \ldots\ldots + b_n z^{-n} \end{aligned} \tag{1.2}$$

in which the delay operator is such that $z^{-i}\ y(t) = y(t - i)$. However, if the parameters are written in the vector form

$$\theta^T = (a_1, \ldots\ldots, a_n : b_1, \ldots\ldots, b_n) \tag{1.3}$$

whilst the regressors are written in the form of

$$x^T(t) = (-y(t-1), \ldots, -y(t-n);\ u(t-1), \ldots, u(t-n)) \tag{1.4}$$

then the system equation (1.1) can be rewritten as:

$$y(t) = \theta^T x(t) + \bar{e}(t) \tag{1.5}$$

Finally, for an ARMAX, or CARMA model, the disturbance is considered to be a coloured noise sequence with zero mean value, such that

$$\bar{e}(t) = C(z^{-1})\ e(t) \tag{1.6}$$

where $C(z^{-1})$ is the monic colouring polynomial of a similar form to $A(z^{-1})$ and $\{e(t)\}$ is a white noise sequence. Hence for $\{\bar{e}(t)\}$ to be a white noise sequence it is simply required that $C(z^{-1}) = 1$, such that $\bar{e}(t) = e(t)$.

2. Recursive Forms of Standard Off-Line Techniques

System Identification can be considered as being the process of modelling systems mathematically whilst giving due consideration to the problems encountered. Problems are found in, for example, filtering noisy data to make it directly useful, the choice of model structure, the statistical analysis of the processed data, and testing of the finalised model in order to ascertain how closely it represents the actual system it is wished to identify.

Many methods are available for System Identification, all however are based on plant information which has already been obtained. This usually involves applying statistical tests to a set of plant input-output data in order to make an estimation of the model order and subsequently the values of the parameters within a model of that particular order. For off-line identification procedures, also termed batch processes, a defined finite amount of plant data is first obtained and it is from this set quantity of information that the resulting system estimations are made.

In the previous chapter, off-line identification techniques were considered, in particular the common least squares and maximum-likelihood approaches were discussed. The plethora of available methods have been covered elsewhere in many texts, those by Eykhoff (3), Sage and Melsa (4), Astrom and Eykhoff (6) and Goodwin and Payne (2), being particularly relevant. Although it is not intended in this chapter to go into a great depth with regard to any of these off-line techniques, some of the most commonly encountered Recursive Identification methods are simply modifications of off-line cases. Indeed, the vast majority of Self-Tuning controllers, Harris and Billings (12), are based on recursive parameter estimation schemes which are merely recursive forms of the least squares or maximum-likelihood procedures already discussed.

The first algorithm for Recursive Identification, to be considered here, is that of Recursive Least Squares (RLS) which as well as remaining a popular choice because of its relatively low computational requirements, has retained a simple methodology in which the function of the statistical test applied is straightforward.

Recursive Least Squares

In this section a form of the Least Squares parameter estimation procedure is considered such that the parameters in a model of the plant are recursively updated. A basic requirement for the recursive algorithm is the storage of only a finite amount of data obtained from the system. Once the data is of a certain age therefore, it is automatically discarded. Hence, at any one time instant a window of past and present input-output data is taken into account. This is indeed true of all the recursive forms of off-line identification techniques. As will be seen shortly, the window can take on one of a number of forms dependant on the relative importance of past and present plant information, this is particularly relevant when time-varying systems are encountered.

To consider further the Recursive form of Least Squares(RLS) parameter estimation, the system output at time instant t is assumed to be related to the plant parameters and the set of past data values by the equation

$$y(t) = \theta^T x(t) + e(t) \tag{2.1}$$

in which the parameter vector $\theta^T = (\theta_1, \theta_2, \ldots, \theta_m)$ and the data vector $x^T(t) = (x_1(t), x_2(t), \ldots, x_m(t))$. It is therefore the object of the parameter estimation exercise to calculate estimates of the plant parameters. If the vector of these estimates, obtained at time instant t, is denoted by $\hat{\theta}(t)$, at that instant the model prediction error is

$$\varepsilon(t) = y(t) - \hat{\theta}^T(t-1)x(t) \tag{2.2}$$

Where the estimated parameter vector $\hat{\theta}$ is very close to its true value θ, the error $\varepsilon(t)$ should be very small. A better estimate vector can be obtained though, if the magnitude of the difference between the actual output value, y(t), and its predicted value $\hat{\theta}(t-1)x(t)$, is taken into account. A new version of $\hat{\theta}$ can therefore be found from

$$\hat{\theta}(t) = \hat{\theta}(t-1) + K(t)\varepsilon(t) \tag{2.3}$$

in which K(t) can hopefully be chosen to improve the estimated values. Clearly the choice of K(t) is an important aspect of the updating procedure, a very large set of values within K(t) leads to large changes in the parameter estimates, even though the error value, $\varepsilon(t)$, may be quite small indicating that the estimated parameters are almost correct. The error between the estimated and true parameter values is then defined as:

$$\lambda(t) = \theta - \hat{\theta}(t) \tag{2.4}$$

such that

$$\lambda(t) = \left[I - K(t)x^T(t) \right] \lambda(t-1) - K(t)e(t) \tag{2.5}$$

where $\lambda(t)$ is a column vector. For rapid updating of the parameter estimates, K(t) must contain elements of high magnitude. It can be seen from (2.5) however, that this also leads to large pertubations in the estimates because of the disturbance term e(t). If K(t), the Kalman gain, is chosen to consist of elements with small magnitude though, this results in a higher degree of noise rejection but a much slower response as far as estimate updating is concerned. In practice the gain K(t) is found in terms of the error covariance matrix, as a time varying gain.

Let the covariance matrix formed by the parameter estimates at time instant t, be P(t), where this is the inverse of R(t):

$$P^{-1}(t) = R(t) = \mathcal{E}\left\{x(t),\ x^T(t)\right\} \tag{2.6}$$

in which $\mathcal{E}\{\cdot\}$ signifies the expected value and observation weightings have not been included. It follows that by choosing the Kalman gain vector as

$$K(t) = \frac{P(t-1)x(t)}{\beta + x^T(t)P(t-1)x(t)} \tag{2.7}$$

which means that also

$$P(t) = \left[I - K(t)x^T(t)\right] P(t-1) \qquad (2.8)$$

then the vector of parameter estimates, $\hat{\theta}(t)$, is the recursively calculated least squares estimate, Ljung and Soderstrom (8).

Further, the value β is a scalar gain factor providing weightings on the observations taken, and where the variance of the disturbance affecting the system is known, β can be either made equal to , or for optimal design, be otherwise related to this value.

The Recursive Least Squares (RLS) parameter estimator then consists of three equations (2.3), (2.7) and (2.8) which must all be re-calculated each time the system input and output are sampled.

With adaptive controllers it is often the case that β is employed as a variable forgetting factor which includes exponential weightings such that new data is considered relatively important whilst its importance declines exponentially with age. However the algorithm may 'blow up' when the data is not persistently exciting enough, the value β is therefore set to be variable with respect to time by linking it with the prediction error such that it can be woken up when necessary. This is discussed further in section 6.

Under normal controlled operating conditions the sequence $\{u(t)\}$ will be provided by means of some calculated equation resulting from the desired control action. Under these conditions if $\{e(t)\}$ is a white noise sequence with zero-mean as described in section 1, then it can be expected that the sequence of parameter vector estimates $\{\hat{\theta}(t)\}$ will converge to the actual vector value θ, as time t tends to infinity. A more detailed discussion of the asymptotic properties of the Least Squares estimator can be found in Astrom and Eykhoff (6), although they are considered further in section 7. It must be noted, however, that the equations used for updating $K(t)$ and $P(t)$ may result in divergence of the estimator because of numerical problems alone. It is therefore best to use a numerically stable updating method such as square root extraction or UD factorisation in order to obtain an efficient algorithm.

One problem encountered when attempting to operate an RLS algorithm is that of giving the parameter vector estimates and the covariance matrix some initial values which do not hinder convergence or cause any other undesired features. The effect that a particular set of initial values has on $\hat{\theta}(t)$ and $P(t)$ decreases with time such that as the parameter estimates approach their true values, so the coefficients in the covariance matrix $P(t)$ tend to zero. A common choice is therefore, $\hat{\theta}(0) = 0$ and $P(0) = \bar{P}.I$, where $\bar{P}$ is a large positive finite number, e.g. 1000. Care must be taken when setting all initial parameter estimates to zero in that a poorly programmed RLS algorithm may not operate with this particular set. The problem can be simply avoided by setting all the $\hat{\theta}(0)$ values to a small finite number (e.g. 0.001). Finally, the effect of setting $P(0)$ to $\bar{P}.I.$ is to show a great deal of uncertainty in the parameter estimates; in the first few time steps some of the estimates may well move around considerably until the coefficients of P(t) reduce drastically, although it is usually the case that the elements of P(t) become very small after only a handful of time steps.

A Simplified Least Squares Estimator

Where an RLS parameter estimation scheme is employed within the framework of a closed loop adaptive control system such as a Self-Tuning Controller it is found that in many cases the time taken to compute one recursion is greater than the time taken for all other necessary calculations, these to include any control law equations. This situation is particularly important when the system under control is of high order, hence requiring that a large number of parameters must be estimated. Then, if computing power is limited, or a high sampling frequency is necessary, certain types of control action may not be implementable because of the time taken to obtain the parameter estimates. A sensible approach is therefore to reduce the computing time taken for parameter estimation, when this is possible, on the provision that no loss of convergence rate or parameter accuracy results.

In the algorithm suggested by Farsi, Karam and Warwick (5) the covariance matrix is assumed to be diagonal for all time. This links the matrix with its initialisation values which are, for the standard RLS algorithm, specified such that P(t) is diagonal and of fairly high magnitude. Under normal operating conditions, as the parameters converge so the diagonal terms reduce considerably whilst the off-diagonal terms remain of extremely small magnitude. The approximate RLS algorithm therefore merely assumes that the diagonal terms retain enough information.

The reduction in computational effort found with the simplified RLS algorithm is considerable. Equations (2.3), (2.7) and (2.8) still need to be calculated at each iteration, and no reduction is apparent in (2.3). However, consider the calculation of P(t-1)x(t) in (2.7), where P(t-1) is an M x M matrix and x(t) is an M x 1 vector. For the standard RLS algorithm M^2 multiplications and $M^2 - M$ additions must be made in order to find the result; if M = 6, for example, then 36 multiplications and 30 additions are necessary. For the simplified RLS algorithm however, with P(t-1) diagonal, only M multiplications are necessary and no additions, so for the example M = 6 only 6 multiplications must take place.

Incorporation of this parameter estimation algorithm into a Self-Tuning Controller, Warwick, Farsi and Karam (7) has shown that the simplified form works just as well as the standard RLS estimator with a considerable saving in computational effort being made.

The Method of Instrumental Variables

The Instrumental Variables technique is a method by which problems due to correlated disturbances, leading to biased estimates in the RLS case, can be overcome. That is, whilst e(t) and x(t) are uncorrelated it is expected that the series of RLS parameter estimates will converge to their true set of values. When $e(t)$ and $x(t)$ are correlated however, the expectation is that the series of RLS parameter estimates will converge to a set of values which is not the true set. The Instrumental variables method is intended to produce a series of estimates which converge to the true set, whether or not the two sequences $e(t)$ and $x(t)$ are correlated.

Defining the instrumental vector as $v(t)$, this is considered to have the property that it is uncorrelated with the zero mean disturbance, $e(t)$. Also, define the pseudo output $\bar{y}(t)$ as:

$$\bar{y}(t) = \hat{\theta}^T(t)\ v(t) \tag{2.9}$$

where $$v^T(t) = (-\bar{y}(t-1),\ldots\ldots,-\bar{y}(t-n);\ u(t-1),\ldots,u(t-n)) \tag{2.10}$$

in which $\{u(t)\}$ is the sequence of system control input signals.

The value $\bar{y}(t)$ is therefore the output obtained from a deterministic model of the stochastic plant based on the estimates obtained of the actual plant parameters. The use of $\hat{\theta}(t)$ in (2.9) is just one particular choice though, as an operator selected set of parameters constant with respect to time would also suffice, Young (1).

The main advantage of the vector $v(t)$ is that despite being uncorrelated with $e(t)$, a requirement, it is also loosely correlated with $x(t)$ such that updated parameter estimates can be obtained by means of (2.3) with however,

$$K(t) = \frac{P(t-1)v(t)}{\beta + x^T(t)P(t-1)v(t)} \tag{2.11}$$

and $P(t)$ related to $K(t)$ and $P(t-1)$ by the expression (2.8).

The complete set of parameter updating equations then becomes (2.3), (2.8) and (2.11) for the case of Instrumental Variables.

3. Stochastic Approximation

The prediction error was introduced in the previous section as

$$\varepsilon(t) = y(t) - \hat{\theta}^T(t-1)x(t) \tag{3.1}$$

in which $y(t)$ is the actual system output whilst the remainder of the right hand side is the predicted output value dependant on estimated plant parameters and past data. The choice of $\hat{\theta}$ at any time instant must be based on some selection criterion, for instance the requirement could be to minimize the variance of the prediction error

$$\hat{\theta} = \min_{\theta} \tfrac{1}{2}\mathcal{E}\{\varepsilon(t)\}^2 \tag{3.2}$$

A solution is found by differentiating the right hand side of (3.2) with respect to θ, which gives $\hat{\theta}$ as the vector which causes equality in (3.3)

$$\mathcal{E}\{Q\} = \mathcal{E}\{x(t)\ \varepsilon(t)\} = 0 \tag{3.3}$$

Unfortunately, since not enough information is available, the expectations taken in (3.2) and (3.3) cannot be calculated implicitly and hence a direct solution cannot be evaluated.

However, at time instant t, both x(t) and $\varepsilon(t)$ are available and therefore an exact value for Q can be obtained. At that time instant then, a particular parameter set is selected, $\hat{\theta}(t)$, and this results in a corresponding value Q at that same time. With the general intention of obtaining a solution to (3.3) by means of choosing the correct set $\hat{\theta}$, so the parameter estimates can be updated at each instant by an amount dependant on how far Q is from zero. If Q(t) is the value of Q at time instant t, an updating procedure is:

$$\hat{\theta}(t) = \hat{\theta}(t-1) + \gamma(t)Q(t) \tag{3.4}$$

in which the gain $\{\gamma(t)\}$ is a sequence of positive scalars, such that $\{\gamma(t)\} \to 0$ as $t \to \infty$. The sequence of estimates $\{\hat{\theta}(t)\}$ will then result in $\{Q(t)\}$ converging to a solution for (3.3) providing certain conditions hold, Ljung and Soderstrom (8).

The updating equation (3.4) can though be rewritten in the form of (2.3), i.e.

$$\hat{\theta}(t) = \hat{\theta}(t-1) + K(t)\,\varepsilon(t) \tag{3.5}$$

where now, it follows that

$$K(t) = \gamma(t)x(t) \tag{3.6}$$

thus showing that this method, on condition that a fairly simple value for $\gamma(t)$ is selected, can result in much less of a computational burden than the standard RLS algorithm.

A straightforward choice of a scalar value for $\gamma(t)$ can be made; i.e. $\gamma(t)$ = constant; although in practice a normalised form is useful where, for example

$$\gamma(t)^{-1} = q(t) = q(t-1) + |x(t)|^2 \tag{3.7}$$

Stochastic Newton Algorithm

The problem to be solved is that of finding a solution to (3.3) in terms of stationary points. The method used is in fact directly analogous to steepest descent functions used for numerical minimization. In general this type of minimization procedure is considered inefficient, especially when the approximates within $\hat{\theta}(t)$ are almost equal to their true values. A better approach is perhaps to use Newton methods in an attempt to achieve improved performance.

For a Newton method the second derivative, or Hessian, of our original cost function, including the variance of the prediction error, is required. In fact the Hessian of the function

$$\tfrac{1}{2}\,\mathcal{E}\{\varepsilon(t)\}^2 \tag{3.8}$$

is

$$\mathcal{E}\{x(t)\ x^T(t)\} \tag{3.9}$$

which is independent of the parameter vector, θ, and therefore can be solved by R in (3.10).

$$\mathcal{E}\{x(t)\ x^T(t) - R\} = 0 \tag{3.10}$$

Or, in terms of a similar argument to that given for (3.4), R can be found recursively from:

$$R(t) = R(t-1) + \gamma(t)\ [x(t)x^T(t) - R(t-1)] \tag{3.11}$$

such that at time instant t, R(t) is an estimate of the Hessian (3.9), that is the second derivative of (3.8).

The Newton updating method can be summarized as:

$$\hat{\theta}(t) = \hat{\theta}(t-1) - \left.\frac{\text{value of first derivative of (3.8)}}{\text{value of Hessian of (3.8)}}\right|\ \theta = \hat{\theta}(t-1) \tag{3.12}$$

which results in an efficient algorithm in the neighbourhood of a solution, but can be inefficient or even divergent when a solution is a long distance away. Hence the Hessian in (3.12) is in general approximated by means of a matrix with assured positive definite conditions such that $\{\hat{\theta}(t)\}$ will always head in the correct direction.

Modifying the previous parameter estimate updating equation to include the benefits of the second derivative, a complete stochastic Newton updating algorithm can be described by:

$$\hat{\theta}(t) = \hat{\theta}(t-1) + \gamma(t)\ R^{-1}(t)\ x(t)\,\varepsilon(t) \tag{3.13}$$

which is of the same form as (3.5) except that the Kalman gain has now become:

$$K(t) = \gamma(t)\ R^{-1}(t)\ x(t) \tag{3.14}$$

where R(t) is updated using (3.11)

4. Model Reference Approach

The main idea behind the Model Reference Approach to recursive identification is to compare the actual system output with the output from a model of the system. The estimated parameters contained within the model are then updated in an attempt to reduce the difference between the two outputs to zero.

If the model output, at time instant t, is defined to be $\bar{y}(t)$, then this is assumed to be generated by means of the model equation

$$\bar{y}(t) = \hat{\theta}^T(t-1)\ v(t) \tag{4.1}$$

where $\hat{\theta}(t)$ is the column vector of parameter estimates and

$$v^T(t) = (-\bar{y}(t-1),\ldots,-\bar{y}(t-n);u(t-1),\ldots,u(t-n)) \tag{4.2}$$

also the sequence $\{u(t)\}$ is the set of input signals applied to both the model and the system itself. It must be noted as well that the parameter vector employed in (4.1) is that obtained at the previous time instant. This is because the value of model output, $\bar{y}(t)$, is subsequently used to calculate the new parameter estimate set. By a similar reasoning to that previously employed, an updating procedure for the parameter estimates is:

$$\hat{\theta}(t) = \hat{\theta}(t-1) + K(t)\left[\, y(t)-\bar{y}(t) \,\right] \tag{4.3}$$

where, as suggested by Landau (13), the gain K(t) can be obtained from

$$K(t) = R^{-1}(t)v(t) \tag{4.4}$$

in which

$$R(t) = R(t-1) + v(t)v^T(t) \tag{4.5}$$

In terms of the use of past model outputs in the model equation (4.1), this particular method bears a certain resemblance to the Instrumental Variables techniques. However the resemblance ends when the updating equation (4.3) is considered, in that the error generated from $y(t) - \bar{y}(t)$ is also dependant on past model output values, i.e. $\bar{y}(t-1)$, $\bar{y}(t-2),\ldots$; this is however not the case with any of the methods, including Instrumental Variables, previously described. An advantage of the Model Reference Approach is that as the estimate sequence $\{\hat{\theta}(t)\}$ is less dependant on the system output, it is also less dependant on the disturbance, e(t), affecting the output.

5. Bayesian Methods

The Bayesian approach to parameter estimation can be regarded as a form of Nonlinear filtering and, as will be described shortly, can be viewed in terms of a Kalman filtering problem. In this method the parameter vector is considered as a set of random variables, information on which can be obtained by inspection of other correlated variables, in particular the system input and output. At a particular time instant a probability density function is required for the parameter vector, and this is dependant on input and output information observed up to and including that time instant t. An estimate of the parameter vector can then be obtained in one of many ways, the most usual being in terms of the conditional expectation, such that

$$\hat{\theta}(t) = \mathcal{E}\{\theta/y(t),u(t)\} \tag{5.1}$$

in which $\hat{\theta}(t)$ is the estimated parameter vector at time instant t. This particular value is in fact the same as that obtained by minimizing (5.2).

$$\hat{\theta}(t) = \min_{\theta} \mathcal{E}|\,\theta - \hat{\theta}(t)\,|^2 \tag{5.2}$$

Although the problem of determining the probability density function, especially in terms of its time dependancy, can be a difficult one for general system descriptions, when a linear relationship exists between the system parameters and the data, and the disturbance affecting the system is zero-mean white noise, a simplification results. Such a system is that already considered in this chapter as:

$$y(t) = \theta^T x(t) + e(t) \tag{5.3}$$

where $\{e(t)\}$ is a white noise sequence with zero-mean and finite variance. Also $x(t)$ is a function of past input and output information such that $x(t)$ does not include $y(t)$ and $u(t)$, but does include $y(t-1)$ and $u(t-1)$ etc.

Updated parameter estimate vectors can then be obtained from:

$$\hat{\theta}(t) = \hat{\theta}(t-1) + K(t)\,\varepsilon(t) \tag{5.4}$$

where $$\varepsilon(t) = y(t) - \hat{\theta}^T(t-1)x(t) \tag{5.5}$$

and $$K(t) = \frac{P(t-1)x(t)}{\Omega + x^T(t)P(t-1)x(t)} \tag{5.6}$$

in which
$$P(t) = \left[I - K(t)x^T(t)\right] P(t-1) \tag{5.7}$$

The term Ω shown in (5.6) is the variance of the disturbance $e(t)$, i.e. $\mathcal{E}\{e^2(t)\} = \Omega$. A proof of these particular equations can be found, in terms of induction based on Bayes's rule, in Ljung and Soderstrom (8).

It can be readily observed that the equations (5.4) - (5.7) used for obtaining updated parameter estimates using Bayesian methods are virtually identical to the set (2.3), (2.7) and (2.8) used to obtain updated RLS estimates. The two results are in fact identical if the forgetting factor β in (2.7) is made equal to the variance value Ω in (5.6), this being applicable for white, Gaussian noise.

<u>The Kalman Filter</u>

The parameter vector updating schemes considered can be viewed in terms of a state-space Kalman filtering problem on condition that the state vector is defined correctly. A state-space description with only measurement noise, $e(t)$, can be written as:

$$X(t+1) = X(t)$$

and $$y(t) = H^T(t)X(t) + e(t) \tag{5.8}$$

in which $X(t)$ is the state vector, and $e(t)$ has zero mean and variance Ω.

It must be noted that there are many different state-space realisations possible for any one particular system. Each realisation is characterised by the type of state vector chosen. This particular selection has been made such that a simple relationship exists between the estimation procedures considered and the Kalman filtering problem.

If the system is described by (5.3) along with

$$\theta(t+1) = \theta(t) \tag{5.9}$$

then the state vector X(t) can be estimated by means of the Kalman filtering equations, Kwakernaak and Sivan (14),

$$\hat{X}(t+1) = \hat{X}(t) + K(t)\left[y(t) - H^T(t)\hat{X}(t)\right] \tag{5.10}$$

where
$$K(t) = \frac{P(t-1)H(t)}{\Omega + H^T(t)P(t-1)H(t)} \tag{5.11}$$

and
$$P(t) = \left[I - K(t)H^T(t)\right]P(t-1) \tag{5.12}$$

This set of equations is found to be exactly that of (5.4), (5.6) and (5.7) if $\hat{X}(t+1) = \hat{\theta}(t)$ and $H(t) = x(t)$.

Although introduced through the Bayesian estimation problem it is apparent that the set of equations (5.10) - (5.12) are equally applicable to the other estimation procedures considered in this chapter, the main filtering equation (5.10) being directly used in many estimators, with the parameter vector replacing the state vector. The gain K(t) is therefore referred to as the Kalman gain and all the standard estimators considered can be regarded as merely special cases of the Kalman filter.

6. Problems with Time-Varying Systems

A general assumption has been made in the preceeding discussion that the recursively estimated parameter vector will converge on a vector containing the actual or true parameter values. This assumption implies that the system itself is not varying with respect to time, and this may well be the case. However, in many situations the system is affected by drifting parameters caused by unmodelled ambient conditions, ageing or physical modifications to the system. In these cases, in order to retain an up-to-date picture of the system the recursive identification procedure must be able to track the system parameters as any drift occurs.

In this section the estimation schemes already introduced will be viewed in terms of how they cope with time-varying systems. In certain instances modifications will need to be made, although in other cases it will be seen that special factors already included in the algorithms can be employed to counteract the system variations.

Recursive Least Squares Estimation

The RLS procedure is intended here to act as a representative of the recursive forms of standard off-line procedures.

For the standard RLS method the parameter vector is required to be

$$\hat{\theta} = \min_{\theta} \mathcal{E}\{\varepsilon(t)\}^2 \tag{6.1}$$

which is found in the RLS case by taking the mean of a set of $\varepsilon(t)$ values, hence replacing the expectational operator. That is the choice of parameter vector will, if considered simply, give equal weighting to all the values of error $\varepsilon(t)$ taken over a finite period. With a time-varying system though, the most recent information concerning the system is of much more use as it gives reference to what the characteristics are like now, rather than what they were like some time previously.

There are several RLS algorithms which include weighting factors in order to forget older data, the general formulae being given by:

$$\hat{\theta}(t) = \hat{\theta}(t-1) + K(t)\,\varepsilon(t) \tag{6.2}$$

$$K(t) = \frac{P(t-1)x(t)}{\beta + x^T(t)P(t-1)x(t)} \tag{6.3}$$

and
$$P(t) = \left[I - K(t)x^T(t)\right]\alpha P(t-1) \tag{6.4}$$

where two variables α and β are used to introduce data forgetting. Four cases in particular will be considered further, dependant on the values taken by α and β.

Firstly, when $\alpha = \beta = 1$, this corresponds to a straightforward identification procedure, no different weightings are placed on data obtained at different times.

Secondly, when $\alpha = 1$ and $0 < \beta \leq 1$ this corresponds to the technique described in section 2. β can either be a scalar value, usually in the range $0.95 \leq \beta \leq 0.99$ or can be related to the error $\varepsilon(t)$ at each time instant t, such that if $\varepsilon(t)$ is larger in magnitude so $\beta = \beta(t)$ becomes smaller. The reason for a time dependant forgetting factor is that if the system parameters remain reasonably constant with respect to time a constant low value of β will result in unnecessary variations in parameter estimates. Further, if sudden system parameter alterations occur and β is too high, the RLS estimator may not be able to retune quickly enough, thus when incorporated in an adaptive controller, control of the system may be lost.

A third possibility is for $\beta = 1$ and $0 < \alpha^{-1} \leq 1$, i.e. the factor α is employed as the sole forgetting factor. Similar types of factor α, to those just described for β, can be considered, i.e. a constant scalar value or a time dependant value. By post multiplying both sides of (6.4) by x(t) it can be seen that:

$$K(t) = \frac{P(t)x(t)}{\alpha\beta} \tag{6.5}$$

Hence with $\beta = 1$ the direct method of increasing all the elements in P(t) by a factor α, see (6.4), also has the effect of decreasing all the elements of the Kalman gain, K(t), by the same amount. It can be seen therefore that α and β have the same type of effect on P(t), but from (6.5) have opposite effects on K(t).

The final approach is to make $\beta = \alpha^{-1}$ such that although all the elements of P(t) are increased, the Kalman gain is not immediately affected by either α or β other than through P(t).

In summary, the overall effect of α or β is to keep the elements of the covariance matrix P(t) at some non zero value. These elements will usually tend to zero as the parameter estimates approach the set of actual values. By slightly increasing P(t) elements therefore, the RLS algorithm will always be awake to any change that may take place in the actual system parameters.

Stochastic Approximation

The Stochastic Approximation approach to parameter estimation will now be reviewed in terms of coping with time varying system dynamics.

Considering again equations (3.13) and (3.11), used to obtain parameter estimate updates, then

$$\hat{\theta}(t) = \hat{\theta}(t-1) + \gamma(t)R^{-1}(t)x(t)\varepsilon(t) \quad (6.6)$$

and $$R(t) = R(t-1) + \gamma(t)\left[x(t)x^T(t) - R(t-1)\right] \quad (6.7)$$

where the gain $\gamma(t)$ can take on one of a number of forms as outlined in section 3, a general requirement being that the sequence $\{\gamma(t)\}$ tends to zero as the parameter estimates approach their actual values, thus causing little change in the updated estimate vector found from (6.6).

For time-varying systems the estimator equations (6.6) and (6.7) must always have one eye open such that the parameter estimate vector can track the actual system parameters. This is most simply ensured by means of selecting $\gamma(t)$ to approach a small finite value rather than zero. In fact $\gamma(t)$ can also be made dependant on the modelling error term $\varepsilon(t)$ such that when the estimates are a bad fit to their actual values, $\gamma(t)$ allows for more rapid retuning of the vector $\hat{\theta}(t)$. Reconsider (6.7) in the form:

$$\frac{R(t)}{\gamma(t)} = [1 - \gamma(t)]\ \frac{R(t-1)}{\gamma(t)} + x(t)x^T(t)$$

then if

$$\gamma(t)S(t) = R(t) \quad (6.8)$$

so

$$S(t) = [\ 1 - \gamma(t)\ ]\ \frac{\gamma(t-1)}{\gamma(t)}\ S(t-1) + x(t)x^T(t) \quad (6.9)$$

and

$$\hat{\theta}(t) = \hat{\theta}(t-1) + S^{-1}(t)x(t)\,\varepsilon(t) \quad (6.10)$$

For a constant value of gain $\gamma(t)$ so $\gamma(t) = \gamma(t-1)$ and subject to $0 < \gamma(t) < 1$ an exponentially weighted sequence $\{S(t)\}$ will result. Hence a value of $0 < \gamma(t) < 0.1$ is often acceptable, where $t > 0$. For a more complicated form of $\gamma(t)$, by linking this with $\varepsilon(t)$ it can be ensured that $\gamma(t) \to 0$ when $\varepsilon(t) \to 0$ and $\gamma(t) \to 0.1$ when $\varepsilon(t)$ is large. It is preferable to relate $\gamma(t)$ to a set of $\varepsilon(t)$ values squared however, to avoid any spurious features. Often $\gamma(t)$ can consist of two parts multiplied together, i.e. an exponential part and an error dependant part.

Model Reference

Methods of dealing with time varying systems in terms of the Model Reference approach to Recursive Identification are not essentially different to those already described. In particular the matrix update equation for R(t), see (4.5), can be altered by a scalar divisor in a similar way to that shown in (2.9). As an example reconsider (4.5) as

$$R(t) = \lambda(t)R(t-1) + v(t)v^T(t) \tag{6.11}$$

where $0 < \lambda(t) \leq 1$, although most likely $0.9 < \lambda(t) \leq 1$. Hence the overall updating procedure for parameter estimates involves (6.11) and (4.3) which is restated as:

$$\hat{\theta}(t) = \hat{\theta}(t-1) + R^{-1}(t)v(t)\,[\,y(t) - \bar{y}(t)\,] \tag{6.12}$$

Exponential forgetting is therefore provided if $\lambda(t)$ is held at a constant value for all time, although a better method is to make $\lambda(t)$ dependant on the error $y(t) - \bar{y}(t)$, where $\bar{y}(t)$ is the model output, such that when this error is small $\lambda(t) \to 1$.

Bayesian Methods

The Bayesian approach to time varying systems is best treated by considering the true parameter vector θ to be varied such that it can be described by:

$$\theta(t+1) = \theta(t) + w(t) \tag{6.13}$$

where w(t) is a zero-mean, Gaussian disturbance with covariance $\bar{\Omega}$.

Coupling (6.13) with the output equation;

$$y(t) = \theta^T(t)\,x(t) + e(t) \tag{6.14}$$

means that the Kalman filtering analogy, as employed in section 5, can be used. If it is assumed that the state vector is in this case the vector of parameter estimates, then applying the Kalman filter to (6.13) and (6.14) will result in the equations:

$$\hat{\theta}(t) = \hat{\theta}(t-1) + K(t)\,\varepsilon(t) \tag{6.15}$$

$$K(t) = \frac{P(t-1)x(t)}{\Omega + x^T(t)P(t-1)x(t)} \tag{6.16}$$

and

$$P(t) = P(t-1) + \bar{\Omega} - \frac{P(t-1)x(t)x^T(t)P(t-1)}{\Omega + x^T(t)P(t-1)x(t)} \tag{6.17}$$

in which

$$\varepsilon(t) = y(t) - \hat{\theta}^T(t-1)x(t) \tag{6.18}$$

Also Ω is the variance of the disturbance sequence, $\{e(t)\}$, and $\hat{\theta}(t)$ is still the conditional mean of $\theta(t)$, if it is considered that information is available up to and including time instant t.

The effect of time varying system dynamics is therefore to cause the inclusion of $\bar{\Omega}$ in the covariance matrix, P(t), update, hence ensuring that P(t) cannot tend to zero. In practice both Ω and $\bar{\Omega}$, if not known, which is most likely, can be chosen as small positive numbers, possibly less than unity although this is really system and noise dependant. The larger their values, especially $\bar{\Omega}$, the more awake the estimator will be and hence it will be more robust in terms of coping with varying parameters. However, a large value of $\bar{\Omega}$ will cause the estimations to vary even in the presence of stationary system parameters. Hence when the parameters are not time varying, convergence of estimates and overall estimator performance will deteriorate. The choice of Ω and $\bar{\Omega}$ must therefore be seen as a trade-off between convergence and robustness. The properties that this type of estimator, and the others considered this far, have in terms of convergence and stability are discussed in the following section.

7. Convergence and Stability

One of the most important questions that must be asked when selecting a parameter estimator for Recursive Identification is: 'Will it converge?'. Unfortunately a detailed mathematical analysis can rapidly prove to be too complicated and rather limiting in many cases. In this section therefore, only a brief overview is given concerning the asymptotic properties, with respect to time, of some of the parameter estimators previously considered.

It must be remembered that an alternative approach to mathematically analysing estimator properties is to operate an estimator on a set of data, obtained from the system to be identified, by means of a simulated experiment most likely by means of a computer. This has an added advantage in that it can be very simple to alter a gain setting or value of a forgetting factor in order to find out the effect that that alteration has on estimator performance. A common problem exists though, of reading too much into a particular simulation run, e.g. because the estimator works well on a particular set of data this does not mean it will work well or even work at all on any other set of data. Simulation results can therefore only be used to show that an estimator can work and that more widespread application of the estimator is likely. In order to make any firmer statements concerning an estimation algorithm a certain amount of mathematical analysis is necessary. This generally takes the form of firstly a specification of data and system type on which the algorithm can be employed, secondly the properties of the estimate sequence with respect to time, in particular as time tends to infinity, can be investigated within the confines of that particular data/system framework.

Asymptotic analysis gives an indication of the actual values or at least region of values, to which the sequence of estimates will converge. However, it also can give information as to the convergence rate and the values to which the elements of the covariance matrix will converge. Unfortunately, because the convergence is in terms of time respective asymptotes, no information is given in terms of how long it will be from start up before the estimate vector obtained is suitable, i.e. how many recursions must occur before the parameter estimates can be employed elsewhere?

Another problem with asymptotic convergence analysis crops up when time varying systems are encountered. In order to cope with varying parameters, as was discussed in the previous section, the Kalman gain vector is usually injected on each recursion in order to keep it away from zero, such that the estimator will follow the system parameters. This means that for the case of time varying parameters it is a requirement, to operate effectively, that the estimates will not converge! Note that this does not mean the estimates must diverge, rather that the gain will tend to a small positive value for slowly time varying systems as opposed to the more normally desired zero value.

For both of the problems mentioned, i.e. number of recursions required and time varying estimates, a return to analysis by means of simulation studies is perhaps most sensible. By means of a simulation one would be able to see how many recursions were most likely required for a specific algorithm operating on a certain type of system. Also by means of simulation work it is possible to investigate how rapidly the system parameters can vary, e.g. for non-linear systems, before the estimator gives poor results.

Least Squares Algorithms

Consider the general set of least squares algorithms, in particular that of extended recursive least squares, which is an extension of RLS in order to estimate the parameters within a coloured noise sequence.

Let the system whose parameters are to be estimated, be described by the CARMA equation

$$A(z^{-1})\,y(t) = z^{-k}\,B(z^{-1})\,u(t) + C(z^{-1})e(t) \tag{7.1}$$

then a sufficient condition for the convergence of the ERLS parameter vector estimate sequence $\{\hat{\theta}(t)\}$ to the actual system values is that

$\left[\frac{1}{C(z^{-1})} - \frac{1}{2}\right]$ is a strictly positive real function.

Further to ensure convergence a necessary condition is that $C(z^{-1})$ itself must be strictly positive real. The method by which these results can be obtained and a further discussion regarding their implications can be found in Ljung and Soderstrom (8).

Two further comments must be made at this point however:

1. The convergence results given here apply only if the system under consideration can be exactly modelled by the CARMA description. Although the results most likely apply to a system which can be adequately described by a CARMA model,unless an exact fit is obtained they are not strictly usable.

2. Because a particular system does not comply with certain convergence conditions that have been laid down as hard and fast rules, this does not mean that a particular parameter estimator should not be used on that system. For example, if $C(z^{-1})$ is not strictly positive real this does not mean that a particular estimator will not converge, it simply means that convergence cannot be ensured for all $A(z^{-1})$ and $B(z^{-1})$ polynomials.

Instrumental Variable Algorithms

The Instrumental Variable method of recursive identification, as described in section 2, can be rewritten by means of the update equations:

$$\hat{\theta}(t) = \hat{\theta}(t-1) + K(t)\,\varepsilon(t) \tag{7.2}$$

$$K(t) = \gamma(t)R^{-1}(t)v(t) \tag{7.3}$$

and
$$R(t) = R(t-1) + \gamma(t)\left[v(t)x^T(t) - R(t-1)\right] \tag{7.4}$$

where
$$\varepsilon(t) = y(t) - \hat{\theta}^T(t-1)\,x(t) \tag{7.5}$$

In the above set of equations $v(t)$ denotes the instrumental vector, as defined in (2.10). Then if the actual system to be identified is defined by:

$$y(t) = \theta^T x(t) + e(t) \tag{7.6}$$

in which $\{e(t)\}$ is a zero-mean, not necessarily white noise, sequence. The parameter estimate vector $\hat{\theta}(t)$ will then approach the actual parameter vector θ as time tends to infinity, as long as the following two conditions are satisfied:

$$\mathcal{E}\{\,v(t)\,e(t)\,\} = 0 \tag{7.7}$$

$$\mathcal{E}\{\,v(t)\,x(t)\,\} \quad \text{is nonsingular, i.e. invertible} \tag{7.8}$$

These conditions are dependent on the model structure being identical to the system structure, this includes the orders of system and model being equal, and also the identification procedure is assumed to take place in an open loop mode.

An extension of the above Instrumental Variable method, due to Young (15), is to include the effect of prefiltering both the output signal and regression vector by an operator specified polynomial. Hence the filtered signals can be defined as:

$$y_F(t) = T(z^{-1})y(t) \tag{7.9}$$

and
$$x_F(t) = T(z^{-1})x(t) \tag{7.10}$$

where $T(z^{-1})$ is the chosen filtering polynomial. If the output signal $y(t)$ and regression vector $x(t)$ are then replaced by their filtered versions in the equation set (7.2) - (7.5), the same asymptotic results will apply on condition that the parameters selected for $T(z^{-1})$ are time invariant.

Numerical Stability

The problems with implementing some of the recursive algorithms considered have, thus far in this section, been looked at in terms merely of theoretical or simulation convergence over a period of time. Unfortunately these factors do not take into account some of the more practical implementation problems such as interface design and connection. It is not intended in this discussion to cover hardware aspects of the problem primarily because of the enormity of the topic, each interface being both system and computer dependent. However, it is felt that an important area is that of computation suitability, in particular numerical procedures to ensure stability in a purely algebraic sense.

Factors to be taken into account in terms of algorithm implementation are (a) the size of computer memory required, dependent on the method employed, (b) the time required for one iteration of the algorithm, and (c) the accuracy of the calculations, the extent of numerical errors due to rounding off for example. It is this latter point of numerical accuracy for which the wide selection of methods available can produce quite different results. In particular, routines and approaches to carry out matrix inversion must be considered in depth because of the occurrence of this particular feature in the algorithms considered, see (7.3).

One method for matrix inversion is to use a Matrix Inversion Lemma, by replacing the particular equation including a matrix to be inverted by a difference equation with improved numerical stability properties. An example is the stabilized Kalman Equation, Bierman (16), in which repeated subtraction operations, introducing multiplied errors, are avoided. Unfortunately the Matrix Inversion Lemma does not ensure numerical stability and requires increased computation, other numerical properties are improved though.

A more popular alternative is to use U-D factorization, i.e. to decompose the matrix to be inverted into two upper triangular matrices and one diagonal matrix. This leads to good numerical stability and relative insensitivity to rounding errors, Bierman (16). The matrix to be inverted can however, when positive definite conditions apply as with the covariance matrix P(t), be factorized into two nonsingular matrices by means of a square root algorithm. At the expense of extra computation, constraints such as triangularity of the factored matrices, can be obtained.

A final numerical method which must be considered for gain updating is the use of fast algorithms, such as ladder techniques, in order to reduce the amount of computation per time step, Ljung and Soderstrom (8).

8. Comments

In this chapter several methods have been discussed for the purpose of Recursive Identification of discrete time system models. Hopefully it has highlighted the most commonly encountered methods, although several have had to be overlooked, e.g. Recursive Maximum Likelihood techniques. Very often the final choice of method is though, dependent on the problem for which the model is to be employed.

Topics of interest when looking into ordinary System Identification, such as experiment design and model order testing naturally have their counterparts in terms of recursive identification. In particular, instances occur when sampling periods and presampling filters can be adjusted along with the type of input signals used. Various other points then arise, such as non uniform sampling periods rather than the uniform sampling interval design considered here.

It is generally assumed, in order to obtain parameter estimate convergence, that the structure of the model is identical to that of the system, hence making the inference that there exists a well defined system order. Often however, in order to keep computation low, a second order recursive identifier is used as standard, this is particularly true when the identifier is incorporated within an adaptive controller. Care must therefore be taken, and convergence cannot be assured when the system is of higher order than this. A common approach is to try a different sampling interval if any convergence problems are encountered or when the estimates appear to be converging to wrong values.

References

1. Young P.C.: 'Recursive estimation and time-series analysis', Springer-Verlag, Berlin, 1984.

2. Goodwin G.C. and Payne R.L.: 'Dynamic system identification', Academic Press, 1977.

3. Eykhoff, P.: 'System identification', John Wiley & Sons, 1974.

4. Sage A.P. and Melsa J.L.: 'System identification, Academic Press, 1971.

5. Farsi M., Karam K.Z. and Warwick K.: 'A simplified recursive identifier for ARMA processes', Electronics Letters, Vol. 20, pp. 913-915, 1984.

6. Astrom K.J. and Eykhoff P.: 'System identification - a survey', Automatica, Vol. 7, pp. 123-162, 1971.

7. Warwick K., Farsi M. and Karam K.Z., 'A simplified pole placement self-tuner', Proc. Control 85, Cambridge Univ., pp. 1-6, 1985.

8. Ljung L. and Soderstrom T., 'Theory and practice of recursive identification', MIT Press, 1983.

9. Strejc V., 'Least squares parameter estimation', Automatica, Vol. 16, pp. 535-550, 1980.

10. Soderstrom T., Ljung L. and Gustavsson I., 'A comparative study of recursive Identification methods', report 7427, Dept. of Automatic Control, Lund Institute of Technology, Lund, Sweden, 1974.

11. Solo V., 'Some aspects of recursive parameter estimation', Int. J. Control, Vol. 32, pp. 395-410, 1980.

12. Harris C.J. and Billings S. eds: 'Self-tuning Control: Theory and Applications', Peter Peregrinus for IEE, rev. 2nd edition, 1985.

13. Landau I.D.: 'Unbiased Recursive Identification using Model Reference Techniques', IEEE Trans. on Automatic Control, Vol. AC-21, pp. 194-202, 1976.

14. Kwakernaak. H. and Sivan. R.: 'Linear Optimal Control Systems', Wiley Interscience, 1972.

15. Young P.C.: 'Some observations on Instrumental Variable methods of time-series analysis', Int. J. Control, Vol. 23, pp. 593-612, 1976.

16. Bierman G.J.: 'Factorization methods for discrete sequential estimation', Academic Press, New York, 1977.

SPECTRAL ANALYSIS AND APPLICATIONS

Dr. P.E. Wellstead

1. INTRODUCTION

The behaviour of engineering systems is usually characterised, in the first instance, by signals which are functions of time. Such signals do not usually convey the information to the engineer in the most suitable form. Hence it is necessary to conduct some signal processing to get the requisite information in a form which can be readily interpreted. Very often the data will be random in nature such that information is "hidden" in the random fluctuations of the time data. Correspondingly, data which are deterministic in nature can profitably be processed to bring out certain features.

Spectral analysis is a rapidly evolving area with new results and algorithms being continually developed and superseded. For this reason, the material presented here consists only of basic results which will not change, as they underpin all aspects of spectral and correlation methods. For the reader who wishes to keep abreast of current developments in spectral and related signal processing techniques, the IEEE Transactions on Acoustics, Speech and Signal Processing are recommended. The journal of the Institute of Sound and Vibration is also a good source. As in all things related to applied random signals processing the book *Random Data: Analysis and Measurement Procedures* (Wiley, 1971) by J.S. Bendat and A.G. Piersol is a basic reference and can be read in conjunction with the material presented here.

2. BASIC RELATIONSHIPS

2.1 Signal Spectra

The theory of spectral analysis is built upon the equivalence of functions of time and frequency. Here, we describe the underlying mathematical relationships. Several of these relationships have already been discussed in earlier Lectures at this School, in particular R1, R2 and L3; they are gathered together here to pave the way for material in later Sections of these notes.

These notes are drawn from two Technical Reports: 008/83 "Methods and Applications of Digital Spectral Techniques" and 009/83 "Theory and Statistical Accuracy of Spectral Analysis", prepared by Dr. P.E. Wellstead for Solartron Instruments, Victoria Road, Farnborough, Hampshire, GU14 7PW, England.

If x(t) is a function of time, t, then its corresponding function of frequency, f, is X(f), defined by the Fourier transformation:

$$X(f) = \int_{-\infty}^{+\infty} x(t)\, \exp(-j2\pi ft)dt \qquad (2.1)$$

where the integral will exist if

$$\int_{-\infty}^{+\infty} |x(t)|dt < \infty \qquad (2.2)$$

The inverse transformation is:

$$x(t) = \int_{-\infty}^{+\infty} X(f)\, \exp(j2\pi ft)df \qquad (2.3)$$

The integral condition (equation 2.2) means that Fourier transforms do not generally exist for random signals which persist for all time and are not bounded in the manner of equation 2.2. Standard tables of Fourier transform pairs are available in the various texts on system theory (e.g. Gabel and Roberts, 1980).

The important feature of frequency representation of a signal is that it enables the action of a linear system transfer function upon that system to be described. In particular, suppose that a stable, linear, time-invariant system has input signal x(t), unit impulse response h(t) and output signal y(t) (Fig. 2.1)

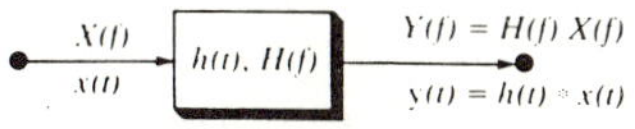

Fig. 2.1 Linear system and its action upon a signal x(t)

The transformed system output Y(f) is

$$Y(f) = H(f).X(f) \qquad (2.4)$$

where H(f) and X(f) are the Fourier transforms of h(t) and x(t), respectively. In the time domain, y(t) is given by the convolution:

$$y(t) = \int_0^{\infty} h(\tau).x(t-\tau)d\tau . \qquad (2.5)$$

2.2 Random Data: Autocorrelation and Power Spectrum

When the output of the system (Fig. 2.1) is corrupted by a random, or noise-like disturbance, or when x(t) is itself a noise process, the simple Fourier relation (equation 2.4) is inapplicable. Specifically, the Fourier transform of a random data record may not exist because it violates the integrability condition (equation 2.2). Moreover, one record of data from a random process only represents one of the many manifestations of the process. Another record of data from the same process, but collected at a different time would be different. For these reasons alternative descriptions of the time and frequency behaviour of random data are used. These are the correlation and spectral density functions.

Consider first a single stationary random process x(t). The power spectrum $S_x(f)$ of x(t) is defined as the Fourier transform of the autocorrelation function $R_x(\tau)$.

$$S_x(f) = \int_{-\infty}^{+\infty} R_x(\tau) \exp(-j2\pi f\tau)d\tau \tag{2.6}$$

For a stationary, ergodic process the autocorrelation can be defined via the time average:

$$R_x(\tau) = \lim_{T\to\infty} \frac{1}{2T} \int_{-T}^{+T} x(t)\, x(t+\tau)dt \tag{2.7}$$

The term 'ergodic' means that the nature of the mechanism which generates a random signal can be obtained by examining one (possibly infinitely long) record of data. Properties of autocorrelation functions have been discussed in Lecture L3 at this School.

In these notes, all signals will be assumed to have zero mean, so that autocorrelation and autocovariance are identical. The normalised autocorrelation (autocovariance) function, given by

$$\rho_x(\tau) = \frac{R_x(\tau)}{\sigma_x^2} \tag{2.8}$$

is used in some applications because it expresses the correlation structure as a proportion of the signal mean squared value, σ_x^2.

The power spectrum is formally defined in terms of equation 2.6, but is in fact, computed in a different manner, which relates to its interpretation as the spectral distribution of average signal power. To see this, consider the autocorrelation function written as the inverse Fourier transform of the power spectrum. Thus,

$$R_x(\tau) = \int_{-\infty}^{+\infty} S_x(f) \exp(j2\pi ft)df \tag{2.9}$$

At $\tau = 0$, this becomes,

$$\sigma_x^{\ 2} = \int_{-\infty}^{+\infty} S_x(f)df \tag{2.10}$$

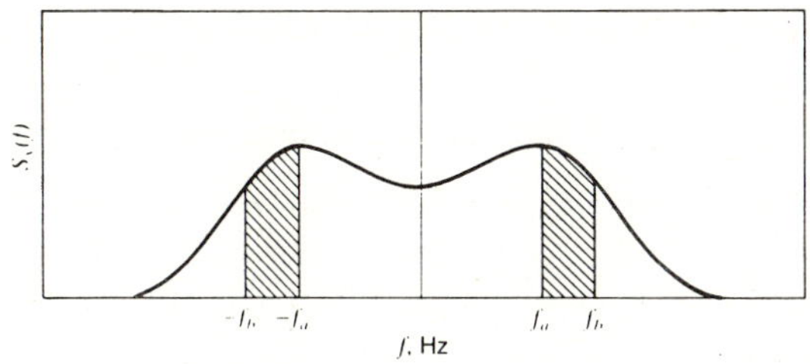

Fig. 2.2 Power spectral density

Equation 2.10 allows $S_x(f)$ to be visualised as a density function for the signal power, expressed as a function of frequency. For example, referring to Fig. 2.2, that fraction of $\sigma_x^{\ 2}$ which is contained in the frequency band f_a to f_b is given by the total shaded area under the power spectral density curve. This interpretation has a clear engineering meaning in terms of signal power and allows a frequency domain quantity to be associated with random signals (albeit in terms of the *average* energy impact of the random process) in the same way that the spectrum is applied with transient and deterministic signals. Moreover, similar relations to equation 2.4 apply to power spectra of signals passed through linear systems. For our linear, time-invariant system (Fig. 2.1)

$$S_y(f) = |H(f)|^2 S_x(f), \tag{2.11}$$

and $|H(f)|^2$ is the modulus squared of the transfer function H(f).

Equation 2.11 is of practical significance because it quantifies how the frequency distribution of power is modified by a system. This important relationship is therefore useful in analysing random signals and synthesising noise controlling filters.

An additional point to note is that power spectra of uncorrelated signals which are added are also additive. Thus, referring to Fig. 2.3, the output power spectrum becomes

$$S_y(f) = S_z(f) + S_n(f) \tag{2.12}$$

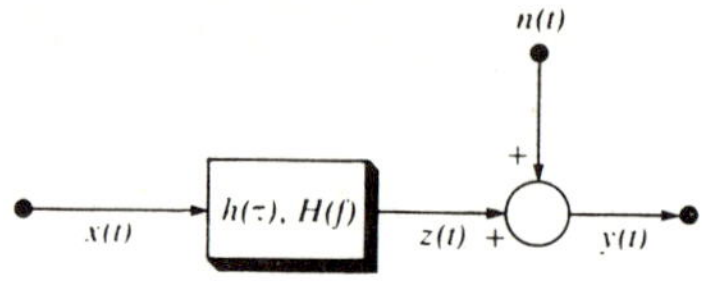

Fig. 2.3 System with noise corrupted output

Related important properties of the power spectrum are:

(i) $S_x(f)$, being a power function, is always real and positive.

(ii) $S_x(f)$ is an even function of frequency such that $S_x(f) = S_x(-f)$.

(iii) The value of the power spectrum at zero frequency is the squared mean of the signal x(t). That is, $S_x(0) = \mu_x^2$ where μ_x is the mean value of x(t).

The latter point has the important practical consequence that significant dc levels are usually removed prior to spectral analysis, otherwise the squared dc component will dominate the display. (This dc suppression is usually achieved by ac coupling of spectral analysis instrumentation, or sample mean removal in digital computer programs.)

For reference purposes some useful autocorrelation functions are given in Table 1 and Table 2 gives the corresponding power spectra.

2.3 Random Data: Crosscorrelation and Cross-Spectrum

In the same manner that the power spectrum relates to the self-power distribution of a random process, the cross-power spectrum relates to the joint power distribution over frequency of two random processes. Formally, cross-spectrum is defined as the Fourier transform of the crosscorrelation function $R_{xy}(\tau)$. Thus

$$S_{xy}(f) = \int_{-\infty}^{+\infty} R_{xy}(\tau)\exp(-j2\pi f\tau)d\tau \tag{2.13}$$

and for a stationary ergodic process, the crosscorrelation function can be defined as a time average:

$$R_{xy}(\tau) = \lim_{T\to\infty} \frac{1}{2T}\int_{-T}^{T} x(t)\, y(t+\tau)dt \tag{2.14}$$

By analogy with the autocorrelation function, the crosscorrelation quantifies the way in which two signals co-vary at different relative points on the time axis. Unlike the autocorrelation function, however, $R_{xy}(\tau)$ does not usually achieve its maximum at $\tau = 0$. For example, if y(t) is a delayed, filtered version of x(t), corresponding to the effects of a time delay system upon x(t) as an input signal,

Type	Autocorrelation Function	
Constant	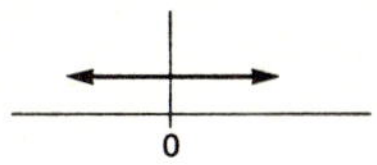	$R_x(\tau) = c^2$
Sine wave	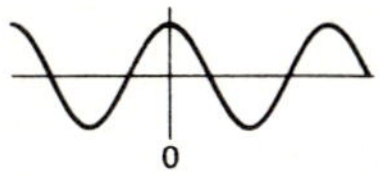	$R_x(\tau) = \frac{X^2}{2} \cos 2\pi f_0 \tau$
White noise	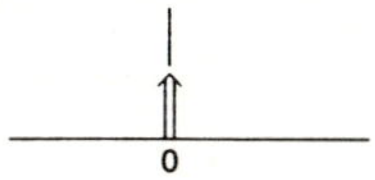	$R_x(\tau) = \alpha\, \delta(\tau)$
First order lag filter	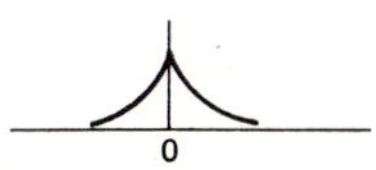	$R_x(\tau) = e^{-\alpha\|\tau\|}$
Low-pass white noise	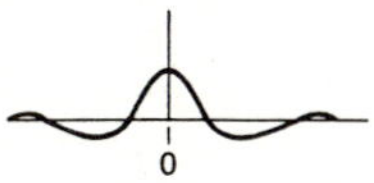	$R_x(\tau) = \alpha\, B \left(\frac{\sin 2\pi B\tau}{2\pi B\tau}\right)$
Band-pass white noise	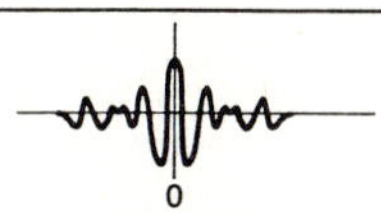	$R_x(\tau) = \alpha\, B \left(\frac{\sin \pi B\tau}{\pi B\tau}\right) \cos 2\pi f_0 \tau$
Exponential cosine	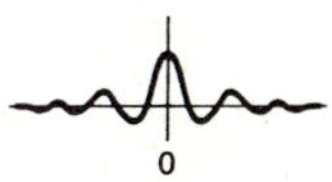	$R_x(\tau) = e^{-\alpha\|\tau\|} \cos 2\pi f_0 \tau$
Exponential cosine, exponential sine	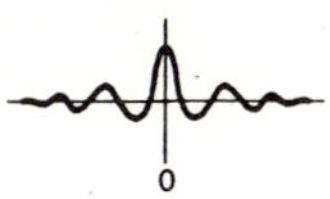	$R_x(\tau) = e^{-\alpha\|\tau\|} (\beta \cos 2\pi f_0 \tau + c \sin 2\pi f_0 \|\tau\|)$

Table 1. Some Autocorrelation Functions

Type	Power Spectral Density Function	
Constant		$G_x(f) = c^2\,\delta(f)$
Sine wave	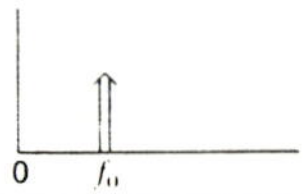	$G_x(f) = \frac{X^2}{2}\,\delta(f - f_0)$
White noise	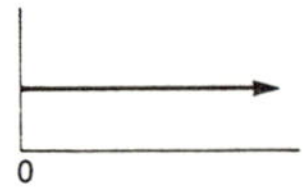	$G_x(f) = \alpha,\ f \geq 0;$ otherwise zero
First order lag filter		$G_x(f) = \frac{4\alpha}{\alpha^2 + 4\pi^2 f^2}$
Low-pass white noise	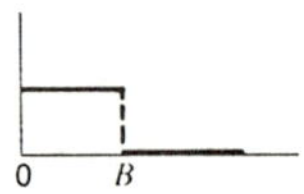	$G_x(f) = \alpha,\quad 0 \leq f \leq B;$ otherwise zero
Band-pass white noise	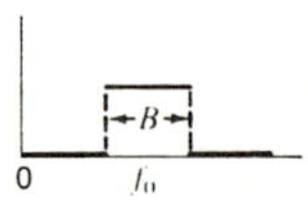	$G_x(f) = \alpha,\quad 0 \leq f_0 - (B/2) \leq f \leq f_0 + (B/2)$ otherwise zero
Exponential cosine	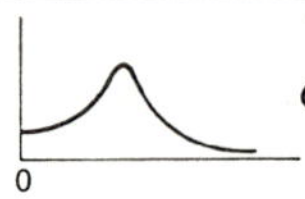	$G_x(f) = 2\alpha\left[\frac{1}{\alpha^2 + 4\pi^2(f + f_0)^2} + \frac{1}{\alpha^2 + 4\pi^2(f - f_0)^2}\right]$
Exponential cosine, exponential sine	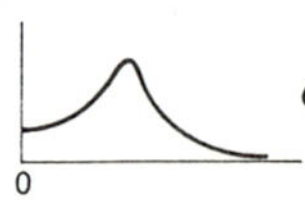	$G_x(f) = \frac{2\alpha\beta + 4\pi c(f + f_0)}{\alpha^2 + 4\pi^2(f + f_0)^2} + \frac{2\alpha\beta - 4\pi c(f - f_0)}{\alpha^2 + 4\pi^2(f - f_0)^2}$

Table 2. Examples of Power Spectra

then $R_{xy}(\tau)$ could resemble Fig. 2.4 in which τ' corresponds to the mean time delay relating the two signals.

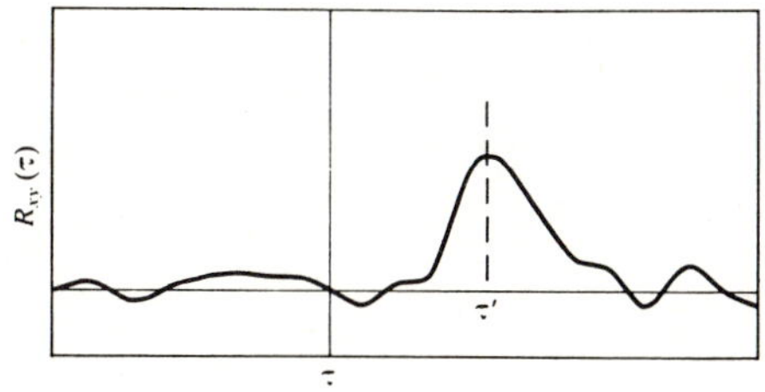

Fig. 2.4 Crosscorrelation function

As a practical aside, it is often worthwhile removing most of the relative time delay between x(t) and y(t) by artificially delaying one of the signals. This 'centres' the display and reduces certain measurement errors associated with signal sampling and finite data records (see Section 3.3).

Important properties of the crosscorrelation function are

(i) $R_{xy}(\tau) = R_{yx}(-\tau)$

(ii) If $R_{xy}(\tau) = 0$ for all τ, the zero mean signals x(t), y(t) are said to be uncorrelated. In the practical sense, this means that they can be considered to arise from independent random mechanisms.

(iii) $|R_{xy}(\tau)|^2 \leq R_x(0)R_y(0)$.

The last property is an inequality which is often used to monitor the maximum possible value of crosscorrelation. The normalised crosscorrelation defined by

$$\rho_{xy}(\tau) = \frac{R_{xy}(\tau)}{\sqrt{R_x(0)R_y(0)}} = \frac{R_{xy}(\tau)}{\sigma_x \sigma_y} \qquad (2.15)$$

is often used in practice, because (recalling the inequality (iii)) it scales the crosscorrelation structure in terms of the root mean square values σ_x,σ_y of the processes x(t) and y(t). Again, it is assumed that both processes are zero mean. The terms crosscorrelation and cross-covariance are sometimes used in the literature. Here the terms crosscorrelation and normalised crosscorrelation are used.

The most important properties of cross-correlation, however, relate to signals filtered by linear systems. For the system of Fig. 2.1,

$$R_{xy}(\tau) = \int_{-\infty}^{+\infty} h(t)R_x(t-\tau)dt \qquad (2.16)$$

This is identical in form to the convolution relating the input-output signals (equation 2.5).

In the frequency domain (obtained by applying the Fourier transform to both sides of equation 2.16), the equivalent relationship is

$$S_{xy}(f) = H(f).S_x(f), \tag{2.17}$$

which is analogous to the relationship between signal spectra in equation 2.4.

A similar frequency domain relation can be derived for the system in Fig. 2.3 which takes the form

$$S_{xy}(f) = H(f).S_x(f) + S_{xn}(f). \tag{2.18}$$

Interestingly, equation 2.18 reduces to equation 2.17 if the noise n(t) and signal x(t) are from independent, uncorrelated sources, since in this case $S_{xn}(f)$ is zero. This result is used extensively in transfer function analysis.

The cross-spectrum is of significant value in testing the joint spectral properties of signals via equations 2.17 and 2.18. It is a complex function, with the following conjugate symmetry

$$S_{xy}(f) = S^*_{yx}(-f) \tag{2.19}$$

From the properties of crosscorrelation functions, if x(t) and y(t) are independent noise processes, then

$$S_{xy}(f) = 0 \text{ for all } f.$$

If the processes are linked by a linear system (as in Fig. 2.1) then the relationship of 2.17 applies. By the same token if both x(t) and y(t) arise from the same process d(t), as indicated in Fig. 2.5, then the cross-spectrum will be:

$$S_{xy}(f) = H_1^*(f)\ H_2(f)\ S_d(f). \tag{2.20}$$

N.B. the superscript * denotes complex conjugation.

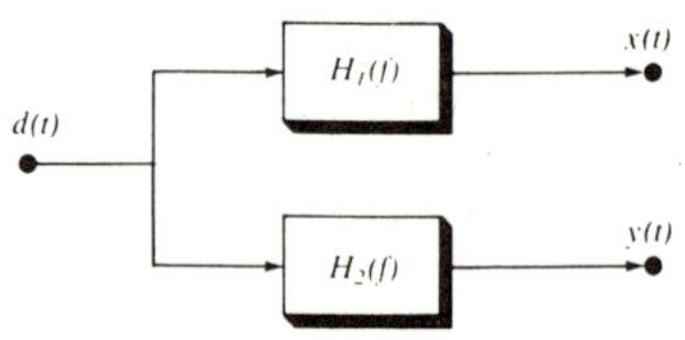

Fig. 2.5 Fully coherent signals

2.4 Spectral Coherency

Expressions 2.17 and 2.20 show that processes x(t) and y(t) are jointly related, but they do not quantify the *degree* of correlation. Such information is provided by the squared coherency function $\gamma^2_{xy}(f)$ defined by

$$\gamma^2_{xy}(f) = \frac{|S_{xy}(f)|^2}{S_x(f)S_y(f)} \tag{2.21}$$

One can visualise $\gamma^2_{xy}(f)$ as being a normalised coefficient of correlation (cf. equation 2.15) between x(t) and y(t) *evaluated at each frequency*. In fact, $\gamma^2_{xy}(f)$ always lies between zero and one. If the processes are independent at a frequency f, then $\gamma^2_{xy}(f)$ is zero at that point. If the processes are linearly related, as in Figs. 2.1 and 2.5 the coherency is one. For intermediate situations as in Figs. 2.3 and 2.6 the coherency will take values between these theoretical limits.

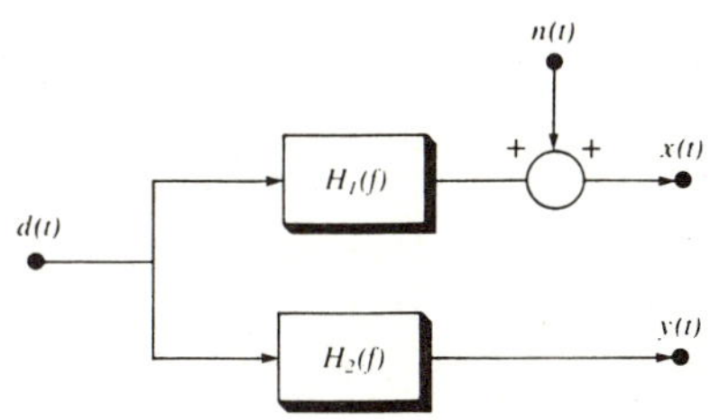

Fig. 2.6 Partially coherent signals

For example, consider the system shown in Fig. 2.3. If n(t) and x(t) are uncorrelated then

$$S_{xy}(f) = H(f).S_x(f) \tag{2.22}$$

$$S_y(f) = |H(f)|^2 S_x(f) + S_n(f) \tag{2.23}$$

$$S_z(f) = |H(f)|^2 S_x(f) \tag{2.24}$$

The squared coherency is therefore expressable as

$$\gamma^2_{xy}(f) = \frac{|S_{xy}(f)|^2}{S_x(f)S_y(f)} \tag{2.25a}$$

$$\gamma^2_{xy}(f) = \frac{1}{1 + \left(\frac{S_n(f)}{S_z(f)}\right)} \tag{2.25b}$$

Because power spectra are positive quantities, equation 2.25b shows that $\gamma^2_{xy}(f)$ must take a value between zero and unity. Equation 2.25 can be re-expressed as

$$\gamma^2_{xy}(f) = 1 - \frac{S_n(f)}{S_y(f)} \tag{2.26}$$

Re-arranging 2.26 gives a convenient formula for computing the noise power spectrum $S_n(f)$, thus

$$S_n(f) = S_y(f)(1 - \gamma^2_{xy}(f)) \tag{2.27}$$

Note again how equations 2.26 and 2.27 show the link between power spectral levels in signal y(t) and noise n(t) and the squared **coherency function.**

2.5 Closed Loop Systems

In practice, systems under study are often inside a feedback loop which cannot be broken. Here we consider some of the spectral relationships which apply in closed loop systems. Consider the system shown in Fig. 2.7, where r(t) is a reference input, possibly corresponding to a measurable control command signal. It is assumed that the signals d(t) and n(t) are independent disturbance signals and, for convenience, the dependence upon frequency will be dropped in equations.

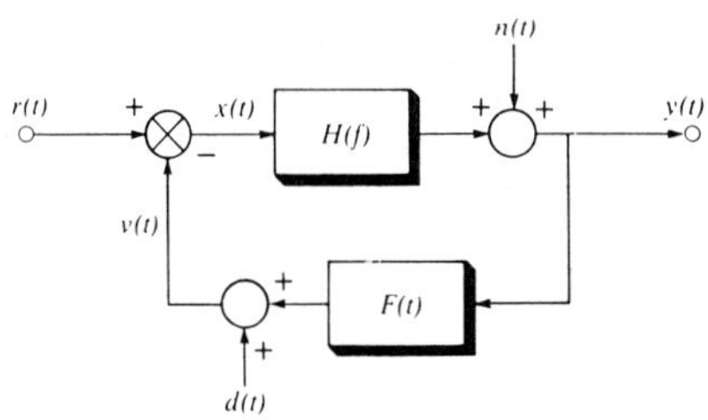

Fig. 2.7 Closed loop system

The power spectra round the loop are

$$S_y|1 + HF|^2 = S_n + |H|^2(S_d + S_r)$$

$$S_x|1 + HF|^2 = |F|^2S_n + S_d + S_r \qquad (2.28)$$

$$S_v|1 + HF|^2 = S_d + |F|^2S_n + |HF|^2S_r$$

Notice that all power spectra are inversely proportional to $|1 + HF|^2$.
Useful cross-spectral relations are

$$|1 + HF|^2S_{xy} = HS_r + HS_d - F^*S_n$$

$$|1 + HF|^2S_{vy} = F^*|H|^2S_r - HS_d + F^*S_n \qquad (2.29)$$

$$|1 + HF|^2S_{vx} = H^*F^*S_r - S_d - |F|^2S_n$$

If r(t) is a measurable signal, possibly even a purposely injected test signal, the following cross-spectral relations are also useful:

$$(1 + HF)S_{vy} = HS_r$$

$$(1 + HF)S_{rx} = S_r \qquad (2.30)$$

$$(1 + HF)S_{rv} = HFS_r$$

The 'normal' approach to frequency response measurement in noise is to apply relations of the form of equation 2.17. Applying this to the system between r(t) and y(t) gives

$$\frac{S_{ry}}{S_r} = \frac{H}{1 + HF} \qquad (2.31)$$

which is the closed loop transfer function between r(t) and y(t).

Applying the same idea between x(t) and y(t) gives

$$\frac{S_{xy}}{S_x} = \frac{HS_r + HS_d - F^*S_n}{|F|^2S_n + S_d + S_r} \qquad (2.32)$$

Note that equation 2.32 does not give the forward path transfer function H(f) expected, because of correlation between y(t) and x(t) due to feedback noise. Certain limiting cases are interesting:

(i) $S_n = 0$; forward path noise free

$\frac{S_{xy}}{S_x} \to H(f)$; the forward transfer function

(ii) $S_r = S_d = 0$; feedback path noise-free

$$\frac{S_{xy}}{S_x} \to \frac{1}{F(f)} \text{ ; the inverse feedback transfer function}$$

In general, the frequency response estimate based on equation 2.32 will depend upon the relative power in n(t), d(t) and r(t) in different frequency bands.

The reference signal r(t) can be used to unscramble correlation effects due to feedback. From equation 2.30 the following equations lead to unbiased estimates of H(f) and F(f):

$$\frac{S_{ry}}{S_{rx}} = H$$

and (2.33)

$$\frac{S_{rv}}{S_{ry}} = F$$

In general, it is difficult to get simple, informative, coherency relations for feedback systems. However, one useful result is

$$\gamma^2_{ry} = \frac{|S_{ry}|^2}{S_r S_y} = \frac{|H|^2 S_r}{S_n + |H|^2(S_d + S_r)} \tag{2.34}$$

which shows that the coherency of the closed loop system from input r(t) to output y(t) is independent of the feedback transfer function F.

3. PRINCIPLES OF DIGITAL SPECTRAL ANALYSIS

3.1 Introduction

Most modern spectral analysis is based upon numerical Fourier transformations derived from sampled versions of continuous signals. This development followed the discovery of the 'fast' Fourier transform algorithms which permit the numerical version of the Fourier integral (equation 2.1) to be computed with great speed and efficiency. However, the application of these digital analysis procedures to continuous phenomena brings certain problems. First, and at an intuitive level, it is not obvious how digital computer based spectral algorithms relate to the traditional hardware scheme of analogue filtering. Second, the use of finite sums of sampled data to represent essentially continuous phenomena introduces an element of approximation which must be accounted for. The following informal discussion is intended to explain the basic computational procedures.

Consider the block diagram (shown in Fig. 3.1) which represents in schematic form the general signal pre-processing associated with all digital spectral analysis. Overall, the pre-processing is intended to determine, from the signal x(t), a discrete facsimile of the continuous Fourier transform X(f), which can be used (in conjunction with the spectra of other signals) to form any required spectral function.

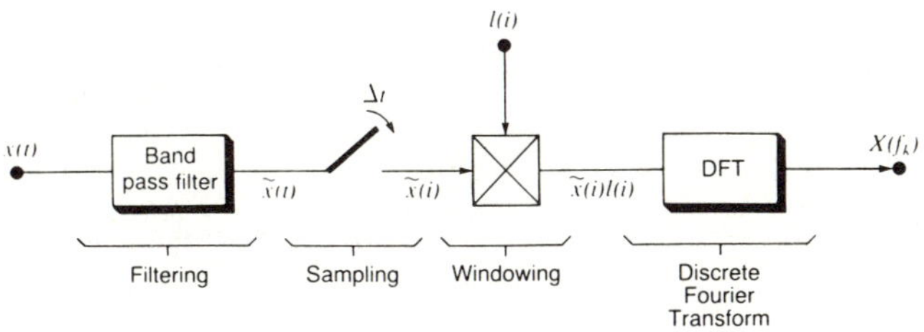

Fig. 3.1 Signal pre-processing for digital spectral analysis

The input filter in Fig. 3.1 is incorporated in order to isolate the frequency range of interest in the analysis. In general, the filtering is band selective and may be achieved by a frequency translation element combined with a high cut-off rate, low pass filter. Moreover, these operations may well be implemented by digital signal processing elements clocking at a suitably higher rate than the spectral sample rate, $f_s = (\Delta t)^{-1}$Hz. For the current discussion, however, it is assumed that the filter is low pass in nature, with cut-off frequency f_c and a high stop band attenuation.

The filtered signal is sampled at a rate of $f_s = (\Delta t)^{-1}$Hz, such that the filtered time signal x(t) is replaced by the sequence x(iΔt), i = 0,1,2,3,... . The act of sampling carries with it a reduction in information which is associated with the lost inter-sample data. This loss is subsequently manifested as an ambiguity between ostensibly unrelated frequency components. The ambiguity (known as aliassing error) arises because the spectrum of a sampled signal is made up of replicas (or aliasses) of the original spectrum X(f) spaced at integer multiples of the sampling frequency, f_s Hz, along the frequency axis (Fig. 3.2). If the signal x(t) has a bandwidth greater than $f_s/2$, the Nyquist frequency, the replicas will overlap and frequency components will be aliassed (Fig. 3.2).

To prevent spectral distortion due to overlapping spectral aliasses, the input must be frequency band-limited to the region 0 to $f_s/2$ Hz. In practice, because of the finite stop-band decay rate in the input filter, it is more common to arrange for the input filter to have a pass-band 0 to f_c, where $f_c \simeq f_s/4$. This ensures that any spectral information at the Nyquist frequency and above is attenuated to insignificant levels. For such an arrangement the distortion introduced by sampling can usually be neglected.

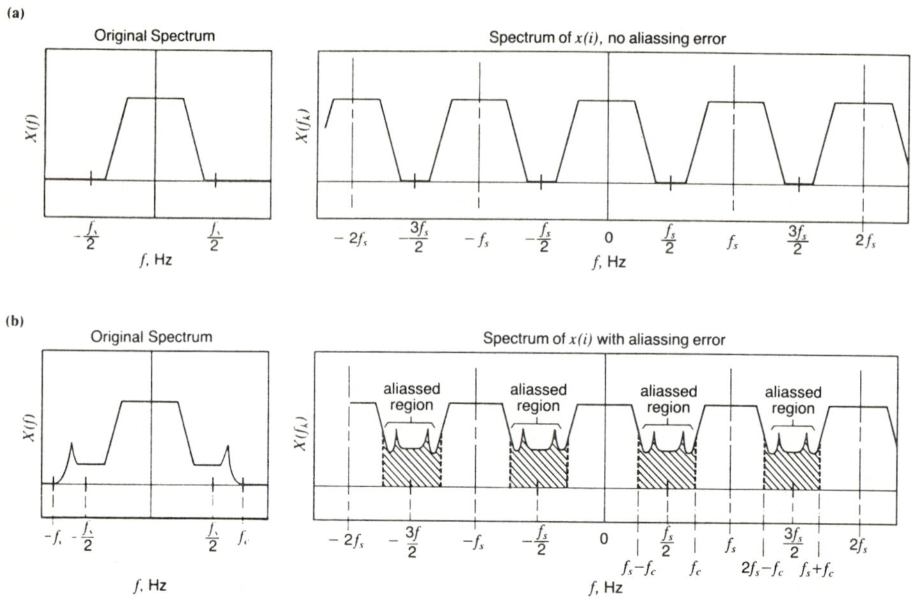

Fig. 3.2 Aliassing error

3.2 Windowing and Spectral Resolution

The data window operation in Fig. 3.1 takes the sequence x(i) and multiplies it by the data window sequence l(i). The object is to select a finite block consisting of T seconds of data, upon which to conduct the discrete Fourier transform defined by the finite summation

$$\tilde{X}(f_k) = \sum_{i=0}^{N-1} x(i)l(i) \exp\left(\frac{-j2\pi ik}{N}\right) \tag{3.1}$$

$$(k = 0,\ldots,N-1)$$

where the frequency ordinates, f_k, are related to the index, k, by

$$f_k = \frac{k}{\Delta tN} \tag{3.2}$$

and T = Δt.N.

The discrete Fourier transform will be discussed later, but for the moment note that the summation in equation 3.1 extends only over the interval 0 to N-1, which corresponds to a time block of $T = N\Delta t$ seconds of the continuous signal $x(t)$. The data window $l(i)$ reflects this feature by formally truncating the data outside the region 0 to T seconds. Specifically, all spectral data windows have the property

$$l(i) = 0 \text{ for } i < 0 \text{ and } i \geq N. \tag{3.3}$$

In between, the data window can take any one of a number of shapes. In order to understand the relevance of data windowing, it is necessary to examine the influence of the windowing on spectral behaviour. To do this, neglect the sampling action (which is in order if the input filter is correctly chosen) and consider the continuous version of the numerical Fourier transform (equation 3.1) based upon a data block of T seconds duration:

$$\tilde{X}(f) = \int_0^T x(t)\, l(t) \exp(-j2\pi ft)dt \tag{3.4}$$

This can be written as:

$$\tilde{X}(f) = \int_{-\infty}^{+\infty} X(\rho)L(f - \rho)d\rho \tag{3.5}$$

where ρ is a dummy frequency variable.

Equation 3.5 is a convolution in the frequency domain and can be written as:

$$\tilde{X}(f) = X(f) * L(f)$$

where * denotes convolution and $L(f)$ is the spectrum of $l(t)$.

In words, equation 3.5 tells us that the spectrum of $x(t)$, calculated using only a finite time block of duration $T = N.\Delta t$ seconds and shaped in that interval using a function $l(t)$, will be the true spectrum $X(f)$ convolved with the spectrum $L(f)$ of $l(t)$. Consider an example, in which $l(t)$ is the rectangular window $l_R(t)$ defined by

$$\begin{aligned} l_R(t) &= 1 \quad 0 \leq t \leq T \\ l_R(t) &= 0 \quad \text{elsewhere} \end{aligned} \tag{3.6}$$

Notice this corresponds to selecting a block of data of duration $T = N\Delta t$ without further modification. The gain spectrum of this window is (Fig. 3.3)

$$|L_R(f)| = T \left| \frac{\sin(\pi fT)}{\pi fT} \right| \tag{3.7}$$

The spectrum of $L_R(f)$ is therefore low-pass with a central lobe of width proportional to $1/T$ and decay rate proportional to $1/f$. Moreover, the area under the window spectrum is unity, because

$$\int_{-\infty}^{+\infty} L_R(f)df = l_R(0) = 1 \tag{3.8}$$

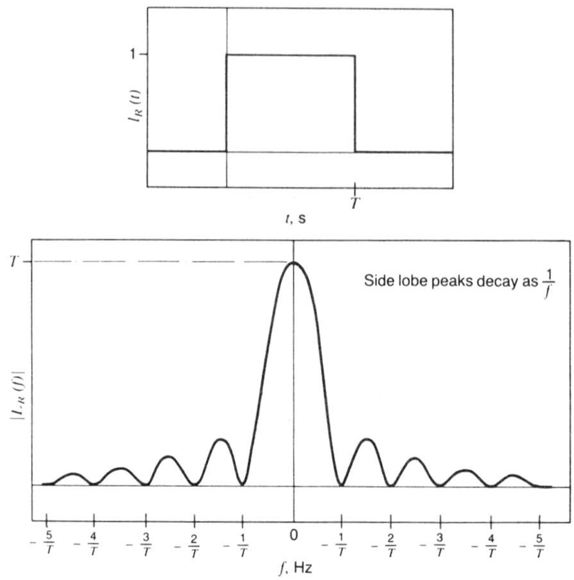

Fig. 3.3 Characteristics of the rectangular window

The data window shape is therefore of the utmost importance since it sets the frequency selective structure of digital spectral analysis. In general, the bandwidth of the resolving filters L(f) is proportional to 1/T, and for greater resolution it is necessary to increase the duration of the data block. However, within the confines of a specific block size it is possible to alter the resolving power by suitable selection of l(t). Specifically, the decay rate of the rectangular window spectrum is relatively slow (Fig. 3.3), so that the interchannel interference of relatively distant spectral components caused by overlapping filter side lobes can be high. It is in an effort to decrease this interference between spectral lines that various data window shapes are used. For example, a commonly used data window is the Hanning window, $l_H(t)$, shown in Fig. 3.4. The Hanning window gain spectrum $|L_H(f)|$ is defined as

$$|L_H(f)| = \frac{T}{2}\left|\frac{\sin(\pi fT)}{\pi fT}\right| \quad \frac{1}{|(1-(fT)^2)|} \tag{3.9}$$

or

$$|L_H(f)| = \frac{1}{2(1 - (fT)^2)} \; |L_R(f)| \qquad (3.10)$$

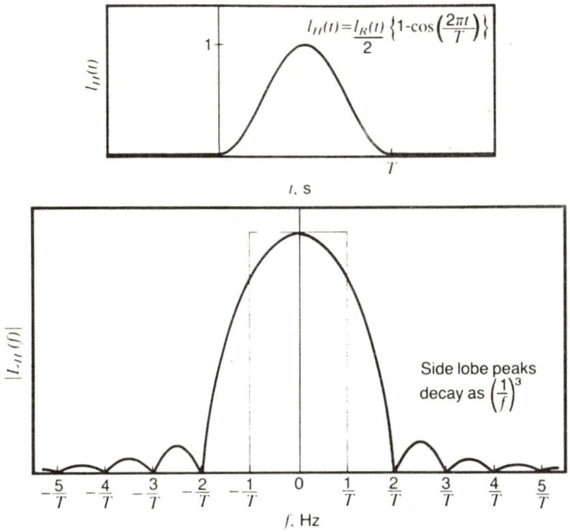

Fig. 3.4 Characteristics of the Hanning Window

Thus, the Hanning window has a stop band decay rate proportional to $(1/f)^3$ and central lobe which is twice as wide as that of the rectangular window. The Hanning window spectrum is also frequently expressed as the weighted sum of three rectangular windows. Specifically, $L_H(f)$ can be written

$$L_H(f) = \frac{1}{2} L_R(f) - \frac{1}{4} L_R(f - 1/T) - \frac{1}{4} L_R(f + 1/T) \qquad (3.11)$$

The merit of the Hanning window is the rapid decay of the side lobes. It is the non-zero side lobes of the spectral window which cause the leakage error to remote spectral regions.

Other choices of window exist which enable a similar increase in stop band attenuation in L(f). However, the most important remaining choice is the 'constant gain', or flat top, window $l_{CG}(t)$, which gives rise to a gain spectrum $|L_{CG}(f)|$ which is nominally constant over the spacing $-\frac{1}{2T} < f < +\frac{1}{2T}$ and yet has acceptable side lobe decay rate. In practice, there is a trade-off between these two desirable properties.

All three above-mentioned windows have their uses, depending on the application, which can be summarised as follows:

(i) Rectangular Window $L_R(f)$

Properties - narrow central lobe, low stop band decay rate.
Uses - analysis of zero mean transient data which are zero at the extremities of the data block. Also used on periodic signals whose period is an *exact* submultiple of the block duration T.

(ii) Hanning Window $L_H(f)$

Properties - relatively narrow central lobe, good stop band decay rate.
Uses - general random signal applications to obtain low spectral leakage, but with good local resolving power (associated with the relatively narrow central lobe).

(iii) Constant Gain (Flat-Top) Window $L_{CG}(f)$

Properties - wide central lobe, good stop band decay rate.
Uses - analysis of periodic signals with a rich line structure. The flat central lobe ensures that all spectral lines in the input signal are captured at their correct gain level and not erroneously attenuated by the skirts of the Hanning or Rectangular window.

3.3 Discrete Fourier Transform

The discrete version of the Fourier transformation is given for a sequence

$a(i)$, $i = 0,1,\ldots, N-1$ as:

$$A(f_k) = \sum_{i=0}^{N-1} a(i) \exp\left(-j\frac{2\pi ik}{N}\right) \tag{3.12}$$

$$k = 0,1,2,\ldots,N-1$$

where the spectral ordinates are positioned at the harmonically related frequencies

$$f_k = k/T,\ k = 0,1,2,\ldots,N-1.$$

If the sequence is real, as is true in most time signal analysis, then the discrete Fourier sequence need only be calculated for $k = 0,1,\ldots,N/2$, since the symmetry

$$A(f_k) = A^*(f_s - f_k) \tag{3.12}$$

applies.

The coefficients for $k > N/2$ can be associated with the first spectral alias. Thus, for a real time signal, the discrete Fourier transform supplies $(N/2 + 1)$ harmonic spectral ordinates spaced at $1/T$ Hz intervals in the region 0 Hz to the Nyquist Frequency $f_N = 1/2\Delta t$.

The discrete Fourier transform (DFT) defined in this way has the advantage that the periodicity in the complex exponent $\{\exp(-j2\pi/N)\}$ can be exploited to reduce the number of multiplications needed to compute the transform from being proportional to N^2, to proportional to N. The resultant 'Fast' Fourier Transform (FFT) algorithms will not be discussed in detail since, for current purposes, they are a means to an end. In fact, the abbreviation FFT has become a generic term for a class of fast numerical discrete Fourier transforms.

3.4 Circular Convolution

An important implication of the complex exponential periodicity is that, *as far as the DFT is concerned*, the finite sequence $a(i), i = 0,\ldots,N-1$, (in equation 3.12) is *assumed* to be periodic in N. Thus, if the input signal does indeed have this period $T = N\Delta t$, then the DFT is exact. In this case no windowing error is involved and no special data window other than the implied rectangular window need be used.

In general, however, the source data will not be periodic in N, and it is important to avoid the errors which arise from the implied periodicity (or circularity) of the finite sequence a(i). In the following, a spectral sampling argument is used to explain the DFT implied periodicity, and to explore its consequences.

The DFT is set up to calculate the spectrum A(f) of the original sequence a(i) at the equally spaced frequency intervals, f_k, given by

$$f_k = k/T, \; k = 0,1,2,3,\ldots \tag{3.14}$$

Thus the DFT actually computes a *sampled* version $A(f_k)$ of the true spectrum A(f). As a result of this implicit sampling, the corresponding time sequences obtained by inverse Discrete Fourier Transformation repeat themselves with period T. In fact, an aliassing phenomenon occurs which is the dual of the frequency spectrum aliassing encountered when sampling a time signal. To be specific, and with respect to the spectrum shown in Fig. 3.5, the time function obtained by inverse Fourier transforming the sampled spectrum $A(f_k)$ is periodic in nature and consists of a series of replicas of the true time function, spaced at integer multiples of $T(= 1/\Delta f)$ seconds. Actually, only the region $-T/2$ to $+T/2$ is computed in the inverse DFT. Nevertheless, if the true time function a(t) is non-zero outside this time band, then overlapping replicas of a(t) will cause aliassing errors to occur (see Fig. 3.6). This time domain aliassing causes errors when DFT spectra are multiplied to achieve a convolution operation in the time domain. In fact, multiplying DFT spectra is equivalent to convolving the aliassed time functions (sometimes called circular convolution) and this may cause the convolved replicas to spread outside the time band $-T/2$ to $+T/2$ and distort the circular convolution.

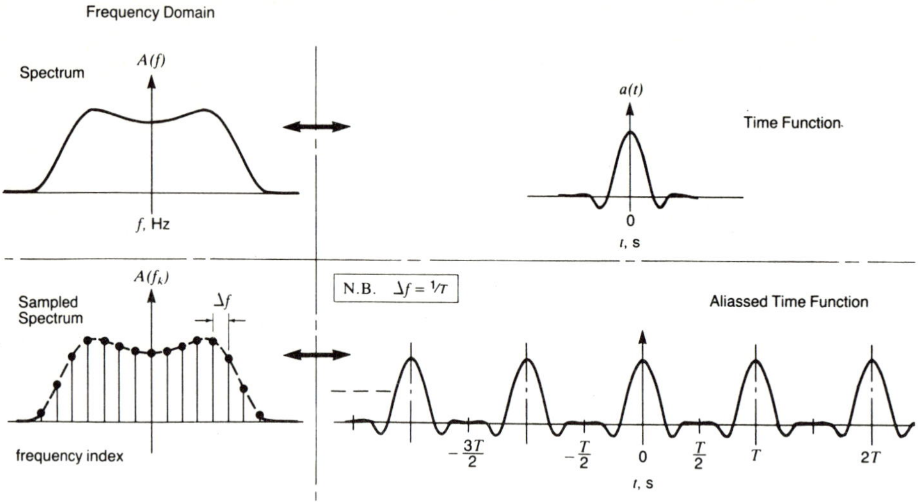

Fig. 3.5 Aliassing of time functions

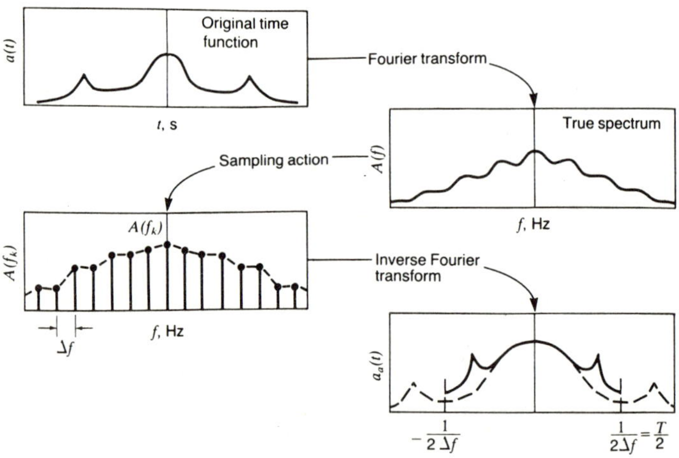

Fig. 3.6 Time domain aliassing

Now, the power spectrum, cross-spectrum and frequency response estimates are obtained by multiplying DFT computed spectra, so that their respective time domain equivalents(autocorrelation, crosscorrelation and impulse response) may suffer from alias overlap. Specifically, if $A(f_k)$ in Fig. 3.5 were a DFT power spectrum estimate, then the autocorrelation estimate obtained by inverse DFT of the sampled power spectrum would be a series of replicas of the true autocorrelation, spaced at integer multiples of $T(= 1/\Delta f)$ seconds along the time axis. Actually, only the region from 0 seconds to T/2 seconds is usually computed. Nevertheless, if the true autocorrelation a(t) is non zero for $t > T/2$ then aliassing of the replicas will occur. In practice this means that the original power spectrum curve should not contain rapid rates of change.

Clearly then, avoiding aliassing errors in the autocorrelation, crosscorrelation and impulse response estimates obtained by inverse DFT is equivalent to the problem of adequate spectral resolution in the spectral estimates. If a spectral estimate is adequately resolved with, say, at least five points on the fastest changing slopes, then time domain aliassing of the inverse function will not occur.

However, the technique should be used with care, in particular, note:

(i) The above reasoning applies to estimates derived for stationary, random processes with no deterministic periodic components.

(ii) If the original data are periodic, or contain a significant periodic component, then the periodic frequency should be equal to $\Delta f = (1/T)$, or a harmonic of Δf and the data window should be rectangular.

(iii) If the original data x(t), y(t) are transient in nature, then to avoid time domain aliassing, the non-zero portions of x(t) and y(t) laid end to end should fit in one block T.

Finally, time domain aliassing can also cause problems when the maximum cross-correlation occurs at a relatively large time lag. Such a time shift may cause overlap of time domain aliases and a concomitant error in the recovered estimate. The possibility of this happening can be diagnosed in the frequency domain by very rapid rates of change in the phase estimates. A practical situation where this might occur is in time delay systems such as testing material transfer and charge transfer devices. If possible, most of the relative time delay in x(t), y(t) should be trimmed out before analysis begins. The same reasoning can be applied to noise-like signals, whereby large relative delays relating two signals can cause errors because their entire joint correlation structure is not contained within the data block span of T seconds. Again, the solution is to trim out relative delays by artificially delaying the advanced signal.

3.5 Summary

(i) The sampling rate $f_s = 1/\Delta t$ sets the maximum bandwidth over which spectral components can be unambiguously distinguished. The upper frequency which can be distinguished is $f_s/2$ (the Nyquist frequency). In practice, the input filter cut-off frequency f_c is set well below the Nyquist frequency.

(ii) The data block duration T sets the limit $\Delta f = 1/T$ within which spectral ordinates can be resolved. For sampled data, the block duration is $T = N.\Delta t$ and the Discrete Fourier Transform computes spectral ordinates only at frequencies $f_k = k/T$, $k = 0,1,2,\ldots,N/2$.

(iii) The implied sampling of the frequency spectrum at Δf intervals means that the inverse Fourier transform is an aliassed time function, consisting of replicas of the true time function spaced at integer multiples of $T = 1/\Delta f$. To avoid time domain aliassing errors the true time function should be non-zero only in the interval 0 to T/2 which by analogy with the frequency sampling theorem, is called the Co-Nyquist period.

(iv) The sequence $X(f_k)$, $k = 0,1,2,\ldots,N/2$ is a useful representation of the underlying spectrum X(f) provided the spacing of the spectral ordinates Δf is sufficient to resolve the components of X(f) and the filter and sampling rate are selected so as to ensure that the sampled signal is band-limited to between 0 and $\frac{1}{2} f_s$ Hz.

4. STATISTICAL PROPERTIES OF SPECTRAL ESTIMATES

4.1 Introduction

Using the digital pre-processing routines outlined in Section 3 (and summarised in Fig. 3.1) the Discrete Fourier Transform of a data block is obtained. The data block is T seconds in duration and corresponds to the span of data selected by the window function. In a digital spectral analysis procedure the windowing function is preceded by a data block gathering function such that sequential blocks of data are drawn in and processed to form a set of m block spectra $X_r(f_k)$, $r = 1,2,3,\ldots,m$. Then, as indicated in Fig. 4.1 the power spectrum estimate $\hat{S}_x(f_k)$ is determined as

$$\hat{S}_x(f_k) = \frac{v}{m} \sum_{r=1}^{m} |X_r(f_k)|^2 \tag{4.1}$$

$$f_k = \frac{k}{T}, \; k = 0,1,2,3,\ldots$$

where the averaging of several data blocks spectra is necessary to smooth certain estimation errors, and v is a scaling function discussed later.

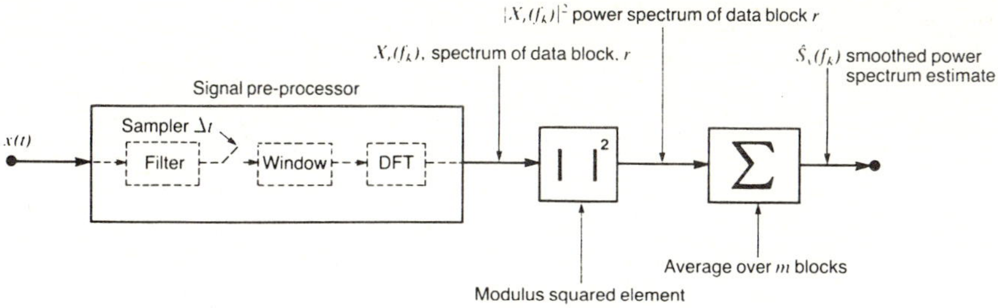

Fig. 4.1 Power spectrum estimation scheme

With the provision of two signal processing channels the cross-spectrum estimate is formed thus:

$$\hat{S}_{xy}(f_k) = \frac{v}{m} \sum_{r=1}^{m} X_r^*(f_k) Y_r(f_k) \tag{4.2}$$

These spectral estimates together with $\hat{S}_y(f_k)$, the power spectrum of y(t), are then used as the basis for all subsequent estimated quantities.

4.2 Power Spectrum

4.2.1 Scaling and Bias

The power spectral estimates computed according to equation 4.1 are subject to both systematic and random errors. Consider first the systematic, or bias, errors in the estimator (equation 4.1). As a preliminary, it is assumed that the input filter in Fig. 4.1 is such that all frequency aliasing errors can be neglected. Then, the systematic error in the power spectrum estimate is associated only with windowing errors. However, certain scaling factors are also required to deal with the transition from continuous signals to discrete computations. Therefore, consider first the block spectrum $X_c(f)$ computed from the continuous time data x(t), windowed by the function l(t) which is zero everywhere except the region 0 to T. It can be shown that the expected value of the quantity $|X_c(f)|^2$ is given by

$$E\{|X_c(f)|^2\} = |L(f)|^2 * S_x(f) \tag{4.3}$$

where $S_x(f)$ is the true power spectrum of the continuous process x(t), and the star * denotes convolution.

If the power spectrum is flat over the pass-band of the window spectrum L(f) then:

$$E\{|X_c(f)|^2\} = S_x(f) \int_{-\infty}^{+\infty} |L(f)|^2 \, df \tag{4.4}$$

where, by Parsevals Theorem:

$$\int_{-\infty}^{+\infty} |L(f)|^2 \, df = \int_0^T l^2(t)dt \tag{4.5}$$

Note that the continuous windowed spectrum $X_c(f)$ is given by

$$X_c(f) = \int_0^T x(t)\ l(t)\ \exp\ (-j2\pi ft)dt \tag{4.6}$$

The discretised version of equation 4.6 is

$$X_c(f) \simeq X_d(f) = \Delta t \sum_{i=0}^{N-1} x(i)\ l(i)\ \exp\ (-j2\pi f\Delta ti) \tag{4.7}$$

where Δt is the sampling interval and i is the integer used to index the sampled sequences $x(i)$, $l(i)$.

The Discrete Fourier Transform is obtained from equation 4.7 by restricting the frequencies at which the discretised transform is computed to the harmonic intervals

$$f_k = \frac{k}{N\Delta t}\ ,\ k = 0,1,\ldots,N \tag{4.8}$$

Thus the DFT of $x(i)\ l(i)$ computed according to

$$X(f_k) = \sum_{i=0}^{N-1} x(i)\ l(i)\ \exp\left(\frac{-j2\pi ki}{N}\right) \tag{4.9}$$

is related to the continuous spectrum $X_c(f)$ by

$$X_c(f)_{f=f_k} \simeq \Delta t\ X(f_k) \tag{4.10}$$

From equations 4.4, 4.5 and 4.10, therefore, the expected value of the rth block power spectrum is related to the true power spectrum of the process $x(t)$ by

$$E\ \{v|X_r(f_k)|^2\} \simeq \frac{v \int_0^T l^2(t)dt}{(\Delta t)^2}\ .\ S_x(f)_{f=f_k} \tag{4.11}$$

Thus, to scale the digital estimates correctly, the scaling factor v is given by

$$v = \frac{(\Delta t)^2}{\int_0^T l^2(t)dt} \tag{4.12}$$

When the window $l(t)$ is defined as the sequence

$l(i)$, $i = 0,\ldots,N-1$ then equation 4.12 becomes

$$v = \frac{\Delta t}{\sum_{i=0}^{N=1} l^2(i)} \tag{4.13}$$

In some texts, the power spectrum is defined in terms of the *sampled* autocorrelation, in which case the term Δt does not appear in the scaling function (equation 4.13). If $x(t)$ is measured in volts (V) then, with the scaling of equation 4.13, the units of the power spectrum are V^2s.

From now on, the fact that the digitally computed estimates are calculated at frequencies f_k will be assumed and the subscript k dropped. The expected value of the smoothed estimate $\hat{S}_x(f)$ (equation 4.1) is, from equation 4.11, given by

$$E\{\hat{S}_x(f)\} = v\{|L(f)|^2 * S_x(f)\} \tag{4.14}$$

In words, the estimated power spectrum has a mean value equal to the convolution of the true power spectrum with the squared gain function of the window spectrum $L(f)$, scaled by v. The bias, or systematic error, in $\hat{S}_x(f)$ is given by

$$B\{\hat{S}_x(f)\} = E\{\hat{S}_x(f)\} - S_x(f) \tag{4.15}$$

Thus, as illustrated in Section 3, for $S_x(f)$ to be undistorted (i.e. unbiassed) it is important that the true spectrum be flat over the bandwidth ($\simeq 1/T$ Hz) of the central lobe of the window spectrum and that the side lobes of $L(f)$ should decay rapidly.

4.2.2 Random Errors

Provided the data blocks used are independent (e.g. non-overlapping) the root mean square value associated with random estimation errors in $\hat{S}_x(f)$ is given by

$$\frac{\text{rmse}\,\{\hat{S}_x(f)\}}{S_x(f)} = \sqrt{\frac{1}{m}} \tag{4.16}$$

If a more precise measure of the quality of estimates is required then confidence intervals must be constructed. Roughly speaking, confidence intervals of say, x%, would specify a band of values within which one is x% certain that the true value lies. Confidence bands for the power spectrum estimator $\hat{S}_x(f)$ are obtained as follows. The estimator (equation 4.1) is distributed as a χ^2, variable with $\nu = 2m$ degrees of freedom. It follows that the $100(1 - \alpha)\%$ confidence limits for an estimate $\hat{S}_x(f)$ based upon m block averages are given by

$$\text{Prob}\left\{\frac{2m\,\hat{S}_x(f)}{\chi_{2m}(1-\alpha/2)} > S_x(f) > \frac{2m\,\hat{S}_x(f)}{\chi_{2m}(\alpha/2)}\right\} = 1-\alpha \tag{4.17}$$

Where $\chi_{2m}(1-\alpha/2)$ and $\chi_{2m}(\alpha/2)$ are the $1-\alpha/2$ and $\alpha/2$ points taken from a tabulated χ^2 distribution with 2m degrees of freedom.

Example

The 95% confidence limits for $\hat{S}_x(f)$ based on m = 12 blocks are

$$\frac{24\,\hat{S}_x(f)}{\chi_{24}(0.975)} > S_x(f) > \frac{24\,S_x(f)}{\chi_{24}(0.025)} \tag{4.18}$$

Now from χ^2 tables, $\chi_{24}(0.975) = 12.4$

and $\chi_{24}(0.025) = 39.36$

Therefore, with 95% probability the true value of $S_x(f)$ lies between 1.935 $\hat{S}_x(f)$ and 0.609 $\hat{S}_x(f)$.

In logarithmic form, the confidence limits are independent of the estimated value, thus, in the example, the 95% confidence limits are 10 log $\hat{S}_x(f)$ + 10 log 1.935 and 10 log $\hat{S}_x(f)$ + 10 log 0.609 (i.e., + 2.9 dB and -2.15 dB of the log estimate).

4.3 Cross Spectrum

4.3.1 Scaling and Bias

For cross-spectral estimates obtained using equation 4.2 the scaling factor v is that given by equations 4.12 and 4.13. The expected value of the estimator is

$$E\,\{\hat{S}_{xy}(f)\} = v\{|L(f)|^2 * S_{xy}(f)\} \tag{4.19}$$

Then, to minimise systematic error (i.e. bias), in cross-spectral estimates it is important that the gain and phase of $S_{xy}(f)$ should vary slowly over the bandwidth of the central lobe of the window spectrum. Also, the side lobes of the window spectrum should decay rapidly.

4.3.2 Random Errors

Provided the data blocks used are independent (e.g. non-overlapping), the root mean square error (rmse) in amplitude and phase of cross-spectral estimates is given by

$$\frac{\text{rmse}\,(|\hat{S}_{xy}(f)|)}{|S_{xy}(f)|} = \left[\frac{1}{2m}\left(1 + \frac{1}{\gamma^2_{xy}(f)}\right)\right]^{1/2} \tag{4.20}$$

$$\text{rmse}(\hat{\phi}_{xy}(f)) = \left[\frac{1}{2m}\left(\frac{1}{\gamma^2_{xy}(f)} - 1\right)\right]^{1/2} \tag{4.21}$$

where $\gamma^2_{xy}(f)$ is the squared coherency function. Normalised rmse curves are given in Wellstead (1984a).

4.4 Coherency

4.4.1 Bias

The systematic error in coherency estimates computed from the power and cross-spectra estimates springs from windowing errors and block averaging errors. Specifically, it can be shown that the expected value of $\hat{\gamma}_{xy}(f)$, estimated from $\hat{S}_{xy}(f)$, $\hat{S}_x(f)$ and $\hat{S}_y(f)$, as given in equations 4.1 and 4.2, is approximately

$$E\,\{\hat{\gamma}^2_{xy}(f)\} \simeq \gamma^2_{xy}(f) + \frac{1}{m}\,(1 - \gamma^2_{xy}(f)) + \frac{A}{N^2}\,(\dot{\phi}_{xy}(f)\gamma_{xy}(f))^2 \tag{4.22}$$

where A is a (small) constant, proportional to the width of the central lobe of the window spectrum and $\dot{\phi}_{xy}(f)$ is the rate of change of phase $\phi_{xy}(f)$ with frequency.

The last term on the right hand side of equation 4.22 is only significant if the cross-spectral phase $\phi_{xy}(f)$ is not relatively constant across the bandwidth of the central lobe of the window spectrum and in high resolution digital analysis it can be neglected.

The second term is associated with the block averaging procedure and explains why, for one average (i.e. m = 1), the estimated coherency is exactly unity. This can also be seen from writing the estimation equation:

$$\hat{\gamma}^2_{xy}(f) = \frac{|\hat{S}_{xy}(f)|^2}{\hat{S}_x(f)\hat{S}_y(f)} \tag{4.23}$$

in terms of the block spectra. For values of $m > 10$ the bias error is usally negligible.

4.4.2 Random Errors

The root mean square error (rmse) associated with coherency estimates is given by

$$\text{rmse}\ \{\hat{\gamma}^2_{xy}(f)\} = \sqrt{\frac{2\gamma^2_{xy}(f)}{m}}\ (1 - \gamma^2_{xy}(f)) \qquad (4.24)$$

Curves of rmse against average number m, for various coherency levels, are plotted in Wellstead (1984b).

Confidence levels for coherency estimates can be obtained by associating a transformed version of coherency with a normally distributed variable. In particular, define the random variable $\eta(f)$ as

$$\eta(f) = \frac{1}{2}\ \ln\left(\frac{1 + \hat{\gamma}_{xy}(f)}{1 - \hat{\gamma}_{xy}(f)}\right) = \tanh^{-1}\ \hat{\gamma}_{xy}(f) \qquad (4.25)$$

For m > 10 and coherency in the range $0.4 \leqq \gamma^2_{xy} \leqq 0.95$, the variable $\eta(f)$ has an approximately normal distribution, with mean value $\mu(f)$ and variance σ^2 given by

$$\mu(f) = \tanh^{-1}\ \gamma_{xy}(f) \qquad (4.26a)$$

$$\sigma^2 = \frac{1}{2(m-1)} \qquad (4.26b)$$

The approximate 100 $(1 - \alpha)$% confidence levels for the transformed variable are given by

$$\text{Prob}\left\{Z(1 - \alpha/2) < \frac{\eta(f) - \mu(f)}{\sigma} < Z(\alpha/2)\right\} = 1 - \alpha \qquad (4.27)$$

where $Z(\alpha/2)$ is the ordinate corresponding to an area $\alpha/2$ under the standard Normal density function and $Z(1 - \alpha/2) = -Z(\alpha/2)$. Rearranging equation 4.27, the 100$(1 - \alpha)$% confidence levels on the coherency function are

$$\left[\tanh\left\{\eta(f) \pm \frac{Z(\alpha/2)}{\sqrt{2m}}\right\}\right]^2 \qquad (4.28)$$

At the 90% confidence level (i.e. $\alpha = 0.1$), $Z(\alpha/2) = 1.645$
At the 95% confidence level (i.e. $\alpha = 0.05$), $Z(\alpha/2) = 1.96$

Example

For an example $\hat{\gamma}^2_{xy}(f) = 0.5$, the transformed variable $\eta(f) = 0.881$. If m = 20, the 90% limits of the transformed variable are 0.881 ± 0.26. Transforming back, the 90% confidence range on the true coherency is 0.305 to 0.664.

A look-up Table, giving 90% confidence limits on coherency for various block average numbers, is given in Wellstead (1984b).

4.5 Discussion

Expressions for bias and random errors in frequency response estimates computed from the power spectrum and cross-spectrum estimates are given in Wellstead (1984b).

The statistical results developed throughout this Section assume that non-overlapping blocks of data, often drawn sequentially from a data record, are used. Where data are limited, it is common practice to overlap data blocks, to get more blocks (and hence more statistical smoothing) from a fixed record length. The effect of this is to correlate the information in a given block with its neighbours, which reduces the smoothing effect of each block. However, because this overlapping gives more blocks to average, the net effect is an improved statistical average.

The technique of overlapping and its effects are discussed in more detail by Welch (1967).

5. CEPSTRAL ANALYSIS

5.1 Introduction

An advantage claimed for spectral analysis is its relatively superior sensitivity when compared with time domain analysis. Cepstral analysis is a means of exploiting and emphasising this sensitivity in order to diagnose the presence of certain features hidden in the original time data. To be more specific, a number of situations arise which cause the spectrum to have periodic components. Using the cepstrum it is often possible to detect these periodicities and relate them back to features of the original data which were not apparent in the time-domain.

When the echo phenomena are separable then an auto and cross-correlation approach is possible. Cases arise, however, where multiple echoes occur at close time intervals and they cannot be separately discriminated in the correlation functions. The cepstrum is a pseudo-correlation function which is aimed at discriminating echo and delay phenomena.

The original application of cepstral techniques was to seismic-signal processing. This area is typified by data which consist of a primary waveform together with filtered, delayed versions (or echoes) of the primary. Frequently however the echoes are buried in noise and as such are indistinguishable in the original time trace. Cepstral analysis can often detect such echoes and their relative positions *vis a vis* the primary waveform.

In practice, any engineering phenomenon which involves multiple transmission paths of a signal to a single measurement point is susceptible to this form of analysis. Potentially much better discriminating power for delay processes can be achieved compared to alternative correlation or cross-spectral phase methods. However, cepstral methods are a subjective tool and can yield good results in one application and inexplicably poor results in another.

Aside from this cautionary note, cepstral analysis is likely to be helpful in any situation involving a basic time function which has been convolved with a sequence of impulse functions. The resulting time data are then repetitive, or reverberant, in a manner which causes periodicities. It is these periodicities which are emphasised and drawn out by cepstral studies.

5.2 Interpretation

The word 'cepstrum' is an anagram of 'spectrum'. The term arose because cepstra are formed by transforming spectra (Bogert et al, 1963). Formally, the power cepstrum $C_x(\tau)$ of a process $x(t)$ is defined as the Fourier transform of the log power spectrum $S_x(f)$ of $x(t)$.

Thus:

$$C_x(\tau) = F[\log (S_x(f))] \tag{5.1}$$

where F is the Fourier transform operator.

The power cepstrum is a function of τ, which has the units of time, but is termed "quefrency" (cf. frequency) to distinguish it from "real" time.

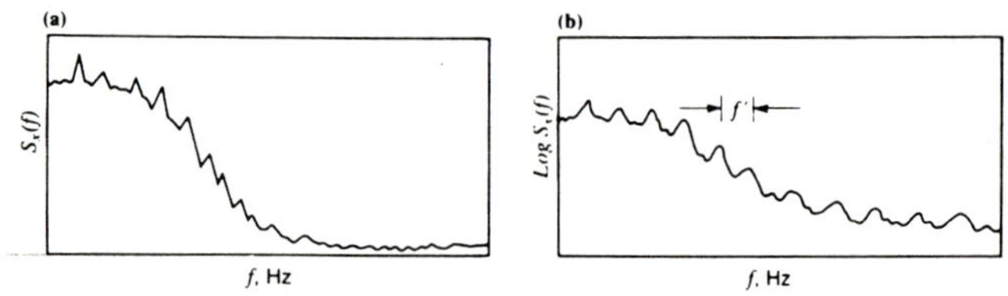

Fig. 5.1 Cepstral analysis of a power spectrum containing periodicities
(a) Original power spectrum
(b) Log power spectrum, emphasising periodic component, period f' Hz.

The technique works as follows. If $S_x(f)$ contains periodic fluctuations as in Fig. 5.1a then taking the logarithm emphasises previously attenuated portions of the spectrum (Fig. 5.1b) such that the cepstrum will contain a peak at the quefrency of the oscillations in $S_x(f)$ (Fig. 5.2)

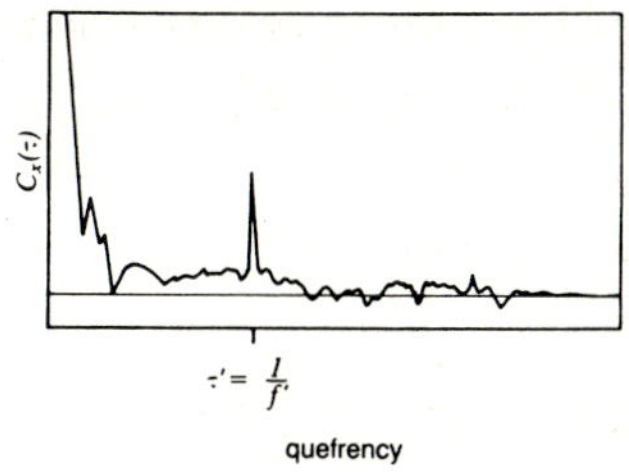

Fig. 5.2 Power cepstrum showing peak at quefrency $\tau' = \frac{1}{f'}$

Note that, in addition to the defining relation equation 5.1, in which the power cepstrum is formed from smoothed power spectrum estimates, it can also be obtained for the energy spectrum, in which case the term power cepstrum is still retained.

A number of phenomena can cause oscillatory patterns in the power spectrum, but chief amongst these are:

(i) non-linearities in the signal generation mechanism, where the periodicities are associated with harmonic components of a fundamental frequency;
(ii) delayed replicas of a signal component added to the original time signal and causing a periodic oscillation in the power spectrum.

Specifically, if a signal x(t) consists of a primary component p(t) plus an echo component p(t - ρ), then,

$$x(t) = p(t) \div p(t - \rho) \tag{5.2}$$

where α is the echo magnitude
and ρ is the echo time delay.

Then the log power spectrum of x(t) can be shown to contain a periodic component of frequency ρ. Thus,

$$\log S_x(f) = \log S_p(f) + 2\alpha \cos 2\pi f\rho + k \tag{5.3}$$

where k is a constant.

When the log power spectrum is Fourier transformed the cosinusoid will show up as an impulse at the quefrency ρ seconds, corresponding to the echo delay time in equation 5.2.

A further useful feature of the cepstrum is its ability (sometimes) to separate various signal components. For example, if a signal y(t) is the result of an unobservable, noise-like signal x(t) passing through some unknown linear system H(f), then the log-power spectrum of y(t) is:

$$\log [S_y(f)] = \log [H(f)] + \log [S_x(f)]. \tag{5.4}$$

That is, the multiplication effect of the transfer function becomes additive in the logarithmic spectrum. Moreover, this additive form is maintained in the cepstrum. Thus:

$$C_y(\tau) = F\,[\log H(f)] + C_x(\tau) \qquad (5.5)$$

Often, the two cepstral functions $F[\log H(f)]$ and $C_x(\tau)$ occupy distinct regions of the quefrency axis, such that they can be separated out. In certain cases the two signal components can be reconstructed by this method.

5.3 Measurement Procedures

The power cepstrum is computed from either a power spectrum or an energy spectrum measurement. Because the aim is to detect peaks, relatively little is known about the systematic and random errors in cepstral estimates. Intuitively, it is clear proper attention should be paid to obtaining correctly resolved power spectral estimates, free from windowing errors, otherwise the spectral periodicity to be diagnosed may be masked from the start.

It must, however, be emphasised that cepstral analysis, as mentioned previously, is a highly subjective tool. Broadly similar applications can occur, with the cepstrum working well in one case but not in another. In general, the quality of cepstral results can be influenced by a number of experimental factors. For example,

(i) Differentiation, or high pass filtering, of the power spectrum or source data can emphasise echo phenomena and lead to more clearly defined cepstral peaks.

(ii) The use of different windows on the time data has been observed to enhance the echo discriminating power of cepstra in certain cases.

It is not yet possible to give general rules on how to improve cepstral analysis and one must, therefore, treat each case separately and be prepared to experiment with different data processing techniques.

5.4 Applications of Cepstral Analysis

The basic application of cepstral methods is the diagnosis of echoes, or reverberation, in time signals, where the power cepstrum of the signal clearly shows a peak at a quefrency corresponding to echo delay time.

An interesting and less obvious example concerns the analysis of random data taken from displacement transducers acting on a moving continuous sheet of paper in a paper making machine (Fig. 5.3). The data record $x(t)$ was apparently random in nature, and the objective was to use power spectral analysis to establish the nature and origin of flexural vibrations in the paper. The cepstrum obtained from the data is shown in Fig. 5.4. Note the cepstral peak at ρ_1 seconds. This suggests that an echo property exists whereby the random vibrations are propagated along the paper both as a flexural wave and by the motion of the paper itself. If this were true then ρ_1 seconds would correspond to the difference between the time taken for the paper to travel from the source point to the measurement point and the flexural wave transmission time. Hence the roller set (Fig. 5.3) responsible for inducing the vibrations can be deduced.

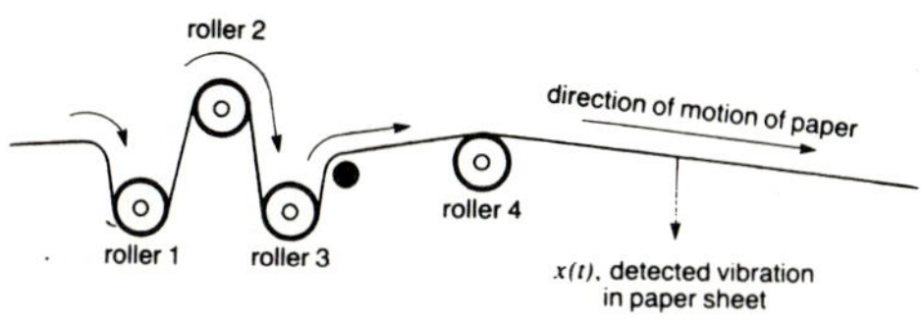

Fig. 5.3 Paper vibration tests

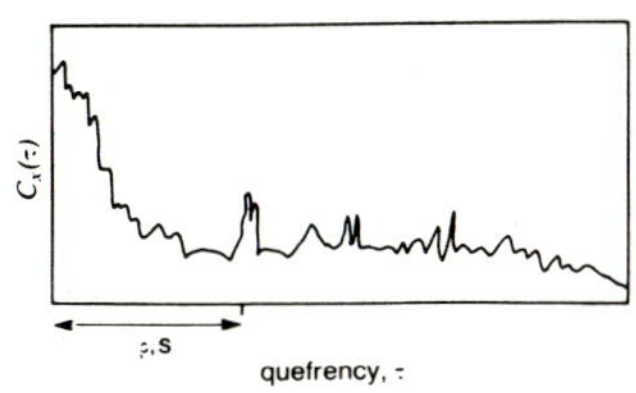

Fig. 5.4 Power cepstrum of paper vibration data

Other applications include voice pitch characterisation (see Wellstead 1984a for an example), machine health monitoring, acoustic testing of rooms and any other process involving echoes, reverberation or modulation.

Few useful results yet exist on the statistical accuracy of cepstral estimates, but the information which is available is discussed by Childers et al. (1977).

6. REFERENCES

J.S. Bendat and A.G. Piersol (1971). "Random Data: Analysis and Measurement Procedures", Wiley.

V. Benignus (1969). "Estimation of the coherence spectrum and its confidence interval using the Fast Fourier Transform", I.E.E.E. Trans., AU-17, pp. 145-150. (See also pp. 198-201). *(Discusses statistical aspects of coherency).*

B.P. Bogert, M.J. Healy and J.W. Tukey (1963). "The **quefrency** analysis of time series of echoes", in "Time-Series Analysis" (M. Rosenblatt, ed.), Wiley. *(The original work on cepstrum analysis).*

D.G. Childers, D.P. Skinner and R.C. Kemerait (1977). "The Cepstrum: a guide to processing", Proc. I.E.E.E., 65, pp. 1428-1443. *(Review paper, including the few useful results existing on statistical analysis of accuracy of cepstral estimates).*

R.A. Gabel and R.A. Roberts (1980). "Signals and Linear Systems", Wiley.

F.J. Harris (1978). "On the use of windows for harmonic analysis with the Discrete Fourier Transform", Proc. I.E.E.E., 66, pp. 51-83. *(Discusses in detail the techniques of windowing in Fourier analysis).*

H.D. Helms (1967). "Fast Fourier Transform method of computing difference equations and simulating filters", I.E.E.E. Trans., AU-15, pp. 85-90. *(Discussion of the methods used for avoiding time domain aliassing in DFT-based convolution procedures).*

L.R. Rabiner and C.M. Rader (eds.). (1972). "Digital Signal Processing", I.E.E.E. Press. *(A collection of papers discussing such topics as the equivalence between the discrete and continuous Fourier transforms, circular convolution and FFT algorithms derived from the work of Cooley and Tukey).*

P.D. Welch (1967). "The use of Fast Fourier Transforms for the estimation of power spectra", I.E.E.E. Trans. AU-15, pp. 70-73. *(Includes discussion of the effects of overlapping data blocks).*

P.E. Wellstead (1975). "Signal aliassing in system identification", Int. Jnl. Control, 22, pp. 363-375. *(Discusses the problems of aliassing and associated spectral estimation errors).*

P.E. Wellstead (1977). "Reference signals in closed loop identification", Int. Jnl. Control, 26, pp. 945-962. *(Discusses the statistical accuracy of the frequency response estimator computed from power and cross-spectrum estimates).*

P.E. Wellstead (1984a). "Methods and applications of digital spectral techniques", Solartron Instruments Technical Report 008/83. *(See footnote on page 210).*

P.E. Wellstead (1984b). "Theory and statistical accuracy of spectral analysis", Solartron Instruments Technical Report 009/83. *(See footnote on page 210).*

OBSERVERS, STATE ESTIMATION AND PREDICTION

Dr. K. Warwick

1. Introduction

System descriptions based on a state-space framework have an extra facility, when compared to polynomial operator descriptions, in that information contained in a model of the system is based not only on input-output data but also on a further variable termed the state of the system. For any one particular system there is a unique set of system inputs and outputs, these are physical impositions and for a real system can only be altered by making actual alterations to the system. Concentrating on a particular output signal, e.g. acceleration rather than velocity of an object, does not in any way alter the available outputs, it merely highlights one of them with respect to the others.

For any one system the system state is certainly not a unique variable, in fact when settling on a specific system description a large number of possible state vectors exist. As an example consider the system to be 'You reading this chapter'. One input could then be effort or energy expended in reading, whereas one output is hopefully, the information taken up. For this system the state could be 'how tired you are', or 'what's on the television whilst you are reading', or 'how bored you are'. All these things will relate to how much information is obtained, i.e. all these things affect the output signal. Similarly they can all be affected by input signal variations. However, very much dependent on the type of person (system) concerned, some states are more useful than others, indeed for certain systems many states are completely irrelevant. The selection of states for a particular system description is therefore extremely important in terms of their usefulness.

Luckily in the mathematical control field the choice of state vectors is perhaps not as haphazard as might be suggested by the previous paragraph. Rather than separate system description frameworks being developed for each individual system, relatively few descriptions are deemed possible and the system is made to fit into the one of these which seems most appropriate. Then, for certain description choices a selection of a specific state vector can result in distinct advantages, such as ease of calculations. Because that particular description is used for many different systems a good idea is held as to which are the best selections to make for a state vector. Having plumped for a state vector and hence a particular state space description of a system the next problem to be faced is how to obtain the values of the parameters within the state vector.

In some cases the state vector may well be measurable either directly or indirectly, this relies on the fact though, that it has some physical meaning, as can be true when certain of the system outputs or inputs are also defined as state variables. However, it is often the case that the state vector parameters are not directly measurable and thus, if it is desired to make any use of the state space description, they must be estimated by employment of the system data that is available.

In the following section systems which are inherently deterministic in nature are considered in terms of a continuous time background. The problem of reconstructing a state vector from system input and output information is then discussed by means of the Luenberger Observer, Luenberger (1). It is also shown how the number of state variables to be estimated can be reduced, hence reducing the amount of necessary computation, when certain of the variables are simply related to known system signals, in particular to one or more output variables. In section 3, state variable estimation is looked at from the point of view of stochastic, discrete-time systems. Hence the Kalman filter is introduced and its importance in terms of computational procedures and on-line real-time controllers is stressed. Various aspects which can be regarded as resulting from the Kalman filter are then considered in the following two sections, with emphasis placed on state vector prediction in section 4 and signal smoothing in section 5. The complete chapter then is intended to act as an introduction to the reasons for using, and the basic equations encountered in, firstly the Luenberger Observer, and secondly the Kalman filter.

2. The Luenberger Observer

When designing a controller for a system which can be described effectively by means of a state space model, it is generally the case that the state vector is, in the first instance, assumed to be either known or directly measurable. The control input can then be formulated as a vector multiple of the state vector, the exact nature of this multiple, termed the state feedback vector, being dependent on the control criterion that it is wished to employ, i.e. optimization, pole placement etc.

For any input-output system many useful state space descriptions can be found, as was pointed out in the introductory section, dependent on the definition given for the state vector in each case. Indeed, the state vector is often chosen merely as a mathematical tool bearing little or no resemblance to any direct physical property of the system. This being so, it is not a common occurrence for the specified state vector to be available as a measured quantity. An obvious exception to this is however encountered when a state vector which comprises simply of system input and output values is used.

If the state vector, in a state space system description, is not directly available as a measurable value, an alternative approach must be employed such that its value can be either calculated or at least an estimate of each state vector element can be made. The estimated vector, as long as it is a reasonably good estimate, can then be employed in place of the unknown actual value within the overall feedback control scheme. Hence the control input is formed as a vector multiple of the state feedback parameters with the estimated state vector. The method of obtaining the state vector estimate is usually referred to as an Observer, in particular for systems which can be considered to be deterministic in nature it is generally known as a Luenberger Observer, Luenberger (1). A requirement of this type of Observer is that the system inputs and outputs that are measurable can be employed to form the estimated state vector such that it is linearly related to the actual state vector.

Any particular choice of Observer can be investigated in terms of its asymptotic properties. Because only an estimate of the state vector is obtained an error will exist, at any time instant, between the estimate and the actual vector. A desirable feature of an Observer is for this error to tend to zero, i.e. for the estimate to become equal to its actual value. However the Observer characteristics can be selected such that the error tends to zero slowly, possibly exponentially, with respect to time. Conversely, a rapid reduction of error could result in undesirable features such as poor noise rejection. Clearly the choice of Observer characteristics is quite an involved procedure and whatever selection is made, because it will most likely be part of a feedback path, the Observer will also affect the properties of the closed-loop system.

Most of the original work on Observer theory was concerned with deterministic, continuous time linear time invariant systems, Luenberger (1), Wolovich (4). In this section therefore, a continuous time approach has been taken, despite the fact that the remainder of the chapter is more concerned with discrete-time systems. Reconsideration of the methods derived in this section in terms of a discrete-time system description is not seen as a difficult task, many authors, see e.g. Tse and Athans (5), having already covered the problem extensively.

Asymptotic Observer

Consider the linear system described by

$$\dot{x}(t) = Ax(t) + Bu(t)$$
$$y(t) = Cx(t) \tag{2.1}$$

in which the system input vector is of dimension m x 1, and the output vector of dimension p x 1. Also the state vector x(t) is of dimension q x 1 such that the system matrix A is of dimension q x q, the distribution matrix B of dimension q x m and the output matrix C of dimension p x q.

It is assumed that all the matrices and vectors, except x(t) and $\dot{x}(t)$, in (2.1) are known. The only thing preventing a direct reconstruction of the system in order to arrive at the state vector itself is then a lack of knowledge of the initial conditions applying to the state vector, although this is based on initial conditions for input and output signals being available, which is usually true. However if an estimate, $\hat{x}(t)$, of the state vector is obtained, a measure of the error between the actual output signal and the output signal that would have been obtained had the estimated state vector been the generating source is given by:

$$l(t) = y(t) - C\hat{x}(t) \tag{2.2}$$

Then the estimated vector can be built up from

$$\dot{\hat{x}}(t) = A\hat{x}(t) + Bu(t) + Gl(t) \tag{2.3}$$

the logic being that the estimated vector is subject to the same system characteristics as the actual state vector, but as an error between x(t) and $\hat{x}(t)$ exists this is accounted for in the term Gl(t). The q x p dimensional matrix G is a feedback gain matrix, the selection of which determines the characteristics of the Observer.

Further consideration of the Observer equations, in particular the choice of G, can be carried out with regard to the error between the actual and estimated state vectors. This error, defined by

$$\varepsilon(t) = x(t) - \hat{x}(t) \tag{2.4}$$

can be calculated by subtracting (2.3) from the state equation in (2.1), such that

$$\dot{x}(t) - \dot{\hat{x}}(t) = Ax(t) + Bu(t) - A\hat{x}(t) - Bu(t) - Gl(t)$$

or, by inclusion of the output equation in (2.1) this becomes

$$\dot{\varepsilon}(t) = (A - GC)\varepsilon(t) \tag{2.5}$$

the solution of which is

$$\varepsilon(t) = \exp\{[A - GC]\,(t - t_o')\}\,\varepsilon(t_o') \tag{2.6}$$

where t_o' signifies the time at which initial conditions are taken. This shows that it is the error due to unknown initial conditions for the actual state vector which produces an error for the rest of time following t_o'. It is apparent that the poles of this error equation are given by the eigenvalues of the matrix (A - GC) and hence if any roots exist in the right half of the S-plane, the error equation will be unstable, thus the error will tend to infinity as time tends to infinity. However, for an observable system all the Observer poles can be arbitrarily chosen by correct selection of the gain matrix, G, Kailath (3). Therefore, by choice of G, the Observer can not only be made stable but also can have a selected rate of convergence, i.e. a selected exponential decay. Hence, asymptotically the error $\varepsilon(t)$ will tend to zero, i.e. the estimated state vector will become equal to the actual state vector, and this will occur, as can be seen from (2.6), irrespective of the input applied and the output produced.

Reduced Order Observer

The estimated state vector $\hat{x}(t)$ is of the same dimension, $q \times 1$, as the actual state vector $x(t)$, this can be seen from (2.3). Hence each of the q components of $\hat{x}(t)$ serves as an approximation to its respective component in $x(t)$. However from the output equation, defined in the system definition (2.1), it is apparent that each of the p outputs is constructed as a linear combination of the q states, most probably unlike the states though, the output variables are directly measurable. This leads one to feel that it might not be absolutely necessary to employ an observer of the same dimension as the actual state vector. In fact an observer of lower dimension can be constructed which removes the redundancy apparent in the full order observer whilst retaining the flexibility in the selection of system dynamics, Luenberger (6), Gopinath (7).

By using the equation set (2.1) as a starting point, assumptions are made firstly that the outputs are linearly independent and secondly that the output matrix C is of the form

$$C = [I \quad 0] \tag{2.7}$$

where I is a $p \times p$ identity matrix, the remainder of C being made up of a $p \times (q - p)$ matrix consisting entirely of zeros. It must be noted that it may well require a change of co-ordinates for the state vector in order to obtain C in such a form, Wolovich (3).

With C in the form of (2.7) it is possible to redefine the state vector $x(t)$ as consisting of two parts, the output vector $y(t)$ and a reduced order state vector, $\bar{x}(t)$, i.e.

$$x(t) = \begin{bmatrix} y(t) \\ \cdots\cdots \\ \bar{x}(t) \end{bmatrix} \tag{2.8}$$

in which $\bar{x}(t)$ is of dimension $(q - p) \times 1$. The system definition (2.1) can then be rewritten as:

$$\begin{aligned} \dot{y}(t) &= A_{11}y(t) + A_{12}\bar{x}(t) + B_1u(t) \\ \dot{\bar{x}}(t) &= A_{21}y(t) + A_{22}\bar{x}(t) + B_2u(t) \end{aligned} \tag{2.9}$$

with the matrix $A = \begin{bmatrix} A_{11} & A_{12} \\ A_{21} & A_{22} \end{bmatrix}$ and $B = \begin{bmatrix} B_1 \\ B_2 \end{bmatrix}$

It can be seen in (2.8) that the top p elements of the full order state vector, namely those of $y(t)$ are directly measurable quantities, merely needing to be differentiated for (2.9). As the input signal $u(t)$ is also directly obtained, this leaves $A_{12}\bar{x}(t)$ as the unknown quantity in the first equation of (2.9). The second equation is then the reduced state equation for $\bar{x}(t)$, as it can be considered that $A_{21}y(t) + B_2u(t)$ is the known input to the equation. By using the same procedure as that for the full order Observer, an Observer can be found for the unknown reduced order state vector $\bar{x}(t)$, which will necessarily be of dimension $(q - p) \times 1$. This will therefore reduce the dimension of the estimated state vector by $p \times 1$, hence resulting in a Reduced Order Observer.

Closed Loop Properties

Whatever the type of control design objective employed, as far as the feedback of a known or measurable state vector is concerned, the closed loop properties of the system can be investigated, with particular emphasis placed on stability. However, once an observer is employed in order to obtain an estimate of the state vector, when this estimated vector is used in place of its actual value the closed loop system is no longer dependent on the true state vector, but rather obtains characteristics from its estimated version.

Where an asymptotic Observer, which reduces the error between actual and estimated state vectors to zero as $t \to \infty$, is used, under steady-state conditions there will be no difference whatsoever. The problem considered in this section is then; if a stable asymptotic Observer is applied to an otherwise stable control system design, will the overall closed loop system remain stable for any type of Observer within this category?

In order to answer the question posed, consider again the state equations (2.1) along with a newly defined control input signal calculation:

$$u(t) = Fx(t) \tag{2.10}$$

where F is a $1 \times q$ row vector containing the q state feedback parameters. Assuming the state vector to be known it follows on substitution of (2.10) into (2.1) that the closed loop eigenvalues are given by

$$\dot{x}(t) = (A + BF)\, x(t) \tag{2.11}$$

However when an observer is used in order to obtain an estimate of the state vector, such that $x(t)$ becomes $\hat{x}(t)$ in (2.10), the closed loop eigenvalues are found by means of the two equations

$$\dot{x}(t) = Ax(t) + BF\hat{x}(t) \tag{2.12}$$

and

$$\dot{\hat{x}}(t) = GCx(t) + (A + BF - GC)\hat{x}(t) \tag{2.13}$$

where the latter is from the general Observer equation (2.3).

It must be noted that due to the control input signal definition (2.10) no external input signal is included in the calculations. As a simple addition to u(t) in (2.10) though, only a trivial extension is required on the equations (2.11)-(2.13) for an external (reference) input to be appended.

In order to obtain the characteristic equation for the complete system described by (2.12) and (2.13), the two equations can be reconsidered in matrix form with the Laplace operator applied, such that the characteristic equation can be calculated from:

$$\text{Char. Eqn.} = \det \begin{bmatrix} sI - A & -BF \\ -GC & sI - (A + BF - GC) \end{bmatrix} \quad (2.14)$$

By adding column 2 to column 1 and subsequently subtracting row 1 from row 2, this determinant is simply reconfigured as:

$$\text{Char. Eqn.} = \det \begin{bmatrix} sI - A - BF & -BF \\ 0 & sI - A + GC \end{bmatrix}$$

$$\text{or Char. Eqn.} = \det(sI - A - BF)\det(sI - A + GC) \quad (2.15)$$

which means that the characteristic polynomial of the complete system is merely the characteristic polynomial of the system assuming the state vector to be known, see (2.11), multiplied by the characteristic equation of the Observer, see (2.5) and (2.6). By selection of F and G, the controller and Observer gain matrices respectively, the eigenvalues can then be chosen to arrive at a stable characteristic polynomial. In fact for a completely controllable and completely observable system the controller and observer poles can be arbitrarily placed.

An important feature results from the characteristic equation being simply a combination of the controller and observer polynomials, in that the two designs can, if deemed appropriate, be carried out independently of each other. Hence the controller can be designed as though the actual state vector is available, and also the Observer can be designed with no regard as to the type of feedback control action, if any, to be applied. This feature is known as the Separation property.

Observer design example

Consider the state space modelled system described by means of the equations, Luenberger (1)

$$A = \begin{bmatrix} -2 & 1 \\ 0 & -1 \end{bmatrix}, \quad B = \begin{bmatrix} 0 \\ 1 \end{bmatrix}, \quad C^T = \begin{bmatrix} 1 \\ 0 \end{bmatrix}$$

An observer equation defined by (2.3) can be used to obtain an estimate of the state vector, the general gain vector being

$$G = \begin{bmatrix} g_1 \\ g_2 \end{bmatrix}$$

The Observer poles are then found from the matrix (A - GC), where in this case

$$A - GC = \begin{bmatrix} -2-g_1 & 1 \\ -g_2 & -1 \end{bmatrix}$$

which has a characteristic equation, calculated from the determinant of $sI - (A - GC)$,

$$s^2 + (g_1 + 3)s + g_1 + g_2 + 2 = 0$$

The gain values g_1 and g_2 can now be selected in order to give the Observer desired pole locations. Suppose it is wished to place the observer poles as $s = -3$ and -4, this would be satisfied by choosing $g_1 = 4$ and $g_2 = 6$ such that the characteristic equation becomes

$$s^2 + 7s + 12 = 0 = (s + 3)(s + 4)$$

Also, the observer itself, as from (2.3), can then be written as

$$\dot{\hat{x}}(t) = \begin{bmatrix} -6 & 1 \\ -6 & -1 \end{bmatrix} \hat{x}(t) + \begin{bmatrix} 0 \\ 1 \end{bmatrix} u(t) + \begin{bmatrix} 4 \\ 6 \end{bmatrix} y(t)$$

3. The Kalman Filter

Many of the ideas developed in the previous section will be carried over into this and the remaining sections. There will however, be two important differences, firstly the following discussion will concern itself with discrete-time rather than continuous-time systems, and secondly the type of systems considered will also directly take into account any noise that affects the system. Indeed this latter element is of significance in that the main object behind Filtering is to reduce the affect that noise or unwanted signals have on measurements obtained or values calculated from those measurements.

An historical account of Filtering in general is given in many works, e.g. see Anderson and Moore (2), and is therefore not summarized here. Suffice it to say that the Kalman filter developed originally in the early 1960's, Kalman and Bucy (9), is particularly useful for discrete-time systems because of its implementability by means of programming tools such as computers. In this section it is intended merely to introduce the basic ideas behind Kalman Filtering and to indicate some of the features obtained.

Consider firstly the discrete-time state space system description

$$\begin{aligned} x(k + 1) &= Ax(k) + Dw(k) \\ \text{and} \quad y(k) &= Cx(k) + v(k) \end{aligned} \tag{3.1}$$

in which the model is assumed to hold for $k \geq 0$. Also $x(k)$ is the state vector at time instant k, $y(k)$ is the output signal at the same instant and is corrupted by measurement noise $v(k)$. The process $w(k)$ is also considered in the form of a noise sequence although it is possible to regard it as a system input. Various assumptions can be made as to the type of noise signal actually encountered, for this section however $v(k)$ and $w(k)$ are assumed to be independent, gaussian white processes with zero mean and finite covariance, equal to R and Q respectively. The filtering problem can then be considered as one in which an estimate of the state vector must be obtained at time instant k using measured values taken up to and including that same time instant. A final limitation must however be imposed in that the state estimate must be available for use also at time instant k.

Firstly though, as the limitation imposed on availability at instant k can appear to be restrictive, assume initially that only measurements taken up to and including time instant (k-1) are at our disposal. The problem is then one of finding the sequence $\mathcal{E}\{x(k/k-1)\} = \hat{x}(k/k-1)$, where $\mathcal{E}\{\cdot\}$ denotes the expected value and $x(k/k-1)$ means the value of state vector at time instant k given measurements taken up to and including time instant (k-1).

The error between actual and estimated state vectors can also be defined:

$$\varepsilon(k/k-1) = x(k) - \hat{x}(k/k-1) \tag{3.2}$$

and further its covariance matrix is then given by:

$$P(k/k-1) = \mathcal{E}\{\varepsilon(\cdot), \varepsilon(\cdot)^T\} \tag{3.3}$$

in which $\varepsilon(\cdot)$ signifies here $\varepsilon(k/k-1)$.

It must also be noted that for the discrete-time model (3.1) the index $k \geq 0$. The initial conditions for the state vector and error covariance matrix must therefore be defined such that

$$\hat{x}(0/-1) = \mathcal{E}\{x_o\} = m_o \tag{3.4}$$

and $$P(0/-1) = P_o \tag{3.5}$$

where $\mathcal{E}\{x_o\}$ is the expected value, given no measurements have been taken, with mean m_o, covariance P_o, is Gaussian and is independent of both v(k) and w(k) sequences.

The Kalman filter can then be defined in terms of the equation

$$\hat{x}(k + 1/k) = [A - KC]\,\hat{x}(k/k-1) + Ky(k) \tag{3.6}$$

in which the Kalman gain matrix is found from

$$K = A\,P(k/k-1)C^T[C\,P(k/k-1)C^T + R]^{-1} \tag{3.7}$$

on condition that $C\,P(k/k-1)C^T + R$ is invertible. Also the error covariance matrix is updated by means of:

$$P(k + 1/k) = [A - KC]\,P(k/k-1)A^T + DQD^T \tag{3.8}$$

Proof of the equation set (3.6) - (3.8) is given in several texts, e.g. Astrom (8), where it is shown that (3.8) depicts the variance of the error $\varepsilon(k/k-1)$, known as the reconstruction error, and that the selection of K as in (3.7) will minimize the mean square reconstruction error, hence leading to an Optimal filter for (3.6)

It was stated initially in this section, following the system definition (3.1), that the state estimate was to be obtained at time instant k by taking into account measurements up to and including those at time instant k. Thus far however, only measurements taken up to and including time instant k - 1 have been incorporated in the estimate of the state vector. Before proceeding with a discussion of the main features of the Kalman filter the effect on the state estimate of including the extra set of measured values will be considered.

Let the alternative state estimate be:

$$\hat{x}(k/k) = \mathcal{E}\{x(k)\} \tag{3.9}$$

with an associated covariance matrix $P(k/k)$. The estimate is then found, in terms of optimal filter design, from the equation:

$$\hat{x}(k/k) = \hat{x}(k/k-1) + \bar{K}(y(k) - C\hat{x}(k/k-1)) \tag{3.10}$$

where $$\bar{K} = P(k/k-1)C^T(CP(k/k-1)C^T + R)^{-1} \tag{3.11}$$

such that $$K = A\bar{K} \tag{3.12}$$

Also, the covariance matrix $P(k/k)$ can be obtained from:

$$P(k/k) = (I - \bar{K}C)\,P(k/k-1) \tag{3.13}$$

So, although the extra measured output value, $y(k)$, appears in the equation (3.10) used to calculate the state estimate $\hat{x}(k/k)$, the covariance equation (3.13) has no such occurrence. $P(k/k)$ can however be calculated from $P(k/k-1)$ just as $\hat{x}(k/k)$ can be from $\hat{x}(k/k-1)$, the latter requiring the inclusion of $y(k)$ though.

It must be noted that throughout this discussion on the Kalman filter the system matrices A,D,C have all been assumed as time invariant, similarly $v(k)$ and $w(k)$ were assumed stationary, resulting in constant R and Q values. However, because $P(k/k-1)$ appears in the equation for K, i.e. (3.7), and K appears in the equation for $P(k+1/k)$, i.e. (3.8), the constant system matrix values do not mean that K and $P(k/k-1)$ will remain constant.

Another important point in the updating procedure for K and $P(k/k-1)$ is the invertibility, i.e. nonsingularity, of the matrix $CP(k/k-1)C^T + R$. Unfortunately there is no guarantee that this matrix will be nonsingular. When the matrix is singular however it can be replaced in (3.7) with its pseudo-inverse, although as $P(k/k-1)$ is time varying this means a nonsingularity check must be applied at each time instant for complete safety.

One of the differences between the Kalman filter and the Luenberger Observer described in section 2, is the allowance for both state and measurement noise. It can be the case however that although the state estimate is built up as a Kalman filter, the system turns out to be noise free. In this case the state and output equations reduce to

$$
\begin{aligned}
x(k+1) &= A\,x(k) \\
\text{and}\quad y(k) &= C\,x(k)
\end{aligned}
\tag{3.14}
$$

such that, due to the output equation, the Kalman filter of (3.6) can be redefined in terms of the state vector estimate as

$$\hat{x}(k+1/k) = [A - KC]\ \hat{x}(k/k-1) + KCx(k)$$

and on subtraction of this equation from the state equation in (3.14), the result is

$$x(k+1) - \hat{x}(k+1/k) = [A - KC]\,[x(k) - \hat{x}(k/k-1)]$$

or in terms of the reconstruction error defined in (3.2)

$$\varepsilon(k+1/k) = [A - KC]\ \varepsilon(k/k-1) \tag{3.15}$$

This equation can then be compared directly with the continuous-time state error of the Luenberger Observer (2.5) and (2.6). The only differences are (a) that the Kalman gain K replaces the Luenberger gain G, and (b) that (3.15) is a discrete-time equation rather than its continuous-time counterpart (2.5). Convergence proofs and properties associated with the Luenberger Observer can therefore be directly translated into discrete-time versions and employed with regard to the noise free Kalman filter.

A final point to note concerning the Kalman filter is its use on systems and controllers represented in terms of polynomials as opposed to the state-space scheme described in (3.1). For any one polynomial description of a system there are numerous possible state space formulations which can be directly employed to describe that same system, the difference between the formulations being essentially due to the choice of state vector. Any polynomial feedback control scheme can thus be reconsidered in terms of an equivalent state-space description. For a stochastic system the polynomial feedback control action will therefore have an equivalent representation in terms of a combination of state feedback parameters and state estimates obtained by means of a Kalman filter. Hence although, on the surface, it may not be obvious, even with a polynomial based feedback control scheme, the equivalent of a Kalman filter is employed. One such example can be found in the field of Self-Tuning Control in which the main bulk of reported schemes are based on a polynomial description. State-space self-tuners have however been developed, Warwick (10), whilst the effective Kalman filtering techniques employed by all of the popular control schemes have been investigated at length, Warwick and Westcott (11).

4. Prediction

In the previous section it was shown how the state estimates $\hat{x}(k/k)$ and $\hat{x}(k/k-1)$ could be obtained by means of filtering. However, although it is common practice to do so, applying the name filter to the updating procedure involved in $\hat{x}(k/k-1)$ leads to something of a misconception. Looked at in one way, an estimate of the state is being calculated at time instant k using only information obtained up to time instant k - 1. Looked at in a slightly different way though, by means of the information available at time instant k - 1 the value of the state vector at time instant k can be predicted. This particular state estimator is thus also referred to as a one-step prediction of the state.

In this section, in the first instance the one-step prediction equations will be briefly reviewed such that a more general N-step ahead predictor can subsequently be considered in the same light. This latter predictor, perhaps obviously, takes its name from its use as a predictor of the state variable vector, N time instants (periods) ahead of the present time instant, N being an integer of value greater than or equal to unity.

Consider a combination of the state equation definitions (3.1) and a discrete version of (2.1), such that both noise and a control input u(k) are allowed for, with time index $k \geq 0$;

$$\begin{aligned} x(k+1) &= Ax(k) + Bu(k) + Dw(k) \\ \text{and} \quad y(k) &= Cx(k) + v(k) \end{aligned} \tag{4.1}$$

In these equations, standard definitions for signals apply, here however it is considered that the noise sequences are related such that a covariance equal to S results. The noise sequences are also assumed to be Gaussian, although this condition can readily be relaxed, resulting in only minor changes in operating definitions for the predictors concerned.

The state estimate update vector can then be written as:

$$\hat{x}(k+1/k) = A\hat{x}(k/k-1) + Bu(k) + K(y(k) - C\hat{x}(k/k-1)) \tag{4.2}$$

which is optimal in the sense of minimum variance estimation, when the Kalman gain is found from,

$$K = (AP(k/k-1)C^T + D\,S)\,(CP(k/k-1)C^T + R)^{-1} \tag{4.3}$$

with covariance matrix:

$$P(k+1/k) = AP(k/k-1)A^T - K(AP(k/k-1)C^T + DS)^T + DQD^T \tag{4.4}$$

Note that these equations are identical to the set (3.6) - (3.8) if the control input u(k) is set equal to zero and the covariance matrix S is zero. Setting u(k) to zero has no effect though on either the Kalman gain K obtained from (4.3) or on the covariance update of P(k/k-1) obtained from (4.4). In fact the control input at instant k is not required until that particular time in order to calculate the state vector estimate for instant $k + 1$. This means that the input sequence does not have to be known apriori for all k, simply knowing the input at the necessary time will suffice. The control input can therefore be obtained in terms of a feedback equation dependent on measurements taken up to and including the same time instant at which the input is required, i.e. $u(k)$ can be generated in terms of measured signals including $y(k)$.

The state prediction considered (4.2) is though just one time instant ahead of the present time. A more general prediction can be made in terms of N time periods ahead. Consider initially a one step ahead version of the state equation given in (4.1).

$$x(k + 2) = Ax(k + 1) + Bu(k + 1) + Dw(k + 1) \tag{4.5}$$

which holds for time invariant A, B and D. Similarly,

$$x(k + 3) = Ax(k + 2) + Bu(k + 2) + Dw(k + 2) \tag{4.6}$$

But a substitution can be made for $x(k + 2)$ in (4.6) from the equation (4.5), and subsequently a substitution can be made for $x(k + 1)$ from (4.1). Hence $x(k + 3)$ can be written in terms of $x(k)$ as:

$$\begin{aligned} x(k + 3) = {} & A^3x(k) + A^2Bu(k) + ABu(k + 1) + Bu(k + 2) \\ & + A^2Dw(k) + ADw(k + 1) + Dw(k + 2) \end{aligned} \tag{4.7}$$

Such that in general:

$$x(k + N) = A^N x(k) + \sum_{i=k}^{\bar{k}} A^{\bar{k}-i} (Bu(i) + Dw(i)) \tag{4.8}$$

where the index $\bar{k} = k + N - 1$.

In the case of a time varying system however, i.e. one in which $A(k + 1)$ is not necessarily equal to $A(k)$ etc., the term A^N becomes $A(k + N - 1)$. $A(k + N - 2) \ldots\ldots A(k)$, with similar effects on the $u(i)$ and $w(i)$ terms, this can perhaps best be seen by means of (4.8).

The random disturbance $w(i)$ is independent, for $i = k$ to $i = \bar{k}$, of the past measurements taken up to and including those at time $t = k - 1$. Taking conditional expectations in order to obtain an estimate of the state vector, it follows that

$$\hat{x}(k + N/k - 1) = A^N \hat{x}(k/k - 1) + \sum_{i=k}^{\bar{k}} A^{\bar{k}-i} Bu(i) \tag{4.9}$$

By subtracting (4.9) from (4.8) the error in the state vector estimate is therefore given by

$$\varepsilon(k + N/k - 1) = A^N \varepsilon(k/k - 1) + \sum_{i=k}^{\bar{k}} A^{\bar{k}-i} Dw(i) \tag{4.10}$$

in which $\varepsilon(k/k-1) = x(k) - \hat{x}(k/k-1)$.

Also the covariance matrix for the N step ahead case is given by:

$$P(k + N/k - 1) = A^N P(k/k-1)(A^N)^T + \sum_{i=k}^{\bar{k}} A^{\bar{k}-i} DQD^T (A^{\bar{k}-i})^T \tag{4.11}$$

But note that for the case N = 1, the state estimate obtained, from (4.9) is

$$\hat{x}(k + 1/k - 1) = A\hat{x}(k/k - 1) + Bu(k) \tag{4.12}$$

A 'better' estimate of the state vector at time k + 1 can however be found by taking into account events up to and including those at time instant k, as is done in the estimate equation (4.2) by including the new output measurement. In order to accommodate this special case one step ahead prediction into the general equation for an N-step ahead predictor, it follows that for an estimate $\hat{x}(k + N)$, when N = 1 equation (4.2) should be used, however when $N > 1$ then a modified form of (4.9) should be employed:

$$\hat{x}(k + N/k) = A^{N-1}\hat{x}(k + 1/k) + \sum_{i=k+1}^{\bar{k}} A^{\bar{k}-i}Bu(i) \tag{4.13}$$

with corresponding revisions being necessary for (4.10) and (4.11). Equation (4.13) is only altered in relation to (4.9) by the removal of the state estimate at time k + 1, hence shifting the indexed dependences by unity, i.e. $k + N/k - 1 \rightarrow k + N/k$.

Because of the output equation defined in (4.1), prediction of the output signal at time instant k + N can be found simply in terms of the state estimate for that time instant, such that

$$\hat{y}(k + N/k) = C\hat{x}(k + N/k)$$

in which it must be remembered that the matrix C is considered to be time invariant.

5. Smoothing

The filtering problem is one in which an estimate of the state is required at a particular time instant, allowing for all measurements taken up to that time instant to be included in the estimation calculations. The prediction problem meanwhile is one in which an estimate of the state which will occur some time in the future (N - time periods ahead) is required now, hence only measurements taken up to the present time instant can be employed to make the prediction.

A final problem which will be briefly considered is one in which an estimate of the present state vector is required, however the estimate is not actually required for some time (N time steps), i.e. a delay of N time steps occurs before the state estimate must be found. Measurements taken from the present time until N further periods have occurred can in this case be used in the estimator calculations. A smoothed state estimate is then one in which an estimate of the state is required at time instant k, and this is obtained such that it is dependent on measurements taken up to and including time instant k + N.

Although various types of signal smoothing have been developed, e.g. fixed lag smoothing, fixed interval smoothing, only one approach, that of fixed point smoothing, will be considered in this section. The choice of fixed point smoothing for explanation as opposed to any other method arises from its similarities with the Kalman filter, and thus linear discrete-time systems affected by white gaussian signals are once more the class of system concerned.

The object of a fixed point smoother is to provide an estimate of the state at time instant j, i.e. $x(j)$, by taking into account all the measurement information obtained up to time instant k, where $k = j + 1, j + 2$, etc. As time progresses therefore, new estimates of the value of the state vector at the fixed time point j, can be obtained. This approach is thus very helpful in giving an estimate of the initial conditions for a particular experiment set up. It is only to be expected then that for the smoother to be useful the estimate of the state at instant j should improve with the index k, i.e. as k gets larger so $\hat{x}(j/k) \to x(j)$. In particular this means that $\hat{x}(j/k)$ should be a better estimate of the state vector $x(j)$ than is the filtered estimate $\hat{x}(j/j)$. If $\hat{x}(j/k)$ was not a better estimate than $\hat{x}(j/j)$ there would be no point in carrying out smoothing, with the extra calculations involved. Much further discussion of smoothing procedures can be found in Meditch (12) and Sage and Melsa (13).

Consider here the state space description first presented as (3.1).

$$\begin{aligned} x(k+1) &= Ax(k) + Dw(k) \\ y(k) &= Cx(k) + v(k) \end{aligned} \tag{5.1}$$

The fixed point smoothing problem is then to determine an estimate of the state vector at time instant j by taking into account measurements taken up to and including time instant k, where $k > j$ and j is fixed. An updating equation for this estimate can be described by:

$$\hat{x}(j/k) = \hat{x}(j/k-1) + \tilde{K}(y(k) - C\hat{x}(k/k-1)) \tag{5.2}$$

in which the initial state estimate is defined by $\hat{x}(j/j-1)$.

Then, for an optimal smoothing algorithm, the gain $\tilde{K}$ is given by

$$\tilde{K} = \tilde{P}(k/k-1)\, C^T\, (CP(k/k-1)C^T + R)^{-1} \tag{5.3}$$

where $\tilde{P}(k+1/k) = \tilde{P}(k/k-1)\,(A - KC)$ (5.4)

in which K is the Kalman gain obtained from the filtering equations described in section 3, in particular (3.7).

The error between actual state $x(j)$ and the estimated state vector $\hat{x}(j/k)$ is then:

$$\varepsilon(j/k) = x(j) - \hat{x}(j/k) \tag{5.5}$$

with a covariance equal to

$$P(j/k) = P(j/k-1) - \tilde{P}(k/k-1)C^T\tilde{K}^T \tag{5.6}$$

Also the initial conditions for the covariance matrix $\tilde{P}(j/j-1)$, are set by $P(j/j-1)$, because of the initial state estimate value.

It can be noted that measured values do not turn up in either the gain equation (5.3) or the covariance update (5.4), this indeed is something which also occurs in the straightforward Kalman filter. It means that both the gain and covariance matrix can be calculated, at a certain time instant, before the particular measurement is taken.

Simplifications can be made to the gain and covariance matrix calculations if $AP(k/k-1)$ is non-singular because from the equation for gain K, (3.7), a substitution leaving the inverse for $AP(k/k-1)$ results in:

$$\tilde{K} = \tilde{P}(k/k-1)\,[AP(k/k-1)]^{-1}K \tag{5.7}$$

The above equation therefore, directly relates the Kalman gain K with the smoothing gain $\tilde{K}$.

One question which needs to be asked is 'how much improvement in the state estimate has been obtained by smoothing?' In Anderson and Moore (2) this is investigated by considering the term $P(j/k) - P(j/j-1)$, i.e. the difference between the most recent covariance matrix, at time instant k, and the initial covariance matrix. A rough conclusion is drawn in that any improvements obtained from smoothing will decrease as the signal-to-noise ratio decreases. Also, for time invariant systems a monotonic improvement is obtained, as k increases, and just about all the improvement occurs within three time constants of the Kalman filter, which is governed by $A - KC$.

In summary, one particular type of smoothing, namely fixed-point smoothing, has been considered here. With this method the signal-to-noise ratio prevalent in the system governs how much of an improvement can be made in the state estimate, the improvement being largest with high ratios.

6. Comments

The intention of this chapter was to give a basic introduction to the Luenberger Observer and state estimation, considering in particular Kalman filtering and state prediction in terms of linear methods. This has been done generally by stating the equations used rather than by obtaining them from first principles. A more detailed mathematical approach can be found in several works, the Luenberger Observer being discussed in Luenberger (1) whilst the Kalman filter is covered in Anderson and Moore (2).

Several topics such as time invariance, implementation, and in terms of filtering, non-linear and Wiener filtering, are considered to be beyond the scope of this chapter, however the topic of smoothing was included despite its perhaps not so obvious widespread usefulness. In general the aim of the chapter was to discuss the Observers and filters from the point of view of their original development. The Luenberger Observer arose with regard to state estimation for deterministic, continuous-time systems, whereas the Kalman filter is appropriate for stochastic, discrete-time systems. Because of this fact, both types of system were introduced where appropriate and developed here, although rigorous system definitions have not been given.

References

1. Luenberger D.G.: 'An Introduction to Observers', IEEE Trans. on Automatic Control, Vol. AC-16, No. 6, Dec. 1971, pp. 596-602.

2. Anderson B.D.O. and Moore J.B.: 'Optimal Filtering', Prentice-Hall Inc., Englewood Cliffs, 1979.

3. Wolovich W.A.: 'Linear Multivariable Systems', Springer-Verlag, New York, 1974.

4. Kailath T.: 'Linear Systems', Prentice-Hall Inc., 1980.

5. Tse E. and Athans M.: 'Optimal Minimal-Order Observer-Estimators for Discrete Linear Time Varying Systems', IEEE Trans. on Automatic Control, Vol. AC-15, pp. 416-426, 1970.

6. Luenberger D.G.: 'Observing the state of a Linear System', IEEE Trans. Mil. Electron., Vol. MIL-8, pp. 74-80, 1964.

7. Gopinath B.: 'On the control of linear multiple input-output systems', Bell Syst. Tech. Journal. Vol. 50, pp. 1063-1081, 1971.

8. Astrom K.J.: 'Introduction to Stochastic Control Theory', Academic Press, 1970.

9. Kalman R.E. and Bucy R.S.: 'New results in Linear Filtering and Prediction Theory', J. of Basic Eng., Trans. ASME, series D, Vol. 83, No. 3, pp. 95-108, 1961.

10. Warwick K.: 'Self-tuning regulators - a state-space approach', Int. J. Control, Vol. 33, No. 5, pp. 839-858, 1981.

11. Warwick K. and Westcott J.H.: 'Filtering techniques in Self-Tuning', IMA Journal Math. Control and Info., Vol. 1, No. 2, pp. 107-116, 1984.

12. Meditch J.S.: 'A Survey of data smoothing for linear and nonlinear dynamic systems', Automatica, Vol. 9, No. 2, pp. 151-162, 1973.

13. Sage A.P. and Melsa J.L.: 'Estimation theory with applications to communications and control', McGraw-Hill, New York, 1971.

14. Strejc V.: 'State space theory of discrete linear control', John Wiley, 1981.

15. Kwakernaak H. and Sivan R.: 'Linear Optimal Control systems', New York: Wiley, 1972.

SERC Vacation School

"SIGNAL PROCESSING FOR CONTROL"

University of Warwick

15-20 September 1985

Introduction to Nonlinear Systems Analysis and Identification

Dr. S. A. Billings

Department of Control Engineering, University of Sheffield

1. Introduction

Most systems encountered in practice are nonlinear to some extent due to inherent distortion introduced by the components of the system such as saturation or because they include deliberately introduced nonlinear effects (e.g. bang-bang controllers). Any system for which the superposition principle does not hold is defined to be nonlinear. Nonlinear systems exhibit phenomena like jumps, limit cycles, hysteresis and chaotic motions which are not possible in linear systems. It is these characteristics which often dictate that the study of nonlinear systems is restricted to specific system structures.

The statistical analysis of nonlinear systems is in general an extremely difficult problem and a unified theory applicable to a broad class of systems does not exist. Systems which contain two or more single-valued nonlinear elements, multivalued nonlinearities or nonlinear functions of two or more system variables are particularly difficult to analyse and recourse is often made to either simulation or piece-wise linear analysis.

The present study briefly reviews some of the methods which are available for the statistical analysis of static and dynamic nonlinear systems including linearisation methods, system identification algorithms, and stochastic control.

2. Static Nonlinear Systems

Consider the system illustrated in Fig.1 where $u(t)$ is applied as an input to a single-valued instantaneous nonlinear element $N(\cdot)$ to produce an output $y(t)$.

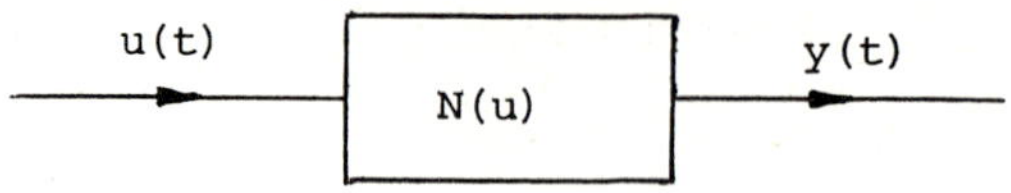

Fig.1

If the input is stationary in the strict sense the k'th order probability density function of y can be obtained from $f_u(u_1,u_2,\ldots u_k; t_1,t_2\ldots t_k)$ the density of u [1]. To determine $f_y(y_1\ldots y_k; t_1,\ldots t_k)$ solve the system of equations $y_1 = N(u_1),\ldots y_k = N(u_k)$ for $u_1,\ldots u_k$ and assuming a unique solution

$$f_y(y_1,\ldots y_k; t_1,\ldots t_k) = \frac{f_u(u_1,\ldots u_k; t_1,\ldots t_k)}{|N'(u_1)|\ldots|N'(u_k)|} \qquad \ldots(1)$$

For example consider the evaluation of the density $f_y(y;t)$ when $y(t) = u^2(t)$. When $y>0$, the eqn $y = x^2$ has two solutions $u_1 = \sqrt{y}$ and $u_2 = -\sqrt{y}$. Further since $|dy/du| = 2\sqrt{y}$, then from eqn (1)

$$f_y(y;t) = \frac{f_u(\sqrt{y};t) + f_u(-\sqrt{y};t)}{2\sqrt{y}}$$

If $y<0$, then $f_y(y) = 0$.

The moments of the output y, Fig.1, can however be expressed directly in terms of the probability density function of the input. The auto-correlation function of the output is for example given by

$$R_{yy}(\tau) = \iint_{-\infty}^{\infty} y_1 y_2\, f(y_1,y_2;\tau)\,dy_1 dy_2 \qquad \ldots(2)$$

or

$$R_{yy}(\tau) = \iint N(u_1)N(u_2)f(u_1,u_2;\tau)\,du_1 du_2 \qquad \ldots(3)$$

An analytical expansion of this integral can be obtained when u is stationary and normally distributed

$$f(u_1,u_2;\tau) = \frac{1}{2\pi\sigma^2(1-\rho^2)^{1/2}} \,\mathrm{Exp}\left\{-\left(\frac{u_1^2+u_2^2-2\rho u_1 u_2}{2\sigma^2(1-\rho^2)}\right)\right\} \qquad \ldots(4)$$

where

$$\begin{aligned} E[u(t)] &= 0 \\ E[u^2(t)] &= \sigma^2 \\ E[u(t)u(t+\tau)] &= \sigma^2\rho(\tau) \end{aligned} \qquad \ldots(5)$$

and $\rho(\tau)$ is the normalised covariance function of the input. Using Mercer's formula [2] eqn (4) can be expanded as

$$\frac{1}{2\pi\sqrt{1-\rho^2}} \exp\left\{-\left(\frac{a_1^2+a_2^2-2\rho a_1 a_2}{2(1-\rho^2)}\right)\right\}$$

$$= \left\{\exp\left\{-\left(\frac{a_1^2}{2}+\frac{a_2^2}{2}\right)\right\}/2\pi\right\} \sum_{n=0}^{\infty} \rho^n Q_n(a_1) Q_n(a_2) \qquad \ldots(6)$$

where $Q_n(a)$ is the n'th order Hermite polynomial

$$Q_n(a) = \exp\left(\frac{a^2}{2}\right) \cdot \frac{(-1)^n}{\sqrt{n!}} \frac{d^n}{da^n}\left\{\exp\left(\frac{a^2}{2}\right)\right\} \qquad \ldots(7)$$

and $a_i = u_i/\sigma$.

Combining eqn's (3), (4) and (6) yields

$$R_{yy}(\tau) = \sum_{n=0}^{\infty} \rho^n k_i^2 \qquad \ldots(8)$$

$$\text{where} \quad k_i = \frac{1}{\sqrt{2\pi}} \int_{-\infty}^{\infty} N(\sigma a) Q_n(a) \exp\left(\frac{a^2}{2}\right) da \qquad \ldots(9)$$

The autocorrelation function of the output eqn (8) is given therefore as a power series of the normalised autocorrelation function of the input where the coefficients k_i, eqn (9), depend on the form of nonlinearity.

For symmetrical non-linearities, $N(u) = -N(-u)$, all even coefficients k_{2j} vanish, and for nonlinearities which can be expressed as truncated power series with zero coefficients for powers greater than j, then $k_i = 0$ for $i>j$. As an example, the output autocorrelation function for the bang-bang nonlinearity

$$y = 1 \;, \quad u>0$$
$$y = -1 \;, \quad u<0$$

is given by

$$R_{yy}(\tau) = \frac{2}{\pi} \operatorname{Sin}^{-1} \rho(\tau) \qquad \ldots(10)$$

Although the input-output cross-correlation function $R_{uy}(\tau)$ can be determined by following a similar procedure to the above a slightly more general result can be obtained by introducing separable processes [3], [23].

Let $f(u_1,u_2;\tau)$ be the second order probability density function of the stationary process u(t) in Fig.1 and define

$$g(u_2,\tau) = \int_{-\infty}^{\infty} u_1 f(u_1,u_2;\tau)du_1 \qquad \text{...(11)}$$

If the g-function separates as

$$g(u_2,\tau) = g_1(u_2)g_2(\tau) \;\forall\; u,\tau \qquad \text{...(12)}$$

then u(t) is said to be a separable process, where

$$g_1(u_2) = \frac{u_2 f(u_2)}{g_2(0)} \qquad \text{...(13)}$$

$$g_2(\tau) = \frac{g_2(0)R_{uu}(\tau)}{R_{uu}(0)} \qquad \text{...(14)}$$

The separable class of random processes is fairly wide and includes the Gaussian process, sine wave process, pseudo-random-binary-sequences etc.

Define the cross-correlation function

$$R_{uy}(\tau) = \iint_{-\infty}^{\infty} u_1 N(u_2) f(u_1,u_2;\tau)du_1 du_2 \qquad \text{...(15)}$$

substituting from eqn's (12) through (14) yields

$$\begin{aligned} R_{uy}(\tau) &= \int N(u_2)g_1(u_2)g_2(\tau)du_2 \\ &= \frac{R_{uu}(\tau)}{R_{uu}(0)} \int N(u_2)u_2 f(u_2)du_2 \\ &= C_F R_{uu}(\tau) \qquad \text{...(16)} \end{aligned}$$

where C_F is a constant scale factor.

Equation (16) which is known as the invariance property, shows that $R_{uy}(\tau)$ is directly proportional to $R_{uu}(\tau)$ for any static nonlinear characteristic $N(\cdot)$ providing u(t) is separable.

3. Nonlinear Systems with Dynamics

3.1 Functional Series Methods

A functional representation of nonlinear systems which is a generalisation of the linear convolution integral was first studied by Volterra early in the twentieth century. Volterra investigated analytic functionals and introduced the representation [4]

$$y(t) = \sum_{n=1}^{\infty} \int_{\Omega} \ldots \int h_n(\tau_1,\ldots\tau_n) \prod_{i=1}^{n} u(t-\tau_i)d\tau_i$$

$$= \sum_{n=1}^{\infty} y_n(t) \qquad \ldots(17)$$

which has become known as the Volterra series. The functions $h_i(\tau_1,\ldots\tau_i)$ in eqn (17) are referred to as Volterra kernels. The kernels are bounded and continuous in each τ_j, symmetric functions of their arguments, and for causal systems $h_i(\tau_1,\ldots\tau_i) = 0$ for any $\tau_j<0$. Systems which contain nonlinear memory elements such as hysteresis or backlash are excluded from the description of eqn (17).

Consider the Volterra series representation of the system illustrated in Fig.2.

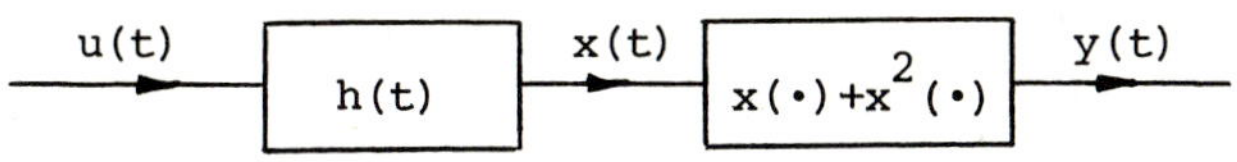

Fig.2

From the convolution integral

$$x(t) = \int h(\tau)u(t-\tau)d\tau$$

But $y(t) = x(t)+x^2(t)$ and therefore

$$y(t) = \int h(\tau_1)u(t-\tau_1)d\tau_1 + \int\int h(\tau_2)h(\tau_3)u(t-\tau_2)u(t-\tau_3)d\tau_2 d\tau_3 \quad \ldots(18)$$

is the Volterra series representation.

Taking the Fourier Transform of the n'th order kernel in eqn (17) yields the n'th order kernel transform [4]

$$H_n(j\omega_1,\ldots j\omega_n) = \int_{-\infty}^{\infty} \ldots \int h_n(\tau_1,\ldots\tau_n)\exp\{-j(\omega_1\tau_1+\ldots+\omega_n\tau_n)\}d\tau_1\ldots d\tau_n \quad \ldots(19)$$

Similarly, considering just the n'th order component of the output $y_n(t)$ and taking Fourier transforms relates the multispectral density

$$Y_n(j\omega_1,\ldots j\omega_n) = H_n(j\omega_1,\ldots j\omega_n)U(j\omega_1)\ldots U(j\omega_n) \qquad \ldots(20)$$

to the input spectrum $U(j\omega)$, from which the output spectrum can be evaluated as

$$Y_n(j\omega) = \frac{1}{(2\pi)^{n-1}}\int_{-\infty}^{\infty}\ldots\int Y_n(j\omega-j\mu_1, j\mu_1-j\mu_2,\ldots j\mu_{n-1})d\mu_1\ldots d\mu_{n-1} \qquad \ldots(21)$$

The statistical analysis of systems described by a Volterra series when the kernels are known has been studied extensively particularly with reference to communication systems [5-8] and Gaussian, sine wave plus Gaussian or random pulse train inputs. An excellent review of the use of functionals in the analysis of nonlinear systems is given by Barnett [9].

Consider the evaluation of the autocorrelation of the output, input-output cross-correlation and associated spectral densities for a system with known Volterra kernels assuming the output is strict-sense stationary. Define the correlation functions

$$R_{uy}(\tau) = E[y(t)u(t-\tau)] = \sum_{n=1}^{\infty} R_{uy_n}(\tau) \qquad \ldots(22)$$

$$R_{yy}(\tau) = E[y(t)y(t-\tau)] = \sum_{m=1}^{\infty}\sum_{n=1}^{\infty} R_{y_n y_m}(\tau) \qquad \ldots(23)$$

To evaluate the above expressions it is necessary to determine the partial correlation functions $R_{uy_n}(\tau)$, $R_{y_n y_m}(\tau)$. Rather than evaluating the general expressions which are given in the literature [5-8] consider the system illustrated in Fig.2 to illustrate the procedure.

Thus from eqn (18)

$$R_{uy}(\tau) = \int h(\tau_1)E\{u(t-\tau_1)u(t-\tau)\}d\tau_1 + \int\int h(\tau_2)h(\tau_3)E\{u(t-\tau_2)u(t-\tau_3)u(t-\tau)\}d\tau_2 d\tau_3 \qquad \ldots(24)$$

$$R_{yy}(\tau) = \iint h(\tau_1)h(\tau_2)E\{u(t-\tau_1)u(t-\tau-\tau_2)\}d\tau_1 d\tau_2$$

$$+ \iiint h(\tau_1)h(\tau_2)h(\tau_3)E\{u(t-\tau_1)u(t-\tau-\tau_2)u(t-\tau-\tau_3)\}d\tau_1 d\tau_2 d\tau_3$$

$$+ \iiint h(\tau_1)h(\tau_2)h(\tau_3)E\{u(t-\tau_1-\tau)u(t-\tau_2)u(t-\tau_3)\}d\tau_1 d\tau_2 d\tau_3$$

$$+ \iiiint h(\tau_1)h(\tau_2)h(\tau_3)h(\tau_4)E\{u(t-\tau_1)u(t-\tau_2)u(t-\tau-\tau_3)u(t-\tau-\tau_4)\} d\tau_1 d\tau_2 d\tau_3 d\tau_4 \quad \ldots(25)$$

In general all the moments of the input process up to order n must be known before eqn's (24) and (25) can be evaluated. When the input is a zero mean white Gaussian process where

$$E\{u(t_1)u(t_2)\ldots u(t_i)\} = 0 \qquad i \text{ odd}$$
$$= \sum_i \prod_{n\neq m} \delta(t_n - t_m) \qquad i \text{ even} \quad \ldots(26)$$

and the summation is over all ways of dividing i objects into pairs, eqn's (24) and (25) reduce to

$$R_{uy}(\tau) = h(\tau) \quad \ldots(27)$$

$$R_{yy}(\tau) = \int h(\tau+\tau_2)h(\tau_2)d\tau_2$$
$$+ 2\iint h(\tau+\tau_3)h(\tau+\tau_4)h(\tau_3)h(\tau_4)d\tau_3 d\tau_4 \quad \ldots(28)$$

Spectral densities are computed by taking Fourier Transforms of the associated correlation functions and using eqn's (19)-(21). An algebra of nonlinear systems based on the Volterra series has been developed by George [10] and this simplifies the notation considerably in many problems.

3.2 The Fokker-Planck-Kolmogorov Equation

Consider the class of dynamic systems which can be represented by the stochastic vector differential equations

$$d\underline{x}/dt = A(\underline{x},t) + C(\underline{x},t)V(t) \quad \ldots(29)$$

where $\underline{x} = \{x_i\}$ are the n-state variables, $A(x,t) = \{a_i\}$ and $C = \{c_{ij}\}$ are coefficient matrices and $\underline{V}(t)$ is an m-dimensional Gaussian white noise vector with the properties $E[\underline{V}(t)] = 0$, $E[\underline{V}(t)\underline{V}(s)^T] = Q\delta(t-s)$, $\underline{Q} = \text{diag}\{\sigma_{ii}^2\}$. $\underline{V}(t)$ can be used to represent random external disturbances, modelling discrepancies and random parametric perturbations.

Because Gaussian white noise is not mathematically meaningful, rewrite eqn (29) in terms of the incremental Wiener process $d\underline{W}(t) = \underline{V}(t)dt$ to yield the Itô [11,12] stochastic differential equation

$$d\underline{x} = \underline{A}(\underline{x},t)dt + \underline{C}(\underline{x},t)d\underline{W}(t) \qquad \ldots(30)$$

where

$$E\{d\underline{W}\} = 0\ , \quad E\{d\underline{W}(t)d\underline{W}(s)^T\} = \underline{Q}\delta(t-s)dt$$

Equation (30) generates a Markov process $\underline{x}(t)$ since $\underline{x}(t)$ and $\underline{w}(t)$ are independent with independent increments. Consequently, the solution of eqn (30) is completely characterized by the first order probability density function $f(\underline{x},t)$ and the transitional probability density function $f(\underline{x},t_2/\underline{x},t_1)$ for $t_2>t_1$, both of which can be shown to satisfy the Fokker-Planck-Kolmogorov equation [1,11,12,13]

$$\frac{\partial f(\cdot)}{\partial t} = -\sum_{i=1}^{n} \frac{\partial}{\partial x_i}(a_i(\underline{x},t)f(\cdot)) + \tfrac{1}{2}\sum_{i=1}^{n}\sum_{j=1}^{n} \frac{\partial^2}{\partial x_i \partial x_j}((\underline{CQC}^T)_{ij}f(\cdot)) \qquad \ldots(31)$$

In general it is of more practical value to determine the moments $m_n = E[x_i^n]$ of the system states rather than the probability density function $f(\underline{x},t)$. Using either the method of moments or Itô's fundamental lemma it can be shown, for example, that the first two moments are given by the solution of the ordinary differential equations [13,14]

$$\frac{dE(x_i)}{dt} = E(a_i(\underline{x},t)) \qquad \ldots(32)$$

$$\frac{dE(x_i x_j)}{dt} = E(a_i x_i + a_j x_j) + E((\underline{CQC}^T)_{ij}) \qquad \ldots(33)$$

for $(i,j) = 1,2\ldots n$, and given initial conditions $\underline{x}(t_o)$.

To illustrate the procedure consider a linear first order system with transfer function $\frac{1}{1+sT}$ driven by unity variance Gaussian white noise. The system model in state-space form is

$$Tdx = -xdx + dw \qquad \ldots(34)$$

and from eqn (31) its associated Fokker-Planck-Kolmogorov equation is

$$\frac{\partial f}{\partial t} = \frac{1}{T}\frac{\partial(xf)}{\partial x} + \frac{1}{2T}\frac{\partial^2 f}{\partial x^2} \qquad \ldots(35)$$

The first two moments are, from eqn's (32) and (33)

$$\frac{dm_1}{dt} = -\frac{1}{T} m_1 \qquad \text{...(36)}$$

$$\frac{dm_2}{dt} = -\frac{2}{T} m_2 + \frac{1}{T^2} \qquad \text{...(37)}$$

with solutions

$$m_1(t) = m_1(0)e^{-t/T} \qquad \text{...(38)}$$

$$m_2(t) = m_2(0)e^{-2t/T} + \frac{1}{2T}(1-e^{-2t/T}) \qquad \text{...(39)}$$

where $m_1(0)$ and $m_2(0)$ are initial conditions.

If the system equations (29) are nonlinear in the states the lower order moment eqn's (32) and (33) are in general functions of higher order moments and closed analytic solutions are not possible. Similarly the probability density function $f(\underline{x},t)$ which satisfies the Fokker-Planck-Kolmogorov equation cannot be found except by linearization or approximation methods. An alternative is to simulate the stochastic difference eqn (29) and evaluate the required moments by averaging over the realisations [13-15].

4. Linearisation Methods

The relative simplicity of the methods of statistical analysis for linear systems compared with the inherent complexity of the analysis outlined above has led to the development of approximation methods based on linearisation techniques [16].

The simplest form of linearisation is based upon the expansion of the nonlinear function in a Taylor series about some operating point and retaining only the linear terms in the analysis. Thus the nonlinear function $g(x_1,\dots x_n)$ is replaced by the approximate expression

$$g(x_1,\dots x_n) \approx g(\bar{x}_1,\dots,\bar{x}_n) + \sum_{i=1} g'_{x_i}(\bar{x}_1,\dots,\bar{x}_n)(x_i-\bar{x}_i) \qquad \text{...(40)}$$

where $\bar{x}_i$ is the mean of x_i and $g'_{x_i} = \partial f/\partial x_i$.

Although eqn (40) is linear with respect to fluctuations it is nonlinear with respect to expectations.

Equation (40) is valid only for continuous functions with continuous first derivatives, and cannot therefore be used to study the characteristics of discontinuous components such as relays or limiters. To linearise such characteristics the method of statistical linearisation was developed.

The earliest method of statistical linearisation was developed by Booton [17] for static nonlinearities and stochastic inputs with zero mean. Booton's method consists of replacing the nonlinearity N(•) by an equivalent gain which is selected so as to minimise the mean square of the difference between the output of the devices.

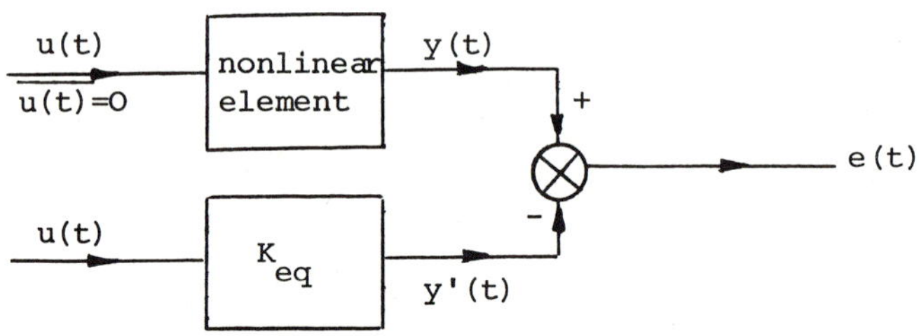

Fig.3

Consider the system illustrated in Fig.3 where

$$e(t) = y(t) - K_{eq} u(t)$$

$$\overline{e^2(t)} = \overline{y^2(t)} - 2K_{eq}\,\overline{u(t)y(t)} + K_{eq}^2\,\overline{u^2(t)} \qquad \ldots(41)$$

Selecting K_{eq} so as to minimise $\overline{e^2(t)}$ yields

$$K_{eq} = \frac{\overline{u(t)y(t)}}{\overline{u^2(t)}} = \frac{\int yuf(u)\,du}{\int u^2 f(u)\,du} \qquad \ldots(42)$$

When the input is Gaussian white $\int u^2 f(u)\,du = \sigma^2$ and hence

$$K_{eq} = \frac{1}{\sigma^2}\int yuf(u)\,du \qquad \ldots(43)$$

It can readily be shown [18] that K_{eq} in eqn (43) is equivalent to the first term in the Wiener series representation eqn (53) of a nonlinear zero memory system.

Somerville and Atherton [19] extended Booton's method to include cases of non-zero mean input signals as illustrated in Fig.4 to yield

$$K_{dc} = \frac{\overline{y(t)}}{\overline{u(t)}}$$

$$K_{eq} = \frac{\overline{u(t)y(t)} - \overline{u(t)}\cdot\overline{y(t)}}{\overline{u^2(t)} - \overline{u(t)}^2} \qquad \ldots(44)$$

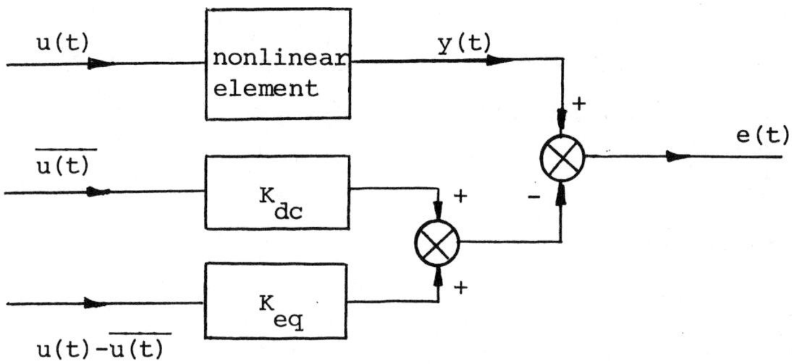

Fig.4

To conserve the spectrum of the output Pupkov [20] proposed replacing the zero memory nonlinear component N(•) by a dynamic stationary linear system

$$N(u(t)) = K_{dc}\overline{u(t)} + \int_{-\infty}^{t} h(t_1)(u(t-t_1)-\overline{u(t-t_1)})dt_1 \qquad \ldots(45)$$

where h(t) is determined to ensure the autocorrelation equivalence of the left and right hand sides of eqn (45)

$$\iint_{-\infty}^{\infty} h(t_1)h(t_2)R_{uu}(\tau-t_1-t_2)dt_1dt_2 = R_{yy}(\tau) \qquad \ldots(46)$$

and K_{dc} is given by eqn (44).

If h(t) in eqn (45) ($\overline{u(t)} = 0$) is selected to minimise the mean squared error this leads to the Wiener-Hopf equation

$$\int h(t_1)R_{uu}(t_1-\tau)d\tau = R_{uy}(\tau) \qquad \ldots(47)$$

Although this equation is in general difficult to solve for h(t), when the input is a separable process [3] from eqn (16)

$$R_{uy}(\tau) = C_F R_{uu}(\tau)$$

and the optimum linear approximation to the nonlinear element is Booton's equivalent gain $C_F = K_{eq}$.

Harmonic linearisation, which yields the describing function [21] consists in the simplest case of replacing the nonlinear element by a linear one whose transfer function is equal to the complex ratio of the fundamental component of the output to the sinusoidal input. The method has been used extensively to obtain mathematical descriptions of a large number of nonlinear systems and several modifications have been developed including describing functions which relate all the output harmonics to the input fundamentals.

5. Identification

Identification algorithms for nonlinear systems can be categorised as functional series methods, algorithms for block oriented systems and parameter estimation techniques [22,23].

5.1 Functional Series and Block Structured Algorithms

Identification using the Volterra series representation eqn (17) involves the measurement of the Volterra kernels. To illustrate the approach consider the identification of a system which can be described by just the first two Volterra kernels

$$y(t) = \int_0^\infty h_1(\tau_1)u(t-\tau_1)d\tau_1 + \iint_0^\infty h_2(\tau_1,\tau_2)u(t-\tau_1)u(t-\tau_2)d\tau_1 d\tau_2 \quad \ldots(48)$$

Defining the mean squared error as $E\{(z(t)-y(t))^2\}$ where z(t) is the measured output and applying calculus of variations yields

$$E\{z(t)\} = \int_0^\infty h_1(\tau_1)E\{u(t-\tau_1)\}d\tau_1 + \iint_0^\infty h_2(\tau_1,\tau_2)E\{u(t-\tau_1)u(t-\tau_2)\}d\tau_1 d\tau_2 \quad \ldots(49)$$

$$E\{z(t)u(t-\sigma)\} = \int_0^\infty h_1(\tau_1)E\{u(t-\tau_1)u(t-\sigma)\}d\tau_1 + \iint_0^\infty h_2(\tau_1,\tau_2)E\{u(t-\tau_1)u(t-\tau_2)u(t-\sigma)\}d\tau_1 d\tau_2 \quad \ldots(50)$$

$$E\{z(t)u(t-\sigma_1)u(t-\sigma_2)\} = \int_0^\infty h_1(\tau_1)E\{u(t-\tau_1)u(t-\sigma_1)u(t-\sigma_2)\}d\tau_1$$

$$+ \iint_0^\infty h_2(\tau_1,\tau_2)E\{u(t-\tau_1)u(t-\tau_2)u(t-\sigma_1)u(t-\sigma_2)\}d\tau_1 d\tau_2 \qquad \ldots(51)$$

The solution of this set of equations for a general stochastic input is extremely difficult. However, if the system input is white Gaussian substituting eqn (26) in eqn's (49)-(51) yields

$$\overline{z(t)} = \int_0^\infty h_2(\tau,\tau)d\tau$$

$$R_{uz}(\sigma_1) = h_1(\sigma_1) \qquad \ldots(52)$$

$$R_{uuz}(\sigma_1,\sigma_2) = \bar{z}\delta(\sigma_1-\sigma_2) + 2h_2(\sigma_1,\sigma_2)$$

and the solution for $h_1(t)$ and $h_2(t_1,t_2)$ is direct providing the mean level $\bar{z}$ is removed. Identification of systems which contain higher than second order kernels is very difficult using this approach. Alternative schemes involve approximating the kernels by an expansion of orthogonal functions and estimating the coefficients [22,23].

Wiener used a Gram-Schmidt orthogonalisation procedure to construct a new functional series where the functionals $\{G_n\}$ are orthogonal for a Gaussian white stimulus. The first two terms in the Wiener series are

$$G_1[k_1,u(t)] = \int_{-\infty}^{\infty} k_1(\tau)u(t-\tau)dt \qquad \ldots(53)$$

$$G_2[k_2,u(t)] = \iint_{-\infty}^{\infty} k_2(\tau_1,\tau_2)u(t-\tau_1)u(t-\tau_2)d\tau_1 d\tau_2$$

$$- P\int_{-\infty}^{\infty} k_2(\tau_1,\tau_1)d\tau_1 \qquad \ldots(54)$$

where P is the power spectral density of the white noise input. In general the Wiener kernels are not equal to the Volterra kernels. Numerous methods have been developed to identify the kernels in Wiener's series [22,23] the most popular being a correlation method by Lee and Schetzen [4]. The procedure consists of computing multidimensional correlation functions between the white Gaussian input and the system output to yield

$$k_n(\tau_1,\ldots\tau_n) = \frac{1}{n!P}\,\overline{\{y(t) - \sum_{m=o}^{n-1} G_m[k_m,u(t)]\}u(t-\tau_1)\ldots u(t-\tau_n)}\quad \tau_1,\tau_2\ldots\tau_n \qquad \ldots(55)$$

In an attempt to reduce the computational burden associated with the functional-series methods various authors have considered the identification of block oriented systems [23,24] which can be represented by interconnections of linear dynamic systems and static nonlinear elements. Consider the system illustrated in Fig.5 to illustrate the approach.

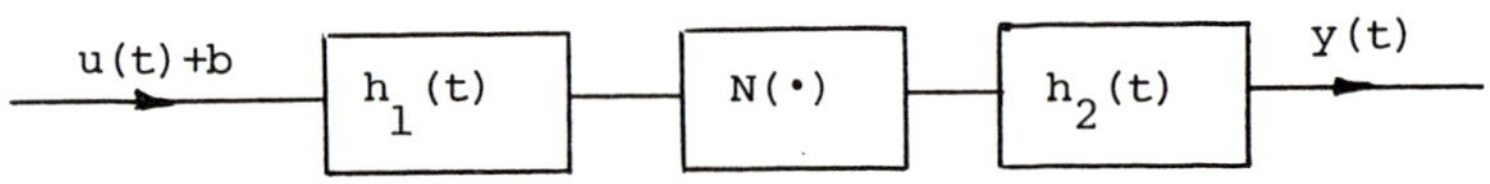

Fig.5

By extending the theory of separable processes [22-24] and using the result of eqn (16) it can readily be shown that for a Gaussian white input with mean level b

$$R_{uy'}(\sigma) = C_{FG}\int h_1(\tau_1)h_2(\sigma-\tau_1)d\tau_1 \qquad \ldots(56)$$

$$R_{u^2y'}(\sigma) = C_{FFG}\int h_1^{\,2}(\sigma-\tau_1)h_2(\tau_1)d\tau_1 \qquad \ldots(57)$$

where providing $h_1(t)$ is stable bounded-inputs bounded outputs C_{FG} and C_{FFG} are constants and the superscript ' indicates that the mean level has been removed from the signal. Estimates of the individual linear subsystems $h_1(t)$ and $h_2(t)$ can be obtained by decomposing eqn's (56) and (57) [24] and the nonlinearity can then be determined. The results of eqn's (56), (57) inherently provide information regarding the structure of the nonlinear system and this can be used to determine the position of the nonlinear element prior to complete identification. Similar results, which provide estimates of the individual component subsystems, are available for feedback, feedforward and multiplicative block oriented systems [24].

5.2 Parameter Estimation Algorithms

Parameter estimation methods for nonlinear systems where the structural form of the describing differential equations are known are now well established [22,23]. When little a priori information is available and the process is treated as a black-box, the usual approach is to expand the input/output using a suitable model representation. Two particular choices of model expansion, the NARMAX model and piecewise linear models will be briefly considered in the present analysis.

5.2.1 The NARMAX Model

If a system is linear then it is finitely realizable and can be represented by the linear difference equation model

$$y(k) = \sum_{i=1}^{n_y} (a_i y(k-i)) + \sum_{i=1}^{n_u} (b_i u(k-i)) \qquad \ldots(58)$$

if the Hankel matrix of the system has finite rank. When the system is nonlinear a similar representation can be derived by utilizing concepts from Nerode realization, multistructural forms and results from differential geometry to yield the nonlinear difference equation model [25]

$$y(k) = F^*[y(k-1),\ldots y(k-n_y),u(k-1),\ldots u(k-n_u)] \qquad \ldots(59)$$

where $F^*[\cdot]$ is some nonlinear function of $u(\cdot)$ and $y(\cdot)$. The model of eqn (59) can be shown [25] to exist whenever

(i) the state-space of the Nerode realization does not have infinite dimensions (i.e. we exclude distributed parameter systems), and

(ii) the linearized system around the origin has a Hankel matrix of maximum rank (i.e. a linearized model would exist if the system were operated close to an equilibrium point).

Equation (59) represents the single-input single-output case but the results have been extended to include multivariable systems. The Hammerstein, Wiener, bilinear, Volterra and other well known nonlinear models can be shown to be special cases of eqn (59).

An equivalent representation for nonlinear stochastic systems can be derived by considering input-output maps based on conditional probability density functions to yield the model

$$z(k) = F\left[z(k-1),\ldots z(k-n_z),u(k-1),\ldots u(k-n_u),\varepsilon(k-1),\ldots\varepsilon(k-n_\varepsilon)\right] +\varepsilon(k) \quad \ldots(60)$$

where $\varepsilon(k)$ is the prediction error. This model is referred to as the Nonlinear AutoRegressive Moving Average model with eXogenous inputs or NARMAX model [25].

A NARMAX model with first order dynamics expanded as a second order polynomial nonlinearity would for example be represented as

$$\begin{aligned} y(k) &= F_2\left[y(k-1),u(k-1)\right] \\ &= C_1y(k-1)+C_2u(k-1)+C_{11}y^2(k-1)+C_{12}y(k-1)u(k-1)+C_{22}u^2(k-1) \end{aligned} \quad \ldots(61)$$

Assuming that the output measurements are corrupted by additive noise

$$z(k) = y(k) + e(k)$$

gives the input-output model

$$\begin{aligned} z(k) = {} & C_1z(k-1)+C_2u(k-1)+C_{11}z^2(k-1)+C_{12}z(k-1)u(k-1) \\ & +C_{22}u^2(k-1)+e(k)-C_1e(k-1)-2C_{11}z(k-1)e(k-1) \\ & +C_{11}e^2(k-1)-C_{12}e(k-1)u(k-1) \end{aligned} \quad \ldots(62)$$

Because the NARMAX model maps the past input and output into the present output multiplicative noise terms are induced in the model even though the noise was additive at the output. In general the noise may enter the system internally and because the system is nonlinear it will not always be possible to translate it to be additive at the output. This situation will again result in multiplicative noise terms in the NARMAX model with the added complication that the noise source and the prediction error will not in general be equal. Since most of the parameter estimation techniques derived for linear systems assume that the noise is independent of the input, biased estimates result when they are applied to nonlinear systems eqn (60).

The recursive extended least squares (RELS) algorithm can however be readily adapted to the NARMAX model, by defining the following vectors

$$Q(k) = [z(k-1), u(k-1), z^2(k-1), z(k-1)u(k-1), u^2(k-1), \varepsilon(k-1),$$
$$\varepsilon(k-1)z(k-1), u(k-1)\varepsilon(k-1), \varepsilon^2(k-1)]^T$$
$$\hat{\theta} = [\hat{c}_1, \hat{c}_2 \ldots \hat{c}_s]^T$$
$$\varepsilon(k+1) = z(k+1) - Q(k+1)^T\hat{\theta}(k) \qquad \ldots(63)$$

for the model of eqn (62) for example. With these definitions the standard RELS algorithm can be applied to yield unbiased parameter estimates. The development of recursive maximum likelihood and instrumental variable algorithms for the NARMAX model is not quite so straightforward [26].

The direct application of an offline maximum likelihood algorithm is not possible because in general the prediction errors will not have a Gaussian distribution. However, by considering the loss function

$$J(\theta) = \frac{1}{2N} \log_e \det \sum_{k=1}^{N} \varepsilon(k;\theta)\varepsilon^T(k;\theta) \qquad \ldots(64)$$

it can be shown that the prediction error estimates obtained by minimising eqn (64) have very similar asymptotic properties to the maximum likelihood estimates even when $\varepsilon(k)$ is non-gaussian. A prediction error algorithm has been developed for the NARMAX model based on this result. This together with least squares derived algorithms [26] have been augmented with a stepwise regression algorithm, a likelihood ratio test and Akaike tests to detect the model structure or significant terms in the model prior to final estimation [28].

Whichever model formulation or identification algorithm is implemented it is important to test that the identified model does adequately describe the data set. When the system is nonlinear the residuals $\zeta(k)$ should be unpredictable from all linear and nonlinear combinations of past inputs and outputs and this condition will hold iff [27]

$$\begin{aligned} R_{\zeta\zeta}(\tau) &= \delta(\tau) \\ R_{u\zeta}(\tau) &= 0 \ \forall\ \tau \\ R_{\zeta\zeta u}(\tau) &= E[\zeta(k)\zeta(k-1-\tau)u(k-1-\tau)] = 0 \ \forall\ \tau \geq 0 \end{aligned} \qquad \ldots(65)$$

Notice that for nonlinear systems the traditional linear tests $R_{\zeta\zeta}(\tau)$ and $R_{u\zeta}(\tau)$ are not sufficient. If instrumental variables or suboptimal least squares are used the residuals may be coloured. It can be shown that in this case the process model is unbiased iff

$$R_{u\zeta}(\tau) = 0 \;\forall\; \tau$$
$$R_{u^{2'}\zeta}(\tau) = E\left[(u^2(k)-\overline{u^2})\,\zeta(k+\tau)\right] = 0 \;\forall\; \tau \qquad \ldots(66)$$
$$R_{u^{2'}\zeta^2}(\tau) = E\left[(u^2(k)-\overline{u^2})\,\zeta^2(k+\tau)\right] = 0 \;\forall\; \tau$$

Experience has shown that when using a prediction error algorithm the tests in both eqn's (65) and (66) often give the experimenter a great deal of information regarding the deficiencies in the fitted model and can indicate which terms should be included in the model to improve the fit [28].

5.2.1.1 <u>An Example</u>

To illustrate some of the ideas associated with parameter estimation based on the NARMAX model consider the identification of a model relating the input volume flow rate u(t) and the level of liquid z(t) in the interconnected tanks illustrated in Fig.6.

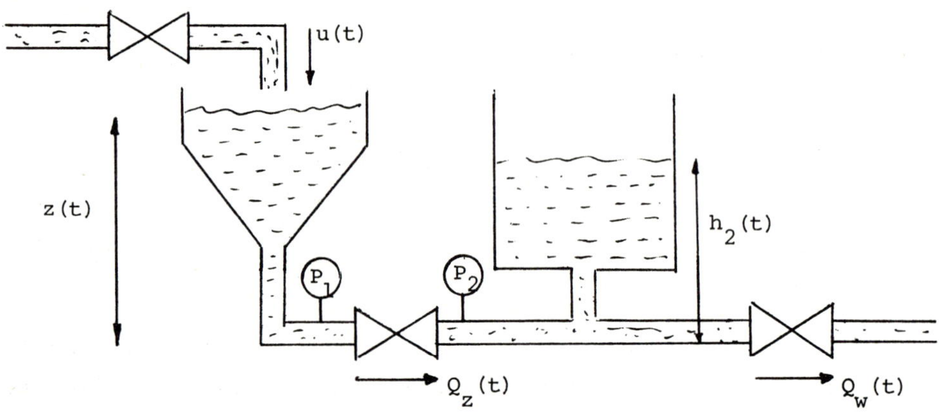

Fig.6. Interconnected Tanks

A zero mean Gaussian signal was used to perturb the input u(t) and 1000 data pairs were recorded by sampling the input and output at 9.6 secs.

In the early stages of any identification procedure it is important to establish if the process under test exhibits nonlinear characteristics which will warrant a nonlinear model. This can readily be achieved using a simple correlation test [23,27]. If the third order moments of the input are zero and all even order moments exist (a sine wave, gaussian or ternary sequence would for example satisfy the properties) then the process is linear iff

$$R_{z'z'^2}(\tau) = E\left[(z(k)-\bar{z})(z(k+\tau)-\bar{z})^2\right] = 0 \;\forall\, \tau \qquad \ldots(67)$$

$R_{z'z'^2}(\tau)$ for the liquid level system is illustrated in Fig.7 and clearly shows that, as expected, the liquid level system is highly nonlinear.

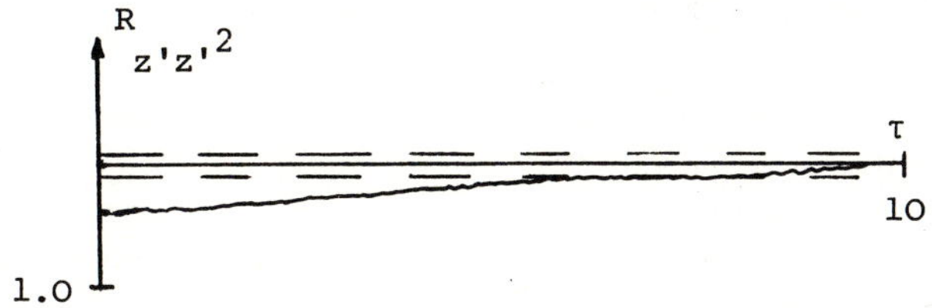

Fig.7. Nonlinear detection test

Initially a linear model was fitted to the data using a maximum likelihood algorithm to give the representation

$$\begin{aligned} z(k) = {} & 0.746z(k-1)+0.340z(k-2)-0.122z(k-3) \\ & +0.471u(k-1)-0.174u(k-2)-0.040u(k-3) \\ & +\varepsilon(k)+0.423\varepsilon(k-1)+0.038\varepsilon(k-2) \end{aligned} \qquad \ldots(68)$$

A comparison of the process and linear model predicted output is illustrated in Fig.8. The model validity tests eqn's (65), (66) for this model are illustrated in Fig.9. Notice that although $R_{\zeta\zeta}(\tau)$ and $R_{u\zeta}(\tau)$ indicate linear adequacy for the model eqn (68), $R_{u^{2\prime}\zeta}(\tau)$ and $R_{u^{2\prime}\zeta^2}(\tau)$ are well outside the 95% confidence bands indicating that nonlinear terms should be included in the model description.

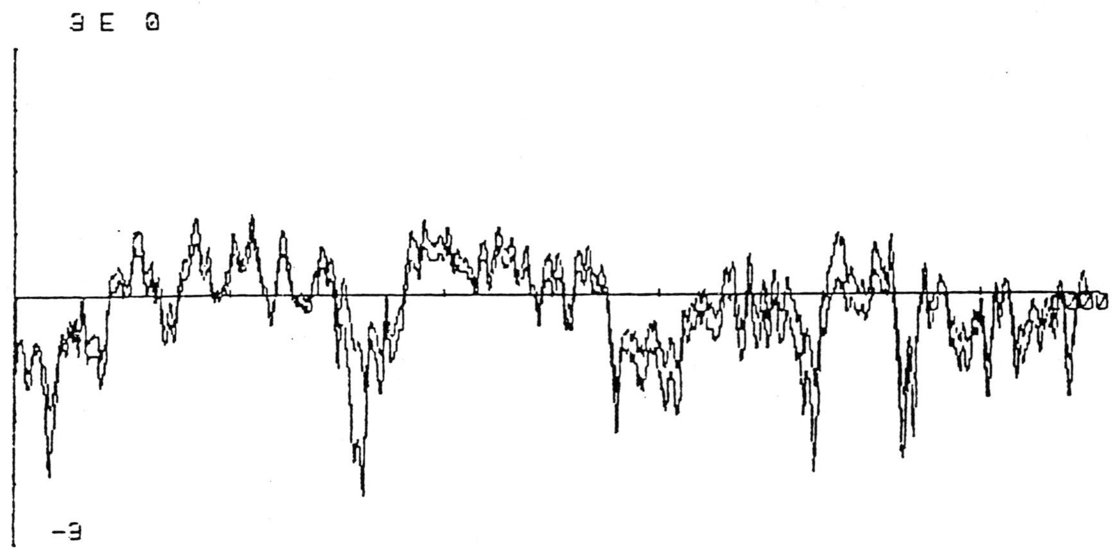

Fig. 8. Process and predicted output : best linear model

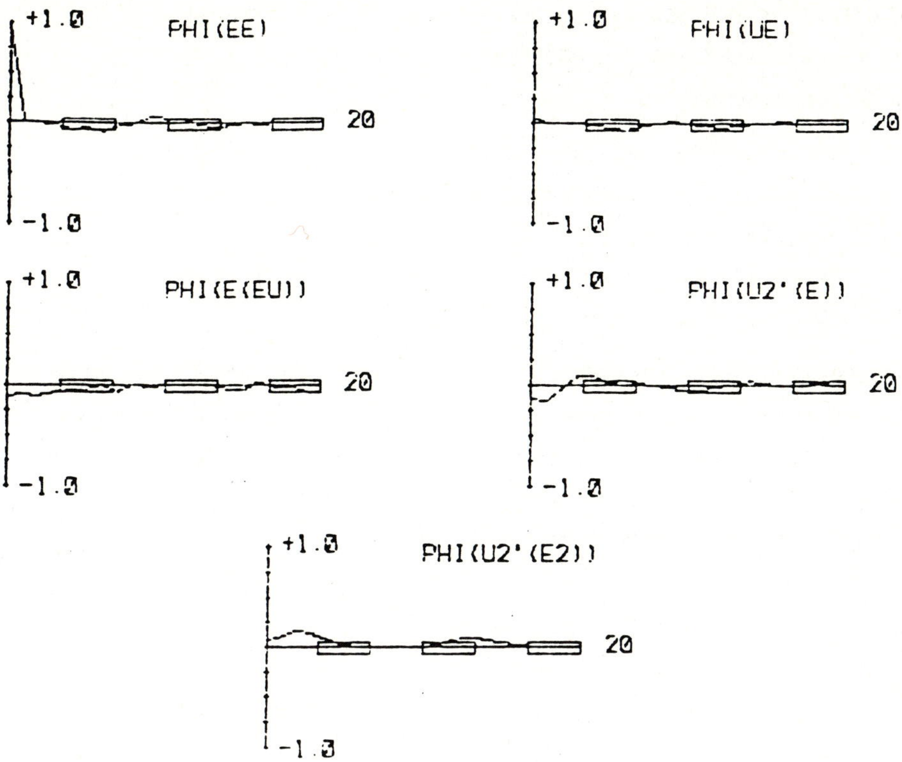

Fig. 9. Model validation : best linear model

The effect of introducing nonlinear terms into the model was therefore investigated and a prediction error algorithm yielded the NARMAX model representation

$$\begin{aligned} z(k) = {} & 0.436z(k-1)+0.681z(k-2)-0.149z(k-3) \\ & +0.396u(k-1)+0.014u(k-2)-0.071u(k-3) \\ & -0.351z(k-1)u(k-1)-0.034z^2(k-2) \\ & -0.135z(k-2)u(k-2)-0.027z^3(k-2)-0.108z^2(k-2)u(k-2) \\ & -0.099u^3(k-2)+\varepsilon(k)+0.344\varepsilon(k-1)-0.201\varepsilon(k-2) \qquad \ldots(69) \end{aligned}$$

The model validity tests for the model of eqn (69) are illustrated in Fig.11 and these together with the comparison of the NARMAX model predicted and process output Fig.10 show the considerable improvement in the prediction capabilities of the estimated NARMAX model eqn (69) compared with the best linear model eqn (68).

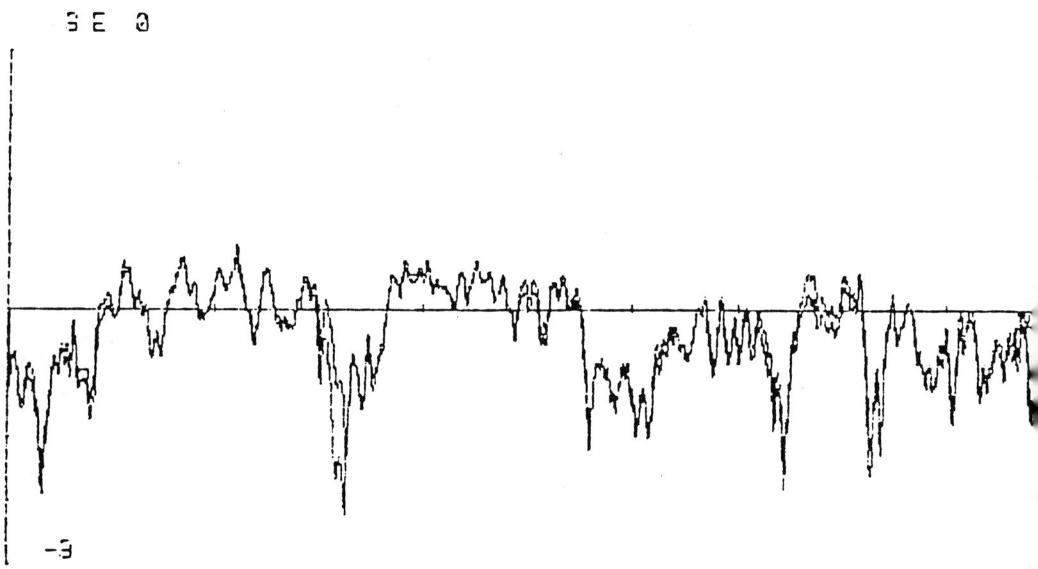

Fig. 10. Process and predicted output for the estimated NARMAX model

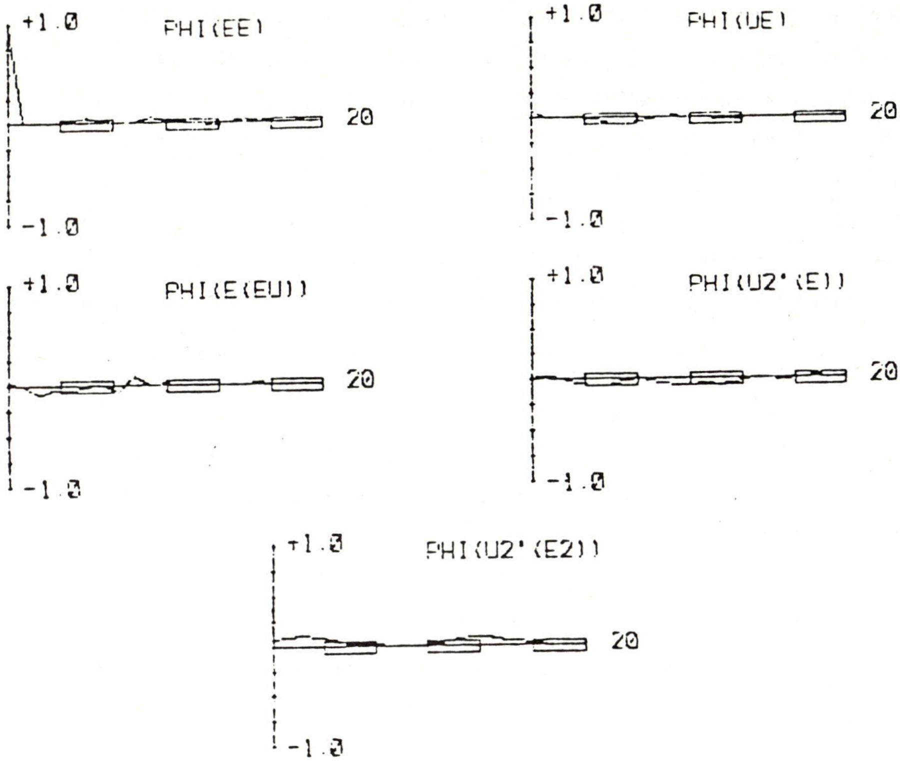

Fig. 11. Model validation : best nonlinear model

5.2.2 Piecewise Linear Modelling

There are several possible ways in which nonlinear systems can be approximated by locally linear models. These involve either the expansions of the NARMAX model eqn (60) using spline functions, spatial piecewise linear models or linear models with signal dependent parameters [29]. The last of these three representations can be fitted by performing a series of linear experiments and repeatedly utilizing linear parameter estimation routines to build up a series of linearized models of the process. This can offer several advantages compared with fitting a global nonlinear model and may be appropriate in situations where this type of approximate system description is adequate. To illustrate the ideas involved we will just consider signal dependent linear modelling.

Linearizing the NARMAX model eqn (60) at a selected operating point $\Delta_k = [z_1, \ldots z_{n_z}, u_1, \ldots u_{n_u}]$, $\Delta_k \varepsilon R$ gives

$$dz(k) = \sum_{i=1}^{n_z} \left.\frac{\partial F[\cdot]}{\partial z(k-i)}\right|_{\Delta_k} dz(k-i) + \sum_{i=1}^{n_u} \left.\frac{\partial F[\cdot]}{\partial u(k-i)}\right|_{\Delta_k} du(k-i) + d\varepsilon(k) \qquad \ldots(70)$$

where for simplicity the noise is assumed to be represented by a single uncorrelated prediction error term $\varepsilon(k)$. Substituting $dz(k) = z(k)-z(k)|_{\Delta_k}$, $du(k) = u(k)-u(k)|_{\Delta_k}$, $d\varepsilon(k) = \varepsilon(k)-\varepsilon(k)|_{\Delta_k}$ into eqn (70) and manipulating gives

$$z(k) = \theta_o|_{\Delta_k} + \sum_{i=1}^{n_z} \theta_i|_{\Delta_k} z(k-i) + \sum_{i=1}^{n_u} \theta_{nz+i}|_{\Delta_k} u(k-i) + \varepsilon(k) \qquad \ldots(71)$$

where

$$\theta_o|_{\Delta_k} = z(k)|_{\Delta_k} - \varepsilon(k)|_{\Delta_k} - \sum_{i=1}^{n_z} \left.\frac{\partial F[\cdot]}{\partial z(k-i)}\right|_{\Delta_k} z(k-i)|_{\Delta_k} - \sum_{i=1}^{n_u} \left.\frac{\partial F[\cdot]}{\partial u(k-i)}\right|_{\Delta_k} u(k-i)|_{\Delta_k}$$

$$\theta_i|_{\Delta_k} = \frac{\partial F[\cdot]}{\partial z(k-i)}$$

$$\theta_{n_z+i}|_{\Delta_k} = \frac{\partial F[\cdot]}{\partial u(k-i)} \qquad \ldots(72)$$

If the process is such that the parameters in the linearized model eqn (71) depend on a signal $\omega(k)$ then eqn (71) can be written as

$$z(k) = \theta_1(\omega(k)) + \sum_{i=1}^{n_z} \theta_i(\omega(k))z(k-i) + \sum_{i=1}^{n_u} \theta_{n_z+i}(\omega(k)).u(k-i) \qquad ...(73)$$

Equation (73) is a signal dependent linear model where $\omega(k)$ is the coefficient dependent signal which may depend on the input, the output or some external variable associated with the operation of the process. In the modelling of a power station for example it may be appropriate to select $\omega(k)$ as the megawatt output. In the liquid level system it would be the level of liquid in the first tank.

Parameter estimation for signal dependent models is usually performed in two stages. Initially the coefficients in a series of linear models as $\omega(k)$ is varied are estimated. These locally linear models are then patched together to form an approximate global nonlinear description of the system under investigation [29]. For example eqn (73) can be expressed as

$$z(k) = \psi^T(k)\underline{\theta}(\omega(k)) + \varepsilon(k) \qquad ...(74)$$

where the definitions of $\psi(\cdot)$ and $\underline{\theta}(\cdot)$ follow directly. Assuming that the signal dependent parameter vector $\underline{\theta}(\omega(k))$ can be approximated by a finite degree polynomial we can write

$$\theta_i(\omega(k)) = \beta_i^T W(k) \quad , \quad i = 0,\dots n_z+n_u \qquad ...(75)$$

where

$$W(k) = [1, \omega(k), \omega^2(k), \dots \omega^\ell(k)]^T$$

$$\beta_i = [\beta_{i0}, \beta_{i1}, \beta_{i2} \dots \beta_{i\ell}]^T$$

Substituting eqn (75) in eqn (74) gives the global nonlinear model

$$z(k) = \psi^T(k)\,[\beta^T W(k)] \qquad ...(76)$$

The advantage of this approach is that it is relatively easy to estimate the model parameters and the results can be readily interpretted using all the well known linear theory. The disadvantage is that the final model will only provide an adequate representation for the system for the particular trajectory of input signal used in the identification unless the process is only mildly nonlinear or the operation of the process moves slowly and smoothly from one operating point to another [29].

5.2.2.1 An Example

The implicit nonlinear model

$$z(k) = 0.5z(k-1)+0.3u(k-1)+0.3z(k-1)u(k-1)+0.5u^2(k-1)+\varepsilon(k) \qquad \text{...(77)}$$

was simulated over the global input range of ±1.0. Eleven first order linearized models with input range ±0.1 were estimated at the different operating levels where $\omega(k) = u(k)$. The final global nonlinear model eqn (76) was estimated as

$$z(k) = [1,z(k-1),u(k-1)] \begin{pmatrix} -0.0019 & 0.0525 & -0.6104 & -0.6156 & -0.3919 \\ 0.4548 & 0.4334 & 0.0580 & -0.1953 & -0.0036 \\ 0.3034 & 1.1151 & 0.3770 & 0.3723 & 0.2348 \end{pmatrix} \times \begin{pmatrix} 1 \\ u(k-1) \\ u^2(k-1) \\ u^3(k-1) \\ u^4(k-1) \end{pmatrix} \qquad \text{...(78)}$$

The final model eqn (78) provided an adequate representation for the system eqn (77) when perturbed by the input used to initially excite the process but significant deficiencies in the model were visible for other rapidly varying inputs.

5.3 Control of Nonlinear Sampled-Data Systems

The choice of model representation for nonlinear systems is vitally important since this will influence its usefulness for both prediction and controller design. In view of the success of the linear difference equation model as a basis for linear controller design procedures it is natural to extend these ideas to the NARMAX model. There are many possibilities here that can be investigated [23,30] and only the simple one-step-ahead adaptive controller based on the NARMAX model will be considered here. This is best illustrated by designing controllers for the liquid level system discussed in section 5.2.1.1.

PI, linear adaptive one-step-ahead and self-tuning regulators were designed for the liquid level system described in section 5.2.1.1. The performance of these controllers is illustrated in Fig.12.

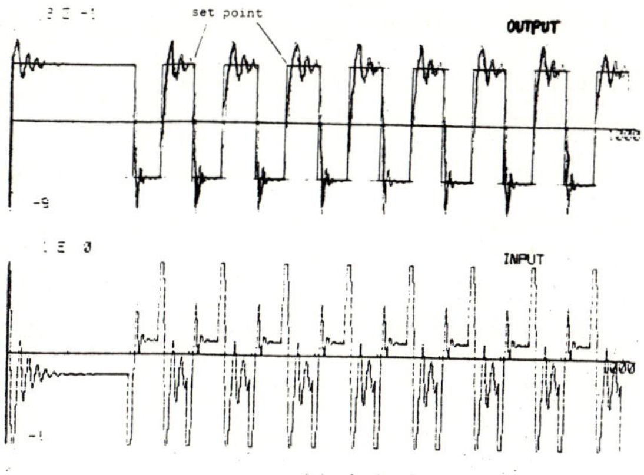

(a) PI Controlled System

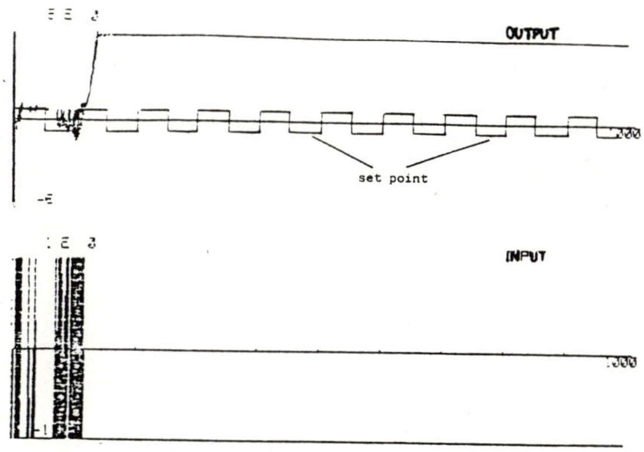

(b) One-step ahead controlled system

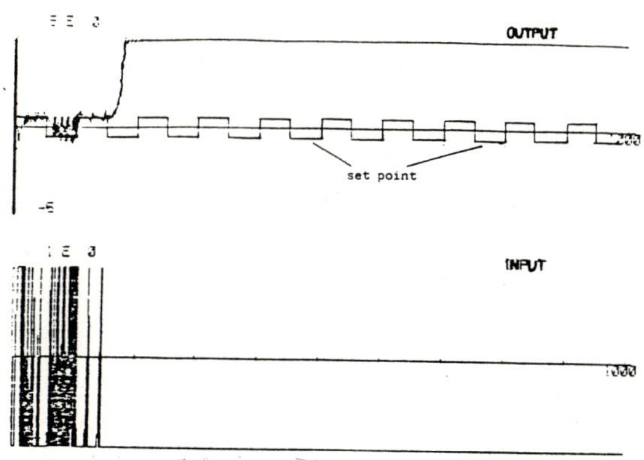

(c) Minimum variance controlled system

Fig.12. Linear controllers applied to the nonlinear liquid level system

Inspection of Fig.12 clearly shows the poor performance of all these linear controllers when applied to the nonlinear liquid level system. The adaptive linear regulators only performed with any degree of satisfaction when the set point signal was slowed right down so that the parameter estimation routines had time to re-adapt to the new operating point.

The adaptive one-step-ahead nonlinear controller was defined, from eqn (69), by the feedback law

$$\begin{aligned} u(k) = \{ & y^* - 0.436z(k) - 0.681z(k-1) + 0.149z(k-2) \\ & - 0.94u(k-1) + 0.071u(k-2) + 0.034z^2(k-1) \\ & + 0.135z(k-1)u(k-1) + 0.027z^3(k-1) + 0.108z^2(k-1)u(k-1) \\ & + 0.099u^3(k-1)\}/\{0.396 - 0.351z(k)\} \end{aligned} \quad \ldots(79)$$

for $[0.396-0.351z(k)] \geq \gamma > 0$

where γ was set to 0.01 and y^* represents the set point. The performance of the controller eqn (79) when applied to the liquid level system is illustrated in Fig.13.

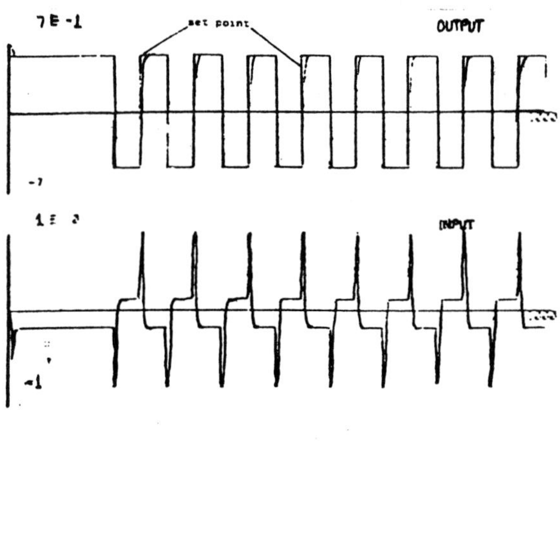

Fig.13. Nonlinear control of the liquid level system

A comparison of the performance of the linear based designs Fig.12 with the nonlinear design Fig.13 clearly shows the excellent response of the system when the nonlinear controller is utilised.

6. Conclusions

The statistical analysis of nonlinear systems is in general a difficult task. Whilst some of the techniques currently available have been briefly described above details of other alternative approaches are readily available in the literature.

Acknowledgements

The author gratefully acknowledges financial support for part of the work presented above from SERC grants GR/B/31163 and GR/D/30587.

References

1. Papoulis, A. 'Probability theory, random variables and stochastic processes', McGraw-Hill, N.Y., 1965.

2. Barrett, J.F., Coales, J.F. 'An introduction to the analysis of nonlinear control systems with random inputs', Proc.IEE, 103C, pp.190-199, 1955.

3. Nuttall, A.H. 'Theory and application of the separable class of random processes', MIT Res. Lab. Elect. Technical Rp 343, 1948.

4. Schetzen, M. 'The Volterra and Wiener theories of nonlinear systems', Wiley, N.Y., 1980.

5. Bedrosian, E., Rice, S.O. 'The output properties of Volterra systems driven by harmonic and Gaussian inputs', Proc.IEEE, 59, pp.1688-1707, 1971.

6. Bussgang, J.J., Ehrman, L., Graham, J.W. 'Analysis of nonlinear systems with multiple inputs', Proc.IEEE, 62, pp.1088-1119, 1974.

7. Rudko, M., Wiener, D.D. 'Volterra systems with random inputs; a formalized approach', IEEE Trans. on Commn., COM-26, pp.217-227, 1978.

8. Barrett, J.F. 'Formula for output autocorrelation and spectrum of a Volterra system with stationary Gaussian input', Proc.IEE, 127, pp.286-289, 1980.

9. Barrett, J.F. 'The use of functionals in the analysis of nonlinear physical systems', J. Elect. & Control, 15, pp.567-615, 1963.

10. George, D.A. 'Continuous nonlinear systems', MIT Res. Lab. Elect., Tech. Report 355, 1959.

11. Jazwinski, A.H. 'Stochastic processes and filtering theory', Academic Press, N.Y., 1970.

12. Doob, J.L. 'Stochastic processes', Wiley, N.Y., 1953.

13. Harris, C.J. 'Stochastic process models and control', SRC Vacation School on Stochastic Processes in Control Systems', Warwick University, April 1978.

14. Harris, C.J. 'Simulation of stochastic processes', ibid.

15. Harris, C.J. 'Modelling, simulation and control of stochastic systems with applications in wastewater treatment', Int. J. Systems Sci., 8, pp.393-411, 1977.

16. Pervozvanskii, A.A. 'Random processes in nonlinear control systems', Academic Press, N.Y., 1965.

17. Booton, R.C. 'The analysis of nonlinear control systems with random inputs', Proc. Symp. on N.L. Ct. Analysis, Brooklyn Poly, 1953.

18. Lawrence, P.J. 'A comparison of nonlinear system approximations', Tech. Note DAG 86, Dynamic Analysis Group, UWIST, 1976.

19. Somerville, M.J., Atherton, D.P. 'Multigain representation for single-valued nonlinearity with several inputs', Proc.IEE, 105C, pp.537-549, 1958.

20. Pupkov, K.A. 'Method of investigating the accuracy of essentially nonlinear automatic control systems by means of equivalent transfer functions', Automn & Remote Contr., 21, pp.126-140, 1960.

21. Atherton, D.P. 'Nonlinear control engineering', Van Nostrand Reinhold, 1975.

22. Billings, S.A. 'Identification of nonlinear systems - a survey', Proc.IEE, 127, pp.272-285, 1980.

23. Billings, S.A., Gray, J.O., Owens, D.H. (Eds). 'Nonlinear System Design', P. Peregrinus, 1984.

24. Billings, S.A., Fakhouri, S.Y. 'Identification of systems composed of linear dynamic and static nonlinear elements', Automatica, 18, pp.15-26, 1982.

25. Leontaritis, I.J., Billings, S.A. 'Input-output parametric models for nonlinear systems, Part I - Deterministic nonlinear systems, Part II - Stochastic nonlinear systems', Int. J. Control, 41, pp.303-344, 1985.

26. Billings, S.A., Voon, W.S.F. 'Least-squares parameter estimation algorithms for nonlinear systems', Int. J. Systems Sci., 15, pp.601-615, 1984.

27. Billings, S.A., Voon, W.S.F. 'Structure detection and model validity tests in the identification of nonlinear systems', Proc.IEE, Part D, 130, pp.193-199, 1983.

28. Billings, S.A., Fadzil, M.B. 'The practical identification of nonlinear systems', 7th IFAC Symp. Ident. Syst. Par. Est., York, 1985.

29. Billings, S.A., Voon, W.S.F. 'Piecewise linear identification of nonlinear systems', (in preparation).

30. Billings, S.A., Tsang, K.M. 'Predictive controller design for nonlinear systems', (in preparation).

AN INTRODUCTION TO DISCRETE-TIME SELF-TUNING CONTROL

Dr. P.J. Gawthrop

1. INTRODUCTION

This chapter provides a tutorial introduction to self-tuning control in its traditional discrete-time setting. We start of with a slightly modified version of the celebrated self-tuning regulator of Astrom and Wittenmark[1]. These modifications include control weighting and set-point following[2]. A weighted model reference controller is then considered which is based on the self-tuning controller of Clarke and Gawthrop[3,2,4,5]. Finally, the pole placement self-tuning controller due to Wellstead, Edmunds, Prager and Zanker[6,7]. is discussed. We attempt to view all three approaches within a common framework: that of emulating unrealisable compensators using a self-tuning emulator.

There are a number of good sources for further reading. The survey paper by Astrom[8] gives a broad overview of adaptive control. The book edited by Harris and Billings (based on an IEE workshop at Oxford)[9] and that edited by Unbehauen[10] (based on a symposium at Bochum) both contain useful tutorial material. The paper by Astrom and Wittenmark[11] gives a very readable tutorial account of pole and pole/zero placement techniques. The book by Goodwin and Sin[12] covers a lot of the background theory.

2. DELAYS AND PREDICTORS

In this section, the control of time-delay systems using a slight modification of the self-tuning regulator of Astrom and Wittenmark[1] is considered. The algorithm is based on the minimum variance controller[13] and its extensions to include control weighting[14,15]

2.1. Systems with delay

Many real systems involve transport processes; examples are fluid flow along a long pipe or material flow along a conveyer. The transport process to which the first minimum variance controller[13] (and later the

self-tuning regulator) was applied was a paper mill.

In discrete time, a pure time delay with output u'(iT) and input u(iT) can be written in the time domain as

$$u'(iT) = u(iT-kT) \tag{2.1.1}$$

where i indexes time and T is the uniform sample interval. In other words, v is u delayed by k sample instants of duration T. Taking the z-transform of u' and u (assuming zero initial conditions) the corresponding z domain equation is

$$U'(z^{-1}) = z^{-k}\, U(z^{-1}) \tag{2.1.2}$$

The rest of the system dynamics are assumed to be given by a linear difference equation leading to the z-domain equation

$$Y(z^{-1}) = \frac{B(z^{-1})}{A(z^{-1})} U'(z^{-1}) \tag{2.1.3}$$

where $U'(z^{-1})$ and $Y(z^{-1})$ are the z-transformed system output and delayed input; $B(z^{-1})$ and $A(z^{-1})$ are polynomials in z^{-1}:

$$A(z^{-1}) = a_0 + a_1 z^{-1} + \dots + a_n z^{-n} \tag{2.1.4}$$

$$B(z^{-1}) = b_0 + b_1 z^{-1} + \dots + b_n z^{-n} \tag{2.1.5}$$

All real systems are subject to random disturbances. One way of modelling such disturbances is as a rational transfer function driven by white noise $\xi(iT)$.

$$V(z^{-1}) = \frac{C(z^{-1})}{A(z^{-1})} \Xi(z^{-1}) \tag{2.1.6}$$

(Note that Ξ is capital ξ)

Putting these three components together gives Figure 2.1.1 and the z-domain equation

$$Y(z^{-1}) = z^{-k} \frac{B(z^{-1})}{A(z^{-1})} U(z^{-1}) + V(z^{-1}) \tag{2.1.7}$$

$$= z^{-k} \frac{B(z^{-1})}{A(z^{-1})} U(z^{-1}) + \frac{C(z^{-1})}{A(z^{-1})} \Xi(z^{-1})$$

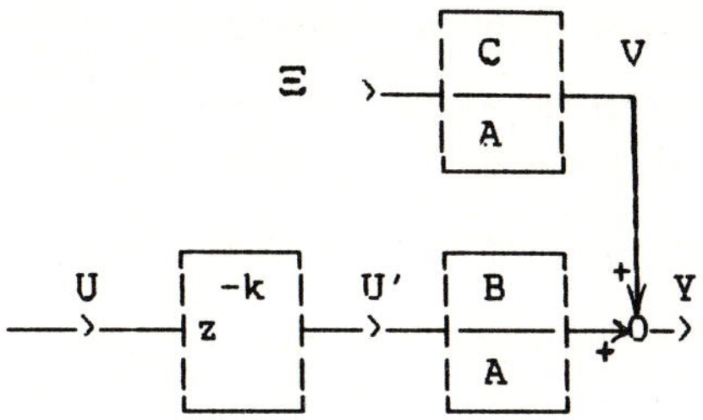

Figure 2.1.1 A time-delay system

There is no loss in generality in assuming common poles for the dynamics and the disturbances as long as pole/zero cancellations are allowed. This system structure is algebraically convenient

2.2. Predicting the future

The key idea in controlling a system with time delay k is that the current control signal is not based on the current system output but on the system output k steps into the future. The purpose of this section is to indicate one way of generating a prediction of the system output.

Given the system dynamics (that is the polynomials A,B and C) there are two sources of uncertainty in predicting the future: the system initial conditions and the unknown disturbance v. The effect of initial conditions will, with correct predictor design, have only transient effect; but the effect of the unknown disturbances will persist. So we will concentrate on minimising the effect of unknown disturbances.

Recalling that the z transform of $y(iT+kT)$ is $z^{-k} Y(z^{-1})$ the value of the system output y k steps into the future is given by:

$$z^k Y(z^{-1}) = \frac{B(z^{-1})}{A(z^{-1})}U(z^{-1}) + z^k \Xi(z^{-1}) \tag{2.2.1}$$

$$= \frac{B(z^{-1})}{A(z^{-1})}U(z^{-1}) + z^k \frac{C(z^{-1})}{A(z^{-1})}\Xi(z^{-1})$$

The first term on the righthand side of this equation $\left(\frac{B(z^{-1})}{A(z^{-1})}U(z^{-1})\right)$ corresponds to past and present values of U, but the other term has a factor z^k indicating future values of V. To make this dependency explicit, $\frac{C(z^{-1})}{A(z^{-1})}$ is expanded by polynomial long division as a power series in terms of z^{-1} as:

$$\frac{C(z^{-1})}{A(z^{-1})} = e_0 + e_1 z^{-1} + \ldots + e_k z^{-k} + \ldots \tag{2.2.2}$$

The details of performing this expansion need not concern us at the moment; but it can always be done. The numbers e_i are called the weighting coefficients or Markov parameters of the transfer function $\frac{C(z^{-1})}{A(z^{-1})}$.

In the time domain, this means that the disturbance v is given by the convolution

$$v(iT) = e_0\xi(iT) + e_1\xi(iT-T) + \ldots + e_k\,\xi(iT-kT) + \ldots \tag{2.2.3}$$

$$= \sum_{j=0}^{\infty} e_j \xi(iT-jT)$$

Thus the future disturbance v(iT+kT) is given by:

$$v(iT+kT) = e_0\xi(iT+kT) + e_1\xi(iT+kT-T) + \ldots + e_k\,\xi(iT) + \ldots \tag{2.2.4}$$

$$= \sum_{j=0}^{\infty} e_j \xi(iT+kT-jT)$$

Typical weighting coefficients are depicted in Fig 2.2.1 where they are plotted against time index j.

Values of $\xi(iT+kT-jT)$ with $j<k$ occur in the future; together with the assumption that the values of $\xi(iT)$ are uncorrelated means that such values cannot be known now (time=iT). However, values with $j\geq k$ can in principle be deduced from measurements of the system output up to the present time by rewriting the system equation as:

$$\Xi(z^{-1}) = \frac{A(z^{-1})}{C(z^{-1})}Y(z^{-1}) - z^{-k}\,\frac{B(z^{-1})}{C(z^{-1})}U(z^{-1}) \tag{2.2.5}$$

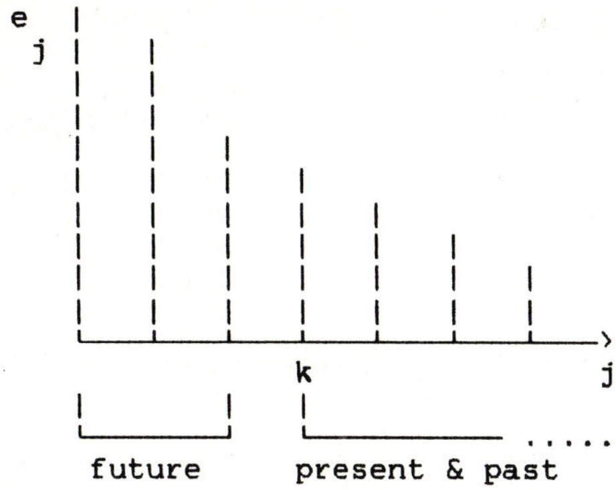

Figure 2.2.1 The weighting sequence expansion

This decomposition into future and past can be captured by rewriting the weighting sequence expansion as:

$$\frac{C(z^{-1})}{A(z^{-1})} = E(z^{-1}) + z^{-k}\frac{F(z^{-1})}{A(z^{-1})} \tag{2.2.6}$$

where

$$E(z^{-1}) \triangleq e_0 + e_1 z^{-1} + .. + e_{k-1} z^{k-1} \tag{2.2.7}$$

$$\frac{F(z^{-1})}{A(z^{-1})} \triangleq e_k + e_{k-1} z^{-1} + e_{k-2} z^{-2} + \ldots \tag{2.2.8}$$

Once again, the exact details of this are not important here; the crucial point is that $z^k \frac{C(z^{-1})}{A(z^{-1})}$ has been divided into a transfer function representing the future ($E(z^{-1})$) and one representing the present and past ($\frac{F(z^{-1})}{A(z^{-1})}$).

Substituting this decomposition into the system equation, the future output may be written as the sum of a prediction $y^\star$ and a prediction error $e^\star$ as

$$y(iT+kT) = y^\star(iT) + e^\star(iT+kT) \tag{2.2.9}$$

where the z-transforms of $y^\star$ and $e^\star$ are

$$Y^\star(z^{-1}) = \frac{B(z^{-1})}{A(z^{-1})}U(z^{-1}) + \frac{F(z^{-1})}{A(z^{-1})}\Xi(z^{-1}) \tag{2.2.10}$$

$$E_k^\star(z^{-1}) = E(z^{-1})\Xi(z^{-1}) \tag{2.2.11}$$

ξ is not directly measurable, but using the deduction equation 2.2.5 it can be eliminated to give:

$$Y^\star(z^{-1}) = \frac{B(z^{-1})}{A(z^{-1})}U(z^{-1}) \tag{2.2.12}$$

$$+ \frac{F(z^{-1})}{A(z^{-1})}\left[\frac{A(z^{-1})}{C(z^{-1})}Y(z^{-1}) - z^{-k}\frac{B(z^{-1})}{C(z^{-1})}U(z^{-1})\right]$$

$$= \frac{F(z^{-1})}{C(z^{-1})}Y(z^{-1})$$

$$+ \frac{B(z^{-1})}{C(z^{-1})}\left[\frac{C(z^{-1})}{A(z^{-1})} - z^{-k}\frac{F(z^{-1})}{A(z^{-1})}\right]$$

Finally using the decomposition identity 2.2.6 the term in [] can be eliminated to give

$$Y^\star(z^{-1}) = \frac{F(z^{-1})}{C(z^{-1})}Y(z^{-1}) + \frac{G(z^{-1})}{C(z^{-1})}U(z^{-1}) \tag{2.2.13}$$

where

$$G(z^{-1}) \triangleq E(z^{-1})B(z^{-1}) \tag{2.2.14}$$

Finally, the z-transformed error equation can be written from 2.2.9 as

$$z^k\ Y(z^{-1}) = Y^\star(z^{-1}) + z^k\ E_k^\star(z^{-1}) \tag{2.2.15}$$

This equation then gives us what we wanted, a prediction of the system output k sample instants into the future.

2.3. Weighted minimum variance control

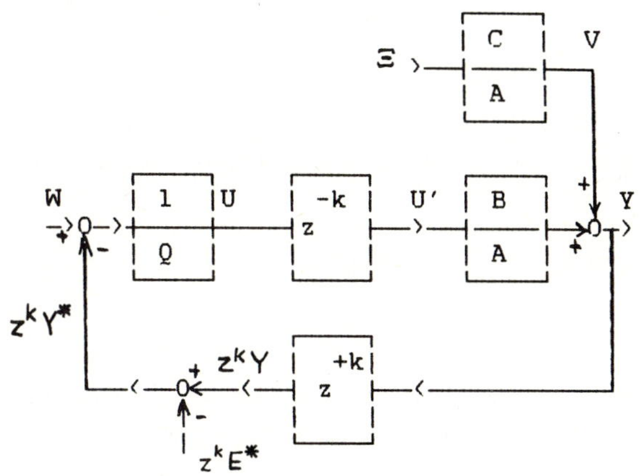

Figure 2.3.1 Weighted minimum variance control

A typical error driven control system may be written as:

$$U(z^{-1}) = \frac{1}{Q(z^{-1})}[W(z^{-1}) - Y(z^{-1})] \qquad (2.3.1)$$

where $\frac{1}{Q(z^{-1})}$ is a transfer function in z^{-1} and W is a suitably filtered setpoint. This may be converted into a predictive controller by replacing $Y(z^{-1})$ by $Y^*(z^{-1})$ generated using equation 2.2.13. Thus the predictive control law is:

$$U(z^{-1}) = \frac{1}{Q(z^{-1})}[W(z^{-1}) - Y^*(z^{-1})] \qquad (2.3.2)$$

Although the prediction $Y^*(z^{-1})$ is generated by the explicit equation 2.2.13, the properties may be conveniently analysed using the error equation 2.2.15. Thus the control equation becomes:

$$U(z^{-1}) = \frac{1}{Q(z^{-1})}[W(z^{-1}) - z^k Y(z^{-1}) + z^k E_k^*(z^{-1})] \qquad (2.3.3)$$

Thus as far as the controller dynamics are concerned, it is as if the controller actually had access to the future output. The important effect of this is that the time delay z^{-k} is eliminated from the closed-loop characteristic equation. Indeed, using the block diagram of Figure 2.3.1 it follows that the closed-loop system is given by:

$$Y(z^{-1}) = \frac{B(z^{-1})}{B(z^{-1}) + Q(z^{-1})A(z^{-1})}[z^{-k} W(z^{-1}) + E_k^{\star}(z^{-1})] \qquad (2.3.4)$$

For stability, it is necessary for the zeros of $B(z^{-1}) + Q(z^{-1})A(z^{-1})$ to lie strictly within the unit disc in the z plane.

A particular case of this controller is the minimum variance controller obtained by setting Q=0. In this case the closed loop system is given by:

$$Y(z^{-1}) = z^{-k} W(z^{-1}) + E_k^{\star}(z^{-1})$$

A stable closed loop system is obtained if all the zeros of $B(z^{-1})$ lie within the unit disc in the z plane. In particular it is necessary (though not sufficient) that $b_0 \neq 0$; this means that the delay k must reflect the true system time delay.

2.4. The self-tuning controller

The weighted minimum variance controller can only be deduced if the system polynomials $A(z^{-1})$, $B(z^{-1})$ and $C(z^{-1})$ are known. This requires two sorts of information: the degrees of the polynomials and the coefficients of the polynomials. If the degrees are known, but the coefficients are unknown, we are lead to the idea of combining recursive least squares with the control law to give a self-tuning algorithm. This idea seems to be due to Kalman[16], but was first presented in a fully developed form based on the minimum variance controller by Astrom and Wittenmark[1]. The extension to weighted minimum variance methods was given by Clarke and Gawthrop[3,2,4,5].

There are two approaches to self-tuning control[11]: the explicit method where the system coefficients are estimated and the corresponding predictor coefficients are computed; and the implicit method where controller parameters are directly estimated.

The implicit method is probably the most appropriate in the context of weighted minimum variance control and is is described here. The key to the implicit approach is to rewrite the system in predictor form in such a way that it is linear in the (predictor) parameters. In this case, we do this as follows:

Choose a stable polynomial $\hat{C}(z^{-1})$ of the same order as $C(z^{-1})$. The predictor 2.2.13 can then be rewritten as:

$$Y^*(z^{-1}) = \frac{F(z^{-1})}{\hat{C}(z^{-1})}Y(z^{-1}) + \frac{G(z^{-1})}{\hat{C}(z^{-1})}U(z^{-1}) + \frac{\hat{C}(z^{-1}) - C(z^{-1})}{\hat{C}(z^{-1})}Y^*(z^{-1}) \quad (2.4.1)$$

Define the data vector x and the corresponding parameter vector θ as

$$x(iT) = \frac{1}{\hat{C}(z^{-1})}\begin{bmatrix} y(iT) \\ y(iT-T) \\ .. \\ y(iT-nT) \\ --- \\ u(iT) \\ u(iT-T) \\ .. \\ u(iT-nT) \\ --- \\ y^*(iT) \\ y^*(iT-T) \\ .. \\ y^*(iT-nT) \end{bmatrix}; \qquad \theta = \begin{bmatrix} f_0 \\ f_1 \\ .. \\ f_n \\ --- \\ g_0 \\ g_1 \\ .. \\ g_n \\ --- \\ \gamma_0 \\ \gamma_1 \\ .. \\ \gamma_n \end{bmatrix} \quad (2.4.2)$$

where $\gamma_i = \hat{c}_i - c_i$. The future system output can then be rewritten as:

$$y(iT+kT) = x^T(iT)\theta + e^*(iT+kT) \quad (2.4.3)$$

This is almost in a form suitable for recursive least-squares estimation; but the elements $y^*(iT)$ are unknown. This may be overcome using the extended least squares method as discussed in many references, for example[8,9] and the relevant chapter of these notes(L7).

3. PHASE LAG AND DERIVATIVES

In this section, we consider the control of systems which have both delay and significant rational dynamics. We concentrate on the

weighted
model-reference method[4] based on the self-tuning controller of Clarke and Gawthrop[3,2,5]. In particular, we demonstrate its intimate relation with the method of the previous section.

3.1. Systems with high relative degree

Time delays are not the only reason why systems are hard to control. Indeed if the underlying continuous system has no delay (giving k=1), the weighted minimum variance control is not interesting as the predictor does not play a significant role having no time delay to predict over apart from that due to the sampling process.

Some continuous-time systems have no pure time delay, but have significant phase lag due to a large excess of poles to zeros (in the s-domain). Such systems could be effectively controlled in the absence of noise by using multiple output derivatives in the feedback loop; but the presence of noise makes such an approach unacceptable. In the previous section, delays were overcome by emulating z^k, in the same way multiple derivatives can be emulated using a similar approach.

3.2. Emulating inverse system models

In the previous section the inverse time delay z^k was emulated using a predictor. Here this approach is extended to emulate an inverse delay in series with an inverse all-pole system given by:

$$\Phi(z^{-1}) \triangleq z^k P(z^{-1})Y(z^{-1}) \tag{3.2.1}$$

$$= p_0 z^k Y(z^{-1}) + p_1 z^{k-1} Y(z^{-1}) + \ldots + p_n z^{k-n} Y(z^{-1})$$

In the previous section, a predictor for $z^k Y(z^{-1})$ was derived. Using a similar approach a predictor $Y_{k-j}{}^*(z^{-1})$ for $z^{k-j}Y(z^{-1})$ can be derived in the form:

$$Y_{k-j}{}^*(z^{-1}) = \frac{F_j(z^{-1})}{C(z^{-1})}Y(z^{-1}) + z^{-j}\frac{G_j(z^{-1})}{C(z^{-1})}U(z^{-1}) \tag{3.2.2}$$

where, if j<k, $F_j(z^{-1})$ and $E_j(z^{-1})$ are given by

$$\frac{C(z^{-1})}{A(z^{-1})} = E_j(z^{-1}) + z^{k-j}\frac{F_j(z^{-1})}{A(z^{-1})} \tag{3.2.3}$$

but if $j \geq k$

$$E_j(z^{-1}) = 0; \; F_j(z^{-1}) = C(z^{-1}) \tag{3.2.4}$$

See[2] for details.

A few moments thought will convince you that a 'predictor' for $\Phi(z^{-1})$ can be written in terms of the $Y_{k-j}{}^{\star}(z^{-1})$ as:

$$\Phi^{\star}(z^{-1}) = p_o Y^{\star}(z^{-1}) + p_1 Y_{k-1}{}^{\star}(z^{-1}) + \ldots + p_n \, Y_{k-n}{}^{\star}(z^{-1}) \tag{3.2.5}$$

$$= \sum_{j=0}^{n} p_j Y_{k-j}{}^{\star}(z^{-1})$$

and

$$\Phi(z^{-1}) = \Phi^{\star}(z^{-1}) + E_{\phi}{}^{\star}(z^{-1}) \tag{3.2.6}$$

where $E_{\phi}{}^{\star}(z^{-1})$ is given by

$$E_{\phi}{}^{\star}(z^{-1}) = p_o E_k{}^{\star}(z^{-1}) + p_1 E_{k-1}{}^{\star}(z^{-1}) + \ldots + p_n \, E_{k-n}{}^{\star}(z^{-1}) \tag{3.2.7}$$

Replacing the $Y_{k-j}{}^{\star}(z^{-1})$ by the explicit equation the $\Phi^{\star}(z^{-1})$ can be written explicitly as:

$$\Phi^{\star}(z^{-1}) = \frac{F(z^{-1})}{C(z^{-1})} Y(z^{-1}) + \frac{G(z^{-1})}{C(z^{-1})} U(z^{-1}) \tag{3.2.8}$$

where

$$F(z^{-1}) \triangleq \sum_{j=0}^{n} p_j F_j(z^{-1}) \tag{3.2.9}$$

$$G(z^{-1}) \triangleq B(z^{-1}) \sum_{j=0}^{n} z^{-j} p_j E_j(z^{-1}) \tag{3.2.10}$$

and the corresponding error equation is

$$\Phi(z^{-1}) = \Phi^{\star}(z^{-1}) + E_{\phi}{}^{\star}(z^{-1}) \tag{3.2.11}$$

3.3. Weighted model-reference control

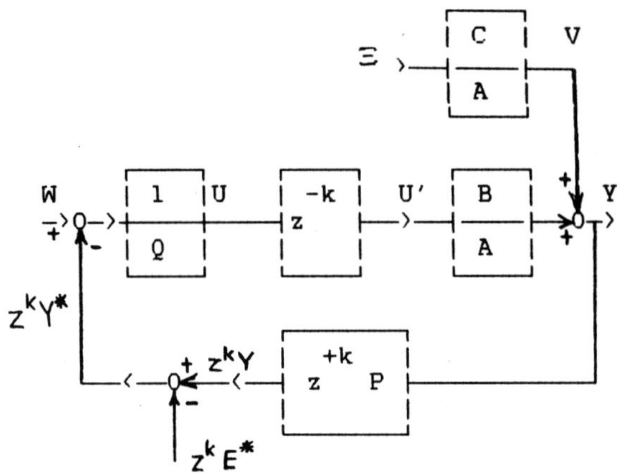

Figure 3.3.1 Weighted model reference control

As in the previous section, a typical error driven control system may be written as:

$$U(z^{-1}) = \frac{1}{Q(z^{-1})}[W(z^{-1}) - Y(z^{-1})] \tag{3.3.1}$$

where $\frac{1}{Q(z^{-1})}$ is a transfer function in z^{-1} and W is a suitably filtered setpoint. This may be converted into a generalised predictive controller by replacing $Y(z^{-1})$ by $\Phi^*(z^{-1})$ generated using equation 3.2.8. Thus the generalised predictive control law is:

$$U(z^{-1}) = \frac{1}{Q(z^{-1})}[W(z^{-1}) - \Phi^*(z^{-1})] \tag{3.3.2}$$

Once again, the properties of this controller may be conveniently analysed using the error equation 3.2.11. Thus the control equation becomes:

$$U(z^{-1}) = \frac{1}{Q(z^{-1})}[W(z^{-1}) - z^k \Phi(z^{-1}) + E_\phi^*(z^{-1})] \tag{3.3.3}$$

Thus as far as the controller dynamics are concerned, it is as if the controller actually had access to the future output multiplied by the inverse all-pole model $P(z^{-1})$. Thus not only is the time delay z^{-k} eliminated from the closed-loop characteristic equation, but also the approximate derivative compensator is effectively included in the loop

dynamics.
Using the block diagram of Figure 3.3.1 it follows that the closed-loop system is given by:

$$Y(z^{-1}) = \frac{B(z^{-1})}{P(z^{-1})B(z^{-1}) + Q(z^{-1})A(z^{-1})}[z^{-k} W(z^{-1}) + E_k^*(z^{-1})] \qquad (3.3.4)$$

For stability, it is necessary for the zeros of $P(z^{-1})B(z^{-1}) + Q(z^{-1})$. $A(z^{-1})$ to lie strictly within the unit disc in the z plane.

A particular case of this controller is the model reference controller obtained by setting Q=0. In this case the closed loop system is given by:

$$Y(z^{-1}) = z^{-k} \frac{1}{P(z^{-1})} W(z^{-1}) + \frac{1}{P(z^{-1})} E_k^*(z^{-1})$$

Thus the closed loop setpoint response is forced to be the same as that of the prespecified closed loop 'model' $\frac{1}{P(z^{-1})}$. This is rather more feasible than the corresponding minimum variance control with $P(z^{-1})=1$. A stable closed loop system is obtained if all the zeros of $B(z^{-1})$ lie within the unit disc in the z plane (it is assumed that $P(z^{-1})$ is chosen to be stable).

3.4. The self-tuning controller

This follows exactly the same pattern as for the minimum variance case. The only difference is that ϕ replaces y in the equation:

$$\phi(iT) = x^T(iT-kT)\theta \qquad (3.4.1)$$

x is the same as before, but the parameters in theta correspond to the parameters in equation (3.2.8).

4. UNSTABLE ZEROS AND THEIR REMOVAL

In this section we consider the control of systems with significant zeros outside the unit circle in the z-plane. A comprehensive survey of this problem is given by Clarke[17]. The basic algorithm is the pole-placement method of Wellstead, Edmunds, Prager and Zanker[6,7].

We present a non-standard approach to this method which provides a link with the methods of the previous two sections. A new algorithm, weighted pole-placement, is introduced.

4.1. Systems with zeros outside the unit circle

It is an unfortunate fact that many discrete-time systems have $B(z^{-1})$ polynomials with zeros outside the unit disc. As discussed in detail by Wellstead et al[6,7] and by Astrom et al[18] this is due to either having continuous-time systems with high relative order, or to having continuous-time systems with a time delay which is not an integer multiple of the sample interval.

Although the two design methods (weighted minimum variance and weighted model reference) discussed in previous sections may give a stable closed-loop system for large control weighting, they will not for sufficiently small weighting. We solve the problem here using a non-standard approach which links pole placement with weighted model reference, based on the idea of emulating an unrealisable system.

4.2. Emulating non-causal inverses

If $B(z^{-1})$ has zeros outside the unit disc, its inverse $\frac{1}{B(z^{-1})}$ may be regarded as being non-causal. Such a non-causal inverse could then be used (in a manner analogous to the non causal z^k) to cancel out the $B(z^{-1})$ polynomial. As with z^k this cannot be implemented directly, but can be emulated using causal transfer functions.

Because we may wish to include the z^k and $P(z^{-1})$ emulation as well we define the auxiliary output $\Psi(z^{-1})$ as:

$$\Psi(z^{-1}) = \frac{1}{B(z^{-1})}\Phi(z^{-1}) = z^k \frac{P(z^{-1})}{B(z^{-1})}Y(z^{-1}) \tag{4.2.1}$$

Recalling that the z transform of $\Psi(iT+kT)$ is $z^{-k} \frac{P(z^{-1})}{B(z^{-1})}Y(z^{-1})$, $\Psi(z^{-1})$ is given by:

$$\Psi(z^{-1}) = z^k \frac{P(z^{-1})}{B(z^{-1})}Y(z^{-1}) = \frac{P(z^{-1})}{A(z^{-1})}U(z^{-1}) + z^k \frac{P(z^{-1})C(z^{-1})}{B(z^{-1})A(z^{-1})}\Xi(z^{-1}) \tag{4.2.2}$$

The first term on the righthand side of this equation $(\frac{P(z^{-1})}{A(z^{-1})}U(z^{-1}))$ corresponds to past and present values of U, but the other term has a factor $\frac{z^k}{B(z^{-1})}$ indicating future values of V. In a similar way to

2.2, this term may be divided into realisable and unrealisable parts via the decomposition:

$$\frac{P(z^{-1})C(z^{-1})}{B(z^{-1})A(z^{-1})} = \frac{E(z^{-1})}{B(z^{-1})} + z^{-k}\frac{F(z^{-1})}{A(z^{-1})} \tag{4.2.3}$$

or, alternatively

$$P(z^{-1})C(z^{-1}) = E(z^{-1})A(z^{-1}) + z^{-k}F(z^{-1})B(z^{-1}) \tag{4.2.4}$$

Once again,the exact details of this are not important here; the crucial point is that $z^{k}\frac{P(z^{-1})C(z^{-1})}{B(z^{-1})A(z^{-1})}$ has been divided into a transfer function representing the future $(\frac{E(z^{-1})}{B(z^{-1})})$ and one representing the present and past $(\frac{F(z^{-1})}{A(z^{-1})})$.

Substituting this decomposition into the system equation, the future output may be written as the sum of an emulating term $\Psi^{\star}(z^{-1})$ and a prediction error $E_{\Psi}^{\star}(z^{-1})$ as

$$\Psi(z^{-1}) = \Psi^{\star}(z^{-1}) + E_{\Psi}^{\star}(z^{-1}) \tag{4.2.5}$$

where

$$\Psi^{\star}(z^{-1}) = \frac{P(z^{-1})}{A(z^{-1})}U(z^{-1}) + \frac{F(z^{-1})}{A(z^{-1})}\Xi(z^{-1}) \tag{4.2.6}$$

$$E_{k}^{\star}(z^{-1}) = E(z^{-1})\Xi(z^{-1}) \tag{4.2.7}$$

ξ is not directly measurable, but using the deduction equation 2.2.5 it can be eliminated to give:

$$\Psi^{\star}(z^{-1}) = \frac{P(z^{-1})}{A(z^{-1})}U(z^{-1}) \tag{4.2.8}$$

$$+ \frac{F(z^{-1})}{A(z^{-1})}[\frac{A(z^{-1})}{C(z^{-1})}Y(z^{-1}) - z^{-k}\frac{B(z^{-1})}{C(z^{-1})}U(z^{-1})]$$

$$= \frac{F(z^{-1})}{C(z^{-1})}Y(z^{-1})$$

$$+ \frac{B(z^{-1})}{C(z^{-1})}[P(z^{-1})\frac{C(z^{-1})}{A(z^{-1})} - z^{-k}\frac{F(z^{-1})}{A(z^{-1})}]$$

Finally using the decomposition identity 4.2.3 the term in [] can be eliminated to give

$$\Psi^{\star}(z^{-1}) = \frac{F(z^{-1})}{C(z^{-1})}Y(z^{-1}) + \frac{G(z^{-1})}{C(z^{-1})}U(z^{-1}) \tag{4.2.9}$$

where

$$G(z^{-1}) = E(z^{-1}) \quad \text{(no B factor !)} \tag{4.2.10}$$

This equation then gives us what we wanted, an emulated value of $\Psi(z^{-1})$

The method can be extended by dividing $B(z^{-1})$ into nice $B^{+}(z^{-1})$ and nasty $B^{-}(z^{-1})$ factors and repeating the above but only cancelling $B^{-}(z^{-1})$. See[11] for details.

4.3. Weighted pole-placement control

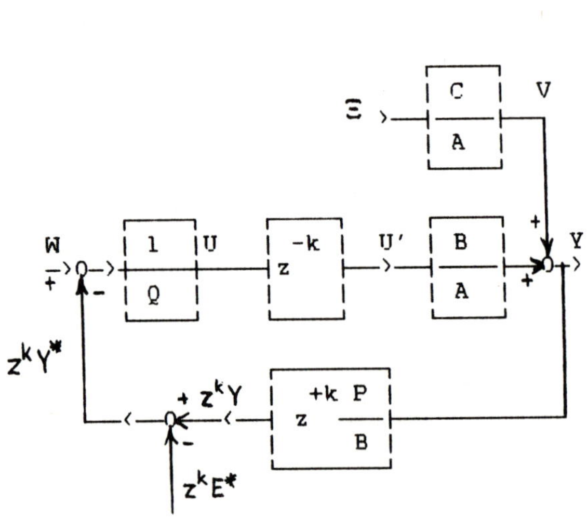

Figure 4.3.1 Weighted pole placement control

As before the emulating control law is given by

$$U(z^{-1}) = \frac{1}{Q(z^{-1})}[W(z^{-1}) - \Psi^{*}(z^{-1})] \tag{4.3.1}$$

Although the prediction $\Psi^{*}(z^{-1})$ is generated by the explicit equation 4.2.9, the properties may be conveniently analysed using the error equation. Thus the control equation becomes:

$$U(z^{-1}) = \frac{1}{Q(z^{-1})}[W(z^{-1}) - z^{k}\frac{P(z^{-1})}{B(z^{-1})}Y(z^{-1}) + E_{\Psi}^{*}(z^{-1})] \tag{4.3.2}$$

Thus as far as the controller dynamics are concerned, it is as if the controller actually had access to the unrealisable signal $\Psi(z^{-1})$. The important effect of this is that the time delay z^{-k} and the zero polynomial $B(z^{-1})$ are eliminated from the loop gain and $P(z^{-1})$ is inserted in series. Indeed, using the block diagram of Figure 4.3.1 it follows that the closed-loop system is given by:

$$Y(z^{-1}) = \frac{B(z^{-1})}{P(z^{-1}) + Q(z^{-1})A(z^{-1})}[z^{-k}\,W(z^{-1}) + E^{*}(z^{-1})] \tag{4.3.3}$$

For stability, it is necessary for the zeros of $P(z^{-1}) + Q(z^{-1})A(z^{-1})$ to lie strictly within the unit disc in the z plane.

A particular case of this controller is the pole-placement controller obtained by setting $Q=0$. In this case the closed loop system is given by:

$$Y(z^{-1}) = z^{-k}\,\frac{B(z^{-1})}{P(z^{-1})}W(z^{-1}) + \frac{B(z^{-1})}{P(z^{-1})}E_{k}^{*}(z^{-1})$$

A stable closed loop system is obtained if all the zeros of $P(z^{-1})$ lie within the unit disc in the z plane. Note that, unlike the other methods, it is not necessary that b_0 should be non-zero; thus the time delay k can be less than the true delay. k is often taken as one and thus the leading coefficients of $B(z^{-1})$ will be zero.

4.4. The self-tuning controller

As it is not possible to generate Ψ, the implicit approach used for the previous two methods is not applicable. It is usual to use an explicit method for self-tuning pole placement[6,7,11].

The system parameters $A(z^{-1})$, $B(z^{-1})$ and $C(z^{-1})$ are estimated using a

suitable
recursive least squares method and the predictor coefficients computed from the identity 4.2.4. This identity may be solved by equating coefficients and solving the resultant linear set of equations[6,7]; or, more neatly, the identity may be solved using Euclids algorithm[19,20].

5. CONTINUOUS-TIME SELF-TUNING

This chapter has concentrated on developing self-tuning methods in their traditional setting: discrete-time equations and the z-domain. However, recently, some authors have suggested that this approach has defects which may be overcome by a continuous-time approach[21,22,23]. One motivation for such an approach is that many discrete-time systems with zeros outside the unit circle arise from continuous-time systems with zeros within the stability region. In such cases, the model reference approach is not applicable in discrete-time but is applicable in continuous-time. Of course the continuous-time algorithms are discretised for implementation, but even so they retain many desirable properties.

Is the effort involved in reading this chapter a waste of time? No, because much of the algebra involved in the discrete approach can be reused in the continuous time approach. From the algebraic point of view, we may as well work in terms of s as in terms of z; though the physical significance is, of course, different.

References

1. Astrom, K.J. and Wittenmark, B., "On self-tuning regulators," Automatica, vol. 9, pp. 185-199, 1973.

2. Clarke, D.W and Gawthrop, P.J., "Self-tuning controller," Proceedings IEE, vol. 122, no. 9, pp. 929-934, 1975.

3. Clarke, D.W. and Gawthrop, P.J., "A generalised self-tuning regulator & Simulation of a generalised self-tuning regulator," Electronics Letters., vol. 11, no. 2, 1975.

4. Gawthrop, P.J., "Some interpretations of the self-tuning controller," Proceedings IEE, vol. 124, no. 10, pp. 889-894, 1977.

5. Clarke, D.W and Gawthrop, P.J., "Self-tuning control," Proceedings IEE, vol. 126, no. 6, pp. 633-640, 1979.

6. Wellstead, P.E., Edmunds, J.E., Prager, D., and Zanker, P., "Self-tuning pole/zero assignment regulators," Int. J. Control., vol. 30, no. 1, pp. 1-26, 1979.

7. Wellstead, P.E., Zanker, P., and Edmunds, J.M., "Pole assignment self-tuning regulator," Proceedings IEE, vol. 126, pp. 781-787..

8. Astrom, K.J., "Theory and applications of adaptive control - A survey," Automatica, vol. 19, no. 5, pp. 471-486, 1983.

9. Harris and Billings (eds.), Self-tuning and adaptive control - theory and applications, Peter Peregrinus, 1981. book

10. H.Unbehauen (ed.), Methods and applications in adaptive control, Springer, 1980. book

11. Astrom, K.J. and Wittenmark, B., "Self-tuning controllers based on pole-zero placement," Proceedings IEE Pt.D., vol. 127, pp. 120-130, 1980.

12. Goodwin, G.C. and Sin, K.S., Adaptive filtering prediction and control, Prentice-Hall, Englewood Cliffs, New Jersey, USA, 1984.

13. Astrom, K.J., Introduction to stochastic control theory, Academic Press, New York, 1970.

14. Young, P.C. and Hastings-James, R., "Indentification and control of discrete-linear systems subject to disturbances with rational spectral density," in Proceedings of 9th IEEE symposium on adaptive processes, CDC, Austin Texas., 1970.

15. Clarke, D.W and Hastings-James, R., "Design of digital controllers for randomly disturbed systems," Proc. IEE, vol. 118, no. 10, pp. 1503-1506, 1971.

16. Kalman, R.E., "Design of a self-optimizing control system," Trans. ASME, vol. 80, p. 468, 1958.

17. Clarke, D.W., "Self-tuning control of nonminimum-phase systems," Automatica, vol. 20, no. 5, pp. 501-518, 1984.

18. Astrom, K.J., Hagander, P., and Sternby, J., "Zeros of sampled systems," Automatica, vol. 20, no. 1, pp. 31-38, 1980.

19. Kucera, V., Discrete linear control: The polynomial equation approach., Wiley, Prague, 1979.

20. MacLane, S. and Birkhoff, G., Algebra, Macmillan, New York, 1967.

21. Gawthrop, P.J., "Hybrid self-tuning control," Proc. IEE, vol. 127, no. 5, pp. 229-236, 1980.

22. Gawthrop, P.J., "A continuous-time approach to discrete-time self-tuning control," Optimal Control: Applications & Methods, vol. 3, no. 4, pp. 399-414, 1982.

23. Goodwin, G.C., "Some observations on robust estimation and control," in Preprints of the 7th IFAC/IFORS symposium on identifictation and system parameter estimation, York, U.K., 1985.

CASE STUDIES

EXPLORING BIOLOGICAL SIGNALS

Prof. D.A. Linkens

1. Introduction

The application of systems techniques to biomedicine is both challenging and stimulating. Major technical challenges are posed by the specific limitations endemic to living systems. These include restricted amount of a-priori structural knowledge, nonlinearities, variability over time, variability between objects, poor measurability of signals, difficult experimental conditions, and limitations in experimental procedure due to ethical considerations. In spite of these massive obstacles, many attempts are being made to explore the biomedical world using an array of systems methods.

This contribution will cover only a small area of applications and methods. Specifically, signal processing and modelling will be confined to one-dimensional time-series considerations. In this way the vast field of medical imaging will be excluded. Motivations for this type of work include a desire to understand better the underlying physiological processes, a movement from qualitative to quantitative knowledge, the possibility of better discrimination between normal and diseased conditions, improved trend prediction relating to health, and on-line therapeutic control. The whole field may be divided into modelling of signals (i.e. signal processing) and modelling of systems (i.e. identification and control). Reviews of modelling and identification in numerous areas of biomedicine can be found in the proceedings of the 7th IFAC Symposium on Identification and System Parameter Estimation, 1985.

The ability to obtain measurements in living systems is very limited, and there is an increasing emphasis on non-invasive techniques. For example, in the respiratory system typical measurements are airflow at the mouth obtained via a pneumotachograph, changes in body volume using a body-plethysmograph, and expired gas analysis. For the cardiovascular system the most common measurement is the ECG obtained using surface electrodes and numerous blood flow techniques. Surface electrical recordings produce information about underlying physiological processes and are referred to as the electrocardiogram (ECG), electro-encephalogram(EEG), electromyogram (EMG) and electrogastrogram (EGG). Those signals are non-specific, however, and tend to give a whole-organ integrated picture rather than details about localised sites. This is very restrictive for bio-systems such as the heart and the gastro-intestinal tract which comprise multiple oscillators co-ordinated together to provide whole organ electrical and mechanical activity. To provide localised information requires invasive techniques. These include indwelling catheters to give pressure measurements at points in the cardio-vascular system, and wire electrodes stitched onto the gastrointestinal tract during abdominal surgery.

The examples described in the following pages are chosen to illustrate the various aspects of physiological understanding, diagnostic tools and therapeutic control. They demonstrate modelling of signals and systems and are aspects in which the author has been involved.

2. Modelling of Signals.

2.1 Gastrointestinal Tract.

It has been known for many years that the stomach, small and large intestines are electrically active having spontaneous oscillations within the frequency range 0.05 to 0.2 Hz.

i) Signal Processing and Analysis.

The clear visual evidence of regular oscillations in electrical potentials recorded from many parts of the gut has prompted attempts to measure the frequency of these 'slow waves' accurately. It is desirable to replace manual measurement of such frequencies by automated methods in order to remove the tedium and time involved. A more important reason for automated methods, however, is to enable precise objective values to be given to measurements which will facilitate discussion between research groups and elucidate fine physiological details in this work. The subject of frequency analysis is of considerable interest in many technological disciplines and there is considerable scope for cross-fertilization between medical and scientific disciplines at this point. In particular one could mention the areas of mechanical vibrations, telecommunications and control system identification. The methods which are now briefly described have largely been developed and applied in such disciplines. Five techniques are summarized, with their relative merits and disadvantages, and some typical results using gut signals are presented.

In assessing the various methods a number of criteria should be applied both to the data which are available and to the information which the technique provides. For the data the number of cycles of oscillation is a very important parameter. Similarly both long and short-term variations in the instantaneous period of slow wave are crucial, as is the amount of noise or artifacts added to the signal. The wave-shape of the data is important and is related to the number of harmonic components present, while the simultaneous existence of more than one rhythm sets a further constraint.

a) Fast Fourier transforms

In recent years the widespread use of digital computers has made this method very popular whenever it is required to measure accurately the frequency components of a signal. In the Fourier transform approach a signal which is characterized by a number of data values versus time is represented by a frequency spectrum comprising a graph of frequency values plotted horizontally and the amplitude of each frequency component plotted vertically. It is calculated via the so-called FFT algorithm developed by Cooley and Tukey.

The FFT method is theoretically based on noise-free data of infinite duration, but in spite of these limitations it has been used successfully to quantify noisy, finite-length medical data in many applications. In terms of accuracy of frequency measurement the crucial factor is the number of cycles of data present. Thus, for n cycles of data recorded and transformed, the discrimination in frequency will be 100/n%. It is also necessary that the frequency should not vary during this number of cycles if a sharp spectral peak is to be obtained in the transform. Frequency components in an FFT having a wide 'spread' rather than a single sharp peak can be caused by factors such as a small number of cycles, variations in instantaneous frequency and/or amplitude, and the presence of large artifacts near the particular frequency.
The presence of large amounts of noise can cause difficulty in determining the significance of peaks in the FFT spectrum. To reduce the spurious peaks caused by artifacts one requires either long time transforms or the averaging of successive transforms, but this aggravates the problems already referred to. Averaging of adjacent peaks in a single spectrum can be used to determine significance of a peak, but this then degrades the frequency accuracy proportionately. The visual detection of significant peaks in the FFT spectrum is easy but subjective. An objective assessment of significance is not straightforward, but two ideas may be useful. First, for pure sinusoidal components in the presence of Gaussian noise one can quantify peak significance simply; particularly if the spectrum is plotted with a log amplitude vertical axis. Medical noise is not, however, Gaussian and in gut recordings there is a preponderance of noise at low frequency values. Another idea therefore is to compare the signal spectrum with the mean and standard deviation of the noise spectrum, and on the basis of this to print out the significant peaks in the signal. It is clear from the basis of the method that multiple rhythms can be detected.

b) Fast Walsh transforms

Fourier transforms assume that a signal can be represented by a summation of sinusoidal components each with a different frequency. In the mathematics of this transformation a large number of multiplications involving sine and cosine functions are required for each FFT. These are time-consuming and would be expensive to implement on a dedicated micro- or mini-computer system.

In contrast, a Walsh transform assumes that a signal can be represented by a summation of discrete functions similar to square waves. When used with a Cooley-Tukey algorithm a fast Walsh transform (FWT) is a very fast process involving simple additions and subtractions instead of multiplications involving sine and cosine expressions. If the main requirement in the data analysis is the measurement of frequency components then a very simple FWT system can be used. If the data are 'squared' using a zero-crossing detector so that logic '1' and

'O' are allocated to positive and negative sections of the signal respectively, then no analogue to digital conversion equipment is required. Figure 1 shows an FWT using this technique on a canine gastric recording. It has given clear peaks at the fundamental and second harmonic frequencies, but two general observations arising from Walsh transforms can also be made. These are that the overall background 'noise' components are increased and that 'spurious' peaks often occur, particularly at the high-frequency end of the spectrum. These differences from the FFT method were clearly visible when using regular data from gastric and small intestinal recordings but were not so apparent when using colonic data. This method has been implemented on a simple microprocessor system using no analogue to digital conversion equipment, but reading out the FWT spectra onto a laboratory oscilloscope via two digital to analogue channels.

c) Auto-regressive (AR) modelling

This technique has been in use for some years for the identification of noisy signals in industrial processes. More recently it has been used by medical researchers in the analysis of EEG, although the emphasis has usually been on the detection of abnormalities in the EEG pattern rather than on the determination of rhythm frequencies. It has been applied in Sheffield to typical gut recordings and is a method that gives accurate estimations of frequencies using only a few cycles of data, unlike the FFT and FWT methods.

The method assumes that the data being analysed can be represented as a Gaussian white noise source operated on by a filter expressed in a digital format referred to as a z-transform notation. Given a specified complexity of filter, referred to as the 'order of the model', the algorithm produces a best least-squares fit to the original data by adjustment of the model coefficients. The amount of error is determined by the difference between the model output and the data, and is referred to as the 'residuals'. The technique is successful if the residuals are substantially white noise, and this is checked by observing the autocorrelation function.

The second part of the method is to determine the frequency components from the model coefficients. This is done by factorizing the z-transform polynomial of the model and plotting the roots on a z-plane diagram. The roots have real and imaginary parts, and a rhythm frequency is indicated by roots which lie near the unit radius circle on the z-plane. Relevant rhythms will give points which are close to, but not exactly on, the unit circle. The distance inside the unit circle is inversely proportional to the amplitude of the rhythm, so that points very close to the circle indicate components with a large amplitude. This provides a simple test of significance for component frequencies, and in fact values of distance from the origin of between 0.95 and 1.0 appear in gut work to be a good indication of significance. Relevant components appear in the right-hand side of the diagram, while high-frequency noise terms are located on the left-hand side. These remarks enable a simple criterion to be applied which gives numerical print-out of relevant frequencies together with an indication of their amplitudes. Points having a magnitude of 0.95-1.0 and located in the right-hand side are read out. This eliminates both high- and low-frequency noise artifact. The order of model must be selected, and this depends on the type of data being processed. Basically the order must be high enough to allow for the probable number of relevant rhythms plus harmonics together with sufficient points for noise representation. This can be seen in Figure 2 which shows the data, z-plane diagram and residuals for a canine duodenal recording. An FFT of 512 points from the same recording gave fundamental and second harmonic components of 0.3 and 0.6 Hz respectively with a frequency discrimination of 0.004 Hz from about 75 cycles of data. The z-plane points 'A' and 'B' represent the equivalent frequencies which gave values of 0.29 and 0.58 Hz using an eigth-order model and 100 data points or 15 cycles only. In this example a low-order model was sufficient to accurately identify the data and give residuals which were substantially white noise as indicated by the autocorrelation function plot of Figure 2c.

It was found that the order of the model and the number of data points had to be increased if there were many harmonics or large noise artifacts. Model orders of about 15 and 250 data points were capable of dealing with the majority of signals from canine and human gastrointestinal recordings.

The autoregressive method appears to give good results on gastrointestinal data, offering numerical print-out of relevant frequencies on short stretches of data covering 5-15 cycles of information. Numerous algorithm extensions have been made recently using Maximum Entropy and Linear Prediction concepts (1).

d) Phase-lock loop

This technique has been widely used in radio communications for the recovery of signals imbedded in noise, and considerable theoretical analysis has been attempted on phase-lock loops (2). The method has been applied to gastrointestinal recordings and has been instrumented using cheap integrated circuit electronics.

A phase-lock loop comprises a feedback control system in which the frequency and phase of a high-accuracy oscillator are adjusted to match the incoming noise-contaminated data. When the loop has locked onto the signal the output from the internal oscillator gives a noise-free signal which can be used to give a measurement of signal frequency plus its variance. A further feature is that a signal can be derived which gives an indication if the system is locked onto a signal. In this way the technique provides objective information on the statistics of the frequency content in the data. Figure 3 shows the dynamic response of a phase-lock loop system used on a human gastric mucosal recording. After a transient disturbance in the incoming data the lock/unlock indicator shows that a satisfactory lock condition can be obtained within a few cycles of the disappearance of the transient disturbance.

The phase-lock loop method is a simple analogue method which gives fast tracking of frequency changes in gastrointestinal rhythms. It should be noted that the internal oscillator frequency range must be limited to one octave (i.e. ratio of maximum frequency to minimum frequency must be less than 2) and similarly the incoming data must be band-pass filtered with the same limits. If this is not done locking can occur into an integer multiple or sub-multiple of the signal frequency. From this it is apparent that the technique cannot deal with the case of multiple rhythm frequencies or harmonics of a single frequency.

e) Raster scanning

This technique is similar to a method used in determining the period of circadian rhythms (3). Electronic ramp generators are used to sweep the beam across the Y-axis of a fibre-optics recording oscilloscope. The signal being analysed is connected to the z-modulation of the oscilloscope so that when a peak occurs a bright spot appears on the screen. If the ramp generator frequency is equal to the data frequency and the X-axis scan is very slow then a series of horizontal dots is recorded. If the data frequency varies then the dots will give an inclined line which eventually disappears off the edge of the screen. The use of nine ramp generators in parallel gives the appearance of continuous lines in this case and makes the visual read-out easier to assimilate. The method requires the ramp generators to be set to the average value of the data frequency, and can clearly only deal with a single frequency. The technique has been transferred to a microcomputer using colour graphics and also applied to heart rate variability signals and described in section 2.2

It is clear, from the description of the five methods which have been applied to gastrointestinal data in Sheffield, that each technique has relative advantages and disadvantages. In choosing the best method for a particular application account must be taken of factors such as the length of data, amount of artifact, presence of harmonic components and multiple rhythms.

ii) Modelling from Signals

It has been suggested that slow-wave oscillations originate in certain 'pacemaker' areas of the gut and are propagated along the organ, in a similar manner to that encountered in cardiac muscle. Various observations have gone against such a 'passive cable' model. For instance, no specialised tissue corresponding to localised pacemaker areas has been identified in the digestive tract. Also, small pieces of excised tissue from various parts of the gut show spontaneous electrical activity, and thus an alternative model comprising a set of mutually-coupled non-linear oscillators has been hypothesized. The small and large intestines have a tubular structure, and show almost complete electrical coordination around the circumference of the organ, in that no phase shift has been recorded in this direction. In contrast, large phase shifts and frequency differences occur along the axis of the organ. As a result, one-dimensional chains of coupled oscillators have been investigated for intestinal modelling (5,6) since phase shifts occur in both directions in this organ.

The basic oscillator equation used in most modelling studies has been the classic van der Pol dynamic:-

$$\ddot{x} - \varepsilon(a^2 - x^2)\dot{x} + \omega^2 x = 0 \tag{1}$$

where the frequency is mainly governed by 'ω', the amplitude by 'a'. and the amount of non-linearity by 'ε'. For intestinal modelling small values of 'ε' have been used, while for the stomach large values have been necessary because of the high harmonic content in the recorded 'slow-waves'. Modifications have been made to this basic equation to allow for damped oscillations and asymmetrical waveforms, following the work of Fitzhugh for neural modelling. Further modifications have been made to make the zero state a stable condition for the modelling of human colonic behaviour.

Van der Pol oscillators may be interconnected in various ways, and one helpful method is based on an equivalent circuit representation of coupled systems, shown in Fig.4. for which the equation for oscillator 1 is:-

$$\ddot{x}_1 - \varepsilon(a_1^2 - x_1^2)\dot{x}_1 + \omega_1^2 x_1 - Lx_2 - R\dot{x}_2 - C\ddot{x}_2 = 0 \tag{2}$$

The three forms of coupling shown in (2) give different phenomena. In the following sections these phenomena will be compared with the physiologically recorded behavioural patterns of frequency entrainment, travelling waves, modulation effects and multimode oscillations.

A major disadvantage of van der Pol dynamics is that they are not directly related to physiological concepts such as ionic conductances and membrane capacitance. The well-known Hodgkin-Huxley equations developed for nerve modelling have been adapted to give auto-rhymic behaviour and used as a basic oscillator unit in the coupled model structure.

a) Analog Computer Simulations

These studies were carried out using an Applied Dynamics AD4 machine capable of giving an 8 oscillator van der Pol structure. Solutions could be run at frequencies of up to 2KHz, except for investigations involving modulation effects which occur over small ranges of parameter settings. It has been found experimentally that intestinal tissue oscillates at a lower frequency when uncoupled. Investigation of various interconnection structures to determine which would give entrained frequencies higher than intrinsic frequencies was easily performed on the AD4. This was found to occur for inductive-type coupling and also when adjacent cell outputs were partially coupled.

In the human small intestine, modulation effects occur and are referred to as 'waxing and waning'. These amplitude and frequency modulation effects have frequently been observed in analog simulations and can be demonstrated clearly using Lissajou figure displays. These waveforms have been recorded on analogue FM tape and subsequently analysed using FFT and shown to have similar spectra to other models using Hodgkin-Huxley dynamics or electronic relaxation oscillators. This has revealed interesting spectral patterns involving injection of adjacent oscillator frequency components and a sideband reflection principle, as can be seen in Fig.5.

In developing a human colonic model which requires three basic phenomena of electrical silence plus two stable thythms, analog simulations were particularly valuable in that various forms of random perturbations could be easily injected to investigate the ease with which the different modes could be excited. Ring structures and modified van der Pol dynamics were simulated in this way. The existence of simultaneous oscillations (called multimodes) has also been investigated, and to obtain these, many simulation runs are required. It has been shown both experimentally, analytically and via simulation that such multimodes can be reproduced using the modified van der Pol dynamics (10). The regions of attraction for the various modes has also been investigated, and for this purpose repetitive operation of the analogue computer is ideal for determining the switching boundary between solutions (11).

b) Digital Computer Simulations

For the coupled oscillator model of the gut the main disadvantage of analog simulation is that sufficient units to model a large portion of an organ cannot be patched. Although the physiological size of a functional tissue oscillator has not yet been determined, it is clear that in an intact organ there must be myriads of oscillators. Unpublished studies have, however, suggested that chain connections of between 16 and 100 units should give representative phenomenological behaviour.

Having determined basic behaviour from analog studies, a 100 unit model using van der Pol dynamics was simulated digitally. Parameters were set up to match the known and estimated human small-intestinal waveshapes and frequencies along the organ. This model has successfully reproduced a number of physiologically observed phenomena (6). The normal read-out from these runs has been a graph of average frequency versus oscillator number. The major phenomenon known to exist in the mammalian duodenum is that of frequency entrainment or the formation of a frequency plateau. This is clearly seen in Fig.6 together with a transient frequency 'dip' near oscillator 25 which travels down the model as time increases. More recent studies into these travelling waves of localised frequency perturbations have shown that the form of coupling plays a major part in their formation. Another physiological effect is that an incision made part way down the small intestine causes the formation of a subsidiary frequency plateau below the cut. This can also be seen to occur in the 100 unit digital model from Fig.7.

As mentioned earlier, an oscillator dynamic more closely related to biochemical concepts is that of the Hodgkin-Huxley equations. These were modified using a conversational digital simulation to give parameters for producing suitable waveforms for smooth muscle modelling. It should be noted that these dynamics are much too complex to make analog simulation a viable method of study. Although the equations are only 4th order, the algebraic relationships are interacting and complex. The coupling together of Hodgkin-Huxley type oscillators has been performed, and the phenomena of entrainment, amplitude and frequency modulation, and harmonic locking have beendemonstrated (12). When coupled together in this way the system equations are very stiff, and hence only a small number of units could be simulated using standard integration. To overcome this problem a method of differential quadrature has been applied to the coupled system and a chain of 30 units has been run successfully (13).

c) Electronic Hardware Simulations

The difficulty encountered in the computer simulation of coupled Hodgkin-Huxley type oscillators suggested the use of simplified electronic hardware as an alternative solution. A number of studies have been made on the basic equations by other workers, varying from complex transistorized circuitry to the use of variable conductance FETs plus operational amplifiers for waveshaping effects. This latter approach has been used to produce a coupled Hodgkin-Huxley type electronic model (14). The electronic implementation has been able to reproduce a number of waveshapes which match the conversational digital simulations referred to previously and which are of interest in smooth muscle studies. When coupled together, the electronic model reproduced the phenomena of frequency entrainment, multiple solutions and multimode behaviour.

The advantages of fast simulation, easy read-out, and quick changes to model structure and parameters have prompted the development of a16-oscillator electronic implementation of van der Pol equations. This hardware model is arranged to allow patching of the oscillators into either chain, matrix, ring, or tubular structures. Each oscillator has an 8-pole DIL switch which enables selection of either basic or modified van der Pol characteristic, passive load resistance, and RLC coupling in both dimensions. This model has given verification of multimode behaviour for coupled fifth power non-linear characteristic, also studied analytically.It has also been used to study the effect of coupling strength on the ability to pace the model using an external pulse stimulus (15). This work has been prompted by the difficulty encountered in attempts to pace the canine colon, in contrast to the comparative ease with which the stomach and small intestine can be paced.

2.2 Cardiovascular System

In this section the analysis and modelling of one component part of the cardiovascular system is described. This comprises the blood pressure reflex control system which involves feedback via the baroreceptors in the circulatory system via the brain to control of peripheral resistance of the blood vessels.

This section is on the analysis methods and simulation studies which have been applied specifically to the elucidation of blood pressure control on a beat-by-beat basis. Raster-scan methods, which have been commonly used in circadian rhythm studies, are particularly suited to pattern recognition investigation of forced oscillatory systems under conditions of partial and complete entrainment. This is done for the blood pressure reflex using an interactive video display technique programmed on a Research Machines 380Z microcomputer. The technique provides accurate calibration of entrainment parameters as well as visual pattern recognition using a simple exclusive-OR mode of screen plotting available to the 380Z. The basic principle of the raster scan is that individual lines of the scan are made up of data points from the time-series being studied. The amplitude of the data is converted to a grey scale of 16 points determining the intensity of each pixel. 80 points per line are used and the sampling frequency is chosen to give about 20 points per cycle of the underlying rhythm. Succeeding lines in the scan are produced from a shifted version of the incoming data. The data shift is chosen interactively to obtain vertical entrainment patterns, and the grey scale values can be similarly altered to visually optimise the contrast of the picture. Calibration of the entrained signal is obtained via an interactively plotted line which gives average frequency and modulation depth and frequency.

Fig.8 shows a raster scan for a thermal stimulus at 0.1 Hz entraining the blood pressure reflex at a frequency of 0.1 Hz. Although the average HRV frequency equals the external stimulus frequency there are regular fluctuations shown by the calibration line, which indicates a condition of almost-entrainment. Simulation studies based on a forced non-linear van der Pol oscillator have shown that this represents a condition of 'beating' (ie. two spectral components). This particular condition has only been achieved for a system with small non-linearity. Ultra-harmonic entrainment has also been achieved, again indicating the existence of a non-linear blood pressure control reflex. The degree of non-linearity can also be inferred from the transient 'capture' time to entrainment upon application of an external stimulus. These non-invasive methods of probing the blood pressure reflex system indicate, therefore, that the control loop comprises a weakly non-linear dynamic system which, for a van der Pol structure, implies a non-linear parameter about 0.1.

Modulation conditions have also been observed when forcing the blood pressure loop via an external stimulus. Theoretical analysis of this condition (16) has shown that

amplitude and frequency modulation can occur, depending again on the degree of non-linearity in the system. The existence of modulation conditions can be detected via conventional FFT spectral analysis, while discrimination between amplitude and frequency modulation can be simply obtained via the raster-scan approach.
The fundamental work using van der Pol modelling has been extended to the study of the dynamic behaviour of the blood pressure control system. In the work a model of the baroreceptor arc associated with changes in peripheral resistance has been developed (17). The physiological phenomenon of respiratory sinus arrhythmia (RSA) has been investigated using the model and compared with results from data published in the physiology literature. Under spontaneous conditions, heart rate variability (HRV) exhibits three main frequency components: a component at the respiratory frequency, a component around 0.1 Hz arising from the baroreceptor arc associated with peripheral resistance, and a region of low frequency activity associated with temperature regulation. The results of the physiological experiments (18) showed that the 0.1 Hz component can be entrained by respiration and that, for a set frequency of breathing, a range of tidal volumes can be established which covers the complete range of entrainment. Simulations were carried out using the computer model of the baroreceptor reflex arc. Under spontaneous conditons the model produced the 0.1 Hz component. For fixed frequencies of breathing the model simulated the effect of the complete physiological range of tidal volumes. Four primary phenomena were observed as tidal volume increased: no interaction, modulation, frequency pulling and finally complete entrainment. Modelling studies have revealed two types of modulation phenomena in the so-called metastable entrainment region. The model of the baroreceptor reflex arc has also been used to study the effect of autonomic neuropathy on heart rate variability. In studies by Watkins and Mackie, the relationship between the amplitude of the HRV signal and tidal volume was plotted as a function of the frequency of breathing. For normal subjects the peak of the characteristic occurs at 0.1 Hz while for autonomic neuropaths a corresponding peak value occurred at approximately 0.08 Hz. These results from the computer simulation showed that the difference in the peak value frequencies probably arises from an increased time delay in the reflex arc in the case of autonomic neuropathy which might arise from a decrease in neural conduction velocity. Both the analysis and modelling studies have shown the existence of specific phenomena which appear in the physiological and simulated data. These findings are seen as illustrations of the ability of such methods to non-invasively study the behaviour of the blood pressure control system on a beat-by-beat basis.

3. Modelling of Systems

Under this heading, two applications are described which use systems identification methods either for obtaining direct diagnostic information or for elucidation of physiological behaviour and drug performance. This is in contrast to the previous section on modelling of signals which uses analysis of signals directly to infer underlying physiological dynamic structure.

3.1 Determination of Lung Mechanics

In this section of study, consideration is given to the mechanics of air-flow in the lungs, rather than the gas exchange between air and blood. To reach the alveoli, air must flow through the branching tree of airways, each of which have resistance and compliance. Although the network is complex and requires a continuous model, it is commonly represented by a simple lumped-parameter model which has clinical relevance and is given by

$$P_m - P_{P\ell} = R_1\dot{V} + R_2\dot{V}^2 + EV \tag{3}$$

where P_m is mouth pressure, $P_{P\ell}$ is pleural pressure, R_1 and R_2 are resistance flow terms, and E is the elastance of the lung and chest-wall. The R_2 term allows for turbulent airflow, when the linear model is inappropriate. The linear model is also poor during forced expiration and during normal expiration of subjects suffering from chronic obstructive lung disorders such as bronchitis, emphysema and asthma. Pleural pressure can be approximated by oesophageal pressure measured via a swallowed balloon catheter. Volume is commonly obtained via integration of the flow signal, or else with the use of a body plethysmograph.

A number of methods have been used for estimating airway resistance R_1 and lung elastance E in (3) and these will be mentioned briefly. The first method uses sampling of $P_{P\ell}$, V and $\dot{V}$ simultaneously. By considering changes in $P_{P\ell}$, and $\dot{V}$ at points of equal V, the pulmonary resistance R_1 can be calculated. Similarly, the elastance E can be calculated from the changes in $P_{P\ell}$ and V between points of equal $\dot{V}$ (19). This method does not allow for noisy measurements and modelling inaccuracies.

A second method solves simultaneous equations involving measurements of P (equal to $P_m - P_{P\ell}$), $\dot{V}$, and V at two or more instants of time. If (3) is satisfied, then

$$\hat{R}_1 = \frac{p_1V_1 - p_2V_1}{\dot{V}_1V_2 - \dot{V}_2V_1} \quad ; \quad \hat{E} = \frac{p_1\dot{V}_2 - p_2\dot{V}_1}{\dot{V}_1V_2 - \dot{V}_2V_1} \tag{4}$$

If there are more than two sets of measurement, a least-squares solution can be obtained, but biased estimates will be given when the measured values are noisy.

The loop-flattening method is basically a visual method using an oscilloscope display. $\dot{V}$ is displayed on the x-axis, while (P-EV) is displayed on the y-axis. $\hat{E}$ is adjusted manually via a potentiometer until a best straight line is obtained, whose slope is proportional to $\hat{R}_1$. If non-linear effects are present then a curve is obtained instead of a straight line, but noisy data can be smoothed by eye. Clearly, the method is not suitable for automated estimation, nor for tracking changes in lung mechanics.

The interrupter method works on the principle that if airflow is occluded at the mouth then P_m will equal alveolar pressure P_A. If the air-flow is measured immediately prior to occlusion, then R_1 can be estimated. The technique does not allow for compressibility of gas, and gives no estimate of lung elastance.

The body-plethysomograph method was introduced by Dubois and others (20). The subject wears a noseclip and breathes rapidly and shallowly through a pneumotachograph. A measurement of the slope λ_1 of a curve of plethysmograph pressure (x-axis) plotted against airflow (y-axis) is obtained from an oscilloscope. A shutter is then closed across the breathing tupe, and a plot of plethysmograph pressure (x-axis) against mouth pressure (y-axis) is obtained and with slope R_2, R_1 is then calculated from

$$R_1 = \frac{\lambda_2}{\lambda_1} \times \frac{\text{pressure calibration}}{\text{flow calibration}} \qquad (5)$$

Objections to this method include a non-physiological breathing regime, and that E is not determined.

A method of forced oscillations at 3 Hz applied to the mouth was developed by Goldman (21). The subject breathes to atmosphere through a long tube which offers a significant impedance to the superimposed 3 Hz signal. From the sinusoidal pressure at the mouth and sinusoidal airflow in the lungs, it is possible to calculate R_1 from peak-to-peak flow and pressure changes at peak flows. The method is impracticable during panting and does not estimate lung elastance.

As an alternative to the above methods an adaptive tracking algorithm based on steepest-descent is described by Nada (22). An error function is defined by

$$\varepsilon = \tilde{R}\dot{V} + \tilde{E}V$$

where $\tilde{R}$ and $\tilde{E}$ are instantaneous errors in tracking R and E.

Using an error performance criterion of $\frac{1}{2}\varepsilon^2$, the steepest descent parameter adjustment equation is given by

$$\begin{bmatrix} \frac{d\hat{E}}{dt} \\ \\ \frac{d\hat{R}}{dt} \end{bmatrix} = \begin{bmatrix} m_1 & 0 \\ \\ 0 & m_2 \end{bmatrix} \begin{bmatrix} V \\ \\ \dot{V} \end{bmatrix} \qquad (6)$$

where m_1 and m_2 are adjustment loop gain coefficients. In this application, Lyapunov theory showed the technique to be asymptotically stable, while m_1 and m_2 were adjusted manually to give fast, but accurate, tracking of $\hat{R}$ and $\hat{E}$.

The adaptive tracker was initially implemented in analogue circuitry, with filters on the pressure and flow signals. The breathing pattern was split into inspiratory and expiratory periods, and volume was obtained by integration of flow, reset at each breathing period. The tracker was applied to data from human volunteers and guinea pig trials. On the human data the tracker was able to follow changes in R and E between periods of normal breathing and panting, and had a switch-on transient response of about 1 breath. On the guinea pig data, the adaptive algorithm was able to perform on-line tracking of changes in airway resistance caused by broncho-constrictor drugs (Fig.9). Comparative studies between the adaptive tracker and the loop-flattening method gave correlation coefficients of 0.97.

The algorithm has also been implemented in microcomputer form, for which it is very suitable particularly when nonlinear effects such as R_2 in (3) are considered. It had already been shown that expiratory data is better fitted using a non-linear model (23). The digital implementation has been applied to neonatal data to investigate changes in lung parameters subsequent to periods of apnoea (cessation of breathing). This implementation also included algorithms to obtain respiratory rate, minute volume and heart rate as well as on-line tracking of inspiration R_1 and E, and expiration R_1 , R_2 and E (24).

As mentioned previously, non-linear identification using the empirical approach of Wiener kernels has been more recently used by Marmarelis & Yamashiro (25) for determining lung mechanics. Their confirmation of the presence of odd-valued non-linearity adds weight to the need to include turbulence effects in modelling studies of the lung gas transport.

3.2 Identification of Muscle Relaxant Drug Dynamics

The work described in this section is part of a series of studies aimed at achieving on-line identification and control of muscle relaxation in operating theatres. In particular, the trials involve operations where abdominal surgery is performed and for which correct levels of muscle relaxation are necessary. The level of muscle relaxation is measured from an evoked EMG obtained at the hand and caused by supramaximal stimulation at a frequency of 0.1 pps applied to the ulnar nerve above the elbow. Drug infusion is via an intravenous cannula supplied from a peristaltic pump driven by a d.c. motor. Using this system successful automatic feedback control on humans in operating theatre has been achieved (26). The initial feedback controller design was based on very little knowledge of the system dynamics. To improve controller design and extend the knowledge of relaxant pharmacokinetics/dynamics a number of identification studies have been undertaken.

There are a number of different relaxant drugs available, and the type considered here are of the non-depolarising form which appear to act by occupying cholingeric receptor sites at the motor end plates. The drugs used in these trials are pancuronium bromide which is in common use and vecuronium which is a newer and faster-acting drug. The identification studies have also been extended to consider the dynamic interactive effect on level of paralysis caused by Halothane and Ethrane which produce unconsciousness.

The conventional physiological method used for identifying the pharmacokinetics relating bloodconcentration levels to drug administration is that of a bolus injection producing the system impulse response. This can produce satisfactory results when the system is linear and provides parameter estimates for compartmental models. When the pharmacological response is also required in the identification (referred to as the pharmaco-dynamics) then bolus dose responses must be interpreted more carefully. Dose-response curves have been inferred from bolus-injections (e.g. (27)) while nonlinear regression methods have been used to attempt simultaneous identification of pharmacokinetics and dynamics (28). Since the main aim of the present work is to obtain regulation of paralysis at a fixed level, the identification has been concentrated on a complete linearised model about an operating point. For such purposes bolus injections are not suitable and hence PRBS signals have been used as the perturbations.

The PRBS trials have been performed on dogs to enable sufficiently long sequences to be used under steady conditions. Initial trials have used open-loop conditions such that a steady level of paralysis was first obtained under closed-loop mode. Subsequently the loop was opened and a PRBS signal superimposed on the last steady-state drug infusion level. Initially a Hewlett-Packard PRBS generator was used with bit intervals of 33.3 seconds or 100 seconds. Currently a Research Machines 380Z microcomputer is being used to produce the PRBS signal with a bit interval equal to any integer multiple of the 10 seconds sampling interval. The 380Z is programmed to provide data logging of the results for off-line identification on another machine. It also provides the controller algorithm software.

Modelling based on bolus injections of pancuronium bromide in humans has produced a 3-compartment structure (29). In contrast, the linearised model about an operating point obtained from PRBS trials has shown two dominant time constants only. This has been confirmed using very small bolus injections superimposed on the steady operating level under open-loop conditions. The identification has been carried out using a generalised least squares algorithm and indicates the presence of a pure time delay in the range of 0.5 - 2.0 minutes. This delay is probably due to effects such as blood perfusion of the response organ, drug diffusion from capillary blood to receptor site and drug receptor association/dissociation responses. The exponential time constants identified have been in the range of 1-3 minutes and 10-40 minutes.

The potentiating effect of Ethrane and Halothane on muscle relaxation has been known for some time. Without steady paralysis levels, however, those effects are difficult to quantify. The closed-loop control scheme being used here and PRBS excitation are well suited to provide identification of these effects. The PRBS signal is provided by mechanically driving the gas vaporiser between levels set up on two cams. Identification of the dynamics shows a pure time delay of about 10 seconds and two exponential time constants similar in magnitude for Ethrane to those for pancuronium bromide.

Instead of off-line identification of open-loop data, two further developments will be described using the blocking drug NC45.These are on-line identification using a recursive least-squares approach,and estimation under closed-loop conditions. Previous studies suggested that the noise contamination in the EMG responses approximates to white noise of a low level, suggesting that recursive least squares identification could give unbiased estimates of parameters. In this way on-line tracking of parameter coefficients related to an underlying compartmental structure should be feasible. This has been demonstrated on open-loop data. On-line identification under closed-loop conditions has been studied by numerous workers and is surveyed by Gustavson (32). The system being studied here is suitable for such identification, and because of the sigmoidal shaped pharmacodynamic characteristic for evoked EMG response the identification is based on a locally linearised approach using differenced input and output data.

The results described above were the necessary precursors to studies involving self-tuning control based on pole-assignment algorithms (30) which give simultaneous parameter estimation and sub-optimal adaptive control. Simulation results indicated that such strategies were feasible under conditions of small noise contamination. Experience in dog trials has shown that good signals can be obtained by very careful electrode placement. Noise components can be further reduced by multirate sampling and filtering since data are collected at 10 second intervals, whereas self-tuning control is better at a sampling period of 1 minute. These algorithms have been applied successfully in animal trials together with Smith predictor methods (31).

4. Conclusions and Acknowledgements

The examples of biomedical signal analysis and systems modelling described in this contribution represent only a small fraction of work being carried out in these areas. It has demonstrated, however, that such techniques are being used increasingly for both diagnostics and elucidation of physiological performance. Such applications must use all the technological systems tools available, since factors such as short time-series, noise contamination and non-linearity are norm in these areas of life science.

All the examples have been taken from research in which the author has partaken. Without the ready and enthusiastic cooperation of numerous clinicians and medical physics staff in several hospitals such work would not have been possible. In particular, grateful acknowledgement of collaboration with the following is given: H.L.Duthie (Welsh National School of Medicine), B.H.Brown (Royal Hallamshire Hospital, Sheffield), A. MacDonald (The London Hospital), R.I.Kitney (Imperial College), A.J.Asbury (Glasgow Royal Infirmary).

5. References

1. Linkens, D.A. (1982) 'Short-time-series spectral analysis of biomedical data.' Proc. IEE Pt. A., 129, p663-672.
2. Gardner, F.M. (1966) 'Phase lock techniques.' John Wiley, New York.
3. Aschoff, J. (1965) 'Circadian Clocks.' Proc. Feldafing Summer School, Sept. 1964, North-Holland Publ., Amsterdam.

4. Nelsen, T.S. & Becker, J.C. (1968) 'Simulation of the electrical and mechanical gradient of the small-intestine.' Am.J.Physiol., 214, p749-757.

5. Sarna, S.K., Daniel, E.E. and Kingma, Y.J. (1971) 'Simulation of slow-wave electrical activity of small-intestine.' Am.J.Physiol., 221, p166-175.

6. Robertson-Dunn, B. & Linkens, D.A. (1974) 'A mathematical model of the slow-wave electrical activity of the human small-intestine.' Med. & Biol. Eng. p750-758.

7. Sarna, S.K., Daniel, E.E. & Kingma, Y.J. (1972) 'Simulation of the electrical control activity of the stomach by an array of relaxation oscillators.' Amer. J. Dig.Dis., 17, p299-310.

8. Linkens, D.A., Taylor I. & Duthie, H.L. (1976) ' Mathematical modelling of the colorectal myoelectrical activity in humans.' IEEE Trans.Bio.Med.Eng. BME-23 p101-110.

9. Patton, R.J. & Linkens, D.A. (1975) 'A conversational program for analytical modelling of action potentials in nerve and muscle.' Comp. Prog. in Biomed., 5, p142-152.

10. Datardina, S.P. & Linkens, D.A. (1978) 'Multimode oscillations in mutually-coupled van der Pol type oscillators with fifth power non-linear characteristics.' IEEE Trans. Cct. & Sys. CAS-25, p308-315.

11. Linkens, D.A. 'Regions of attraction in coupled non-linear oscillators.' IEEE Trans. Cct. & Sys. CAS-26, p663-666.

12. Linkens, D.A. & Datardina, S.P. (1977) 'Frequency entrainment of coupled Hodgkin-Huxley type oscillators for modelling gastro-intestinal electrical activity.' IEEE Trans. Bio.Med.Eng. BME-24, p362-365.

13. Patton, R.J. & Linkens, D.A. (1976) 'A differential quadrature method for simulation of gastro-intestinal signals.' 8th AICA Int. Cong. on 'Simulation of Systems.' Delft, p142-152.

14. Patton, R.J. & Linkens, D.A. (1977) 'Hodgkin-Huxley type electronic modelling of gastro-intestinal electrical activity.' Med. & Biol. Eng., 16, p195-202.

15. Linkens, D.A. (1978) 'Canine colonic pacing & coupled oscillator synchronisation' J. Physiol., 278, p26P.

16. Linkens, D.A.(1979) 'Modulation analysis of forced nonlinear oscillators for biological modelling.' J. Theoret. Biol., 77, p235-251.

17. Kitney, R.I. (1979) 'A nonlinear model for studying oscillations in the blood pressure control system.' J.Biomed.Eng.1, p89-99.

18. Selman, A., MacDonald, A., Kitney,R.I., & Linkens, D.A.(1982) 'The interaction between heart-rate & respiration, Part I : experimental studies in man.' Automedica, 4, p131-139.

19. Neergaard, R. & Wirz, K. (1927) 'Uber eine methode zur messung der lungenelasticitat am lebenden imenschen, insbersondere beim emphysema.' Z Klein. Med., 105, p35-51.

20. Dubois, A.B., Botelho, S.Y. & Comroe, J.H. (1956) 'A new method for measuring airway resistance in man using a body-plethysmograph.' J.Clin.Invest., 35, p327-335.

21. Goldman, M., Knidson, R.J., Mead, J., Peterson, N., Schwaber, R.J. & Wohl,M.E., (1970). 'A simplified measurement of respiratory resistance by forced oscillation.' J. Appl. Physiol. 28, p113-116.

22. Nada, M.D.(1979) 'Lung parameters identification' in 'Biological Systems modelling & control.' Linkens (ed) Peregrinus 1979 p.242-274.

23. Nada, M.D. & Linkens, D.A. (1977) 'An adaptive technique for estimating the parameter of a nonlinear mathematical lung model.' Med. & Biol. Eng. & Comput., 15, p149-154.

24. Linkens, D.A. & Rimmer, S.J. (1982) 'On-line identification of mechanical properties of the lung.' Trans. Inst. M.C., 4, p177-185.

25. Marmarelis, V.Z. & Yamashiro, S.M. (1982) 'Nonparametric modelling of respiratory mechanics and gas exchange.' 6th IFAC Symp. Ident. & Syst. Par. Est. Washington p586-591.

26. Brown, B.H., Asbury, A.J., Linkens, D.A., Perks, R., & Anthony, M., (1980) 'Closed-loop control of muscle relaxation during surgery.' Clin. Phy. Physiol. Meas., 1, p203.

27. Shanks, C.A., Somogyi, A.A. & Triggs, E.J. (1979) 'Dose-response & plasma concentration-response relationships of pancuronium in man.' Anaesthesiology, 55, p111.

28. Sheiner, L.B., Stanski, D.R, Vozeh, S., Miller, R.D. and Ham, J. (1979) 'Simultaneous modelling of pharmacokinetics & pharmacodynamics application to d-tubocurarine.' Clin. Pharmacol. Th., 25, p358.

29. Hull,C.J., Van Beem, H.B.H., McLeod, K. Sibbald, A. & Watson, M.J. (1978) 'A pharmacodynamic model for pancuronium.' Br.J. Anaesth., 50, p1113.

30. Wellstead, P.E., Prager, D. & Zanker, P. (1979) 'Pole assignment self-tuning regulator.' Proc. IEE. Pt. D, 126 p781.

31. Linkens, D.A., Menad, M. & Asbury, A.J. (1985) ' Smith Predictor & Self-tuning control of muscle relaxant drug administration.' Proc. IEE. Pt. D.

32. Gustavsson, I., Ljung,L., & Soderstrom, T. (1977) 'Identification of processes in closed-loop: identifiability and accuracy aspects.' Automatica, 13, p59.

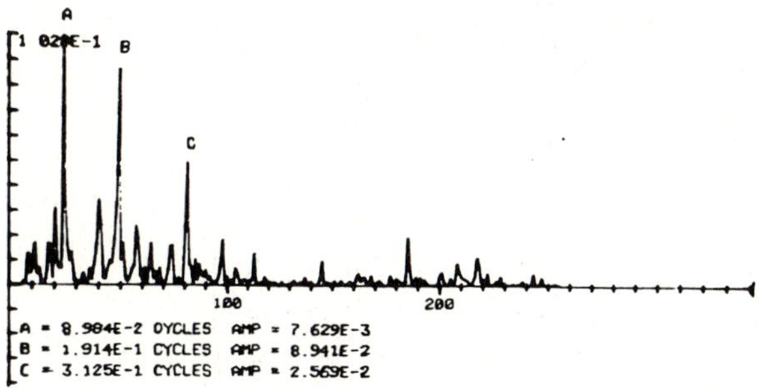

Fig. 1. FWT of a canine gastric signal showing fundamental (A) and second harmonic (B) peaks.

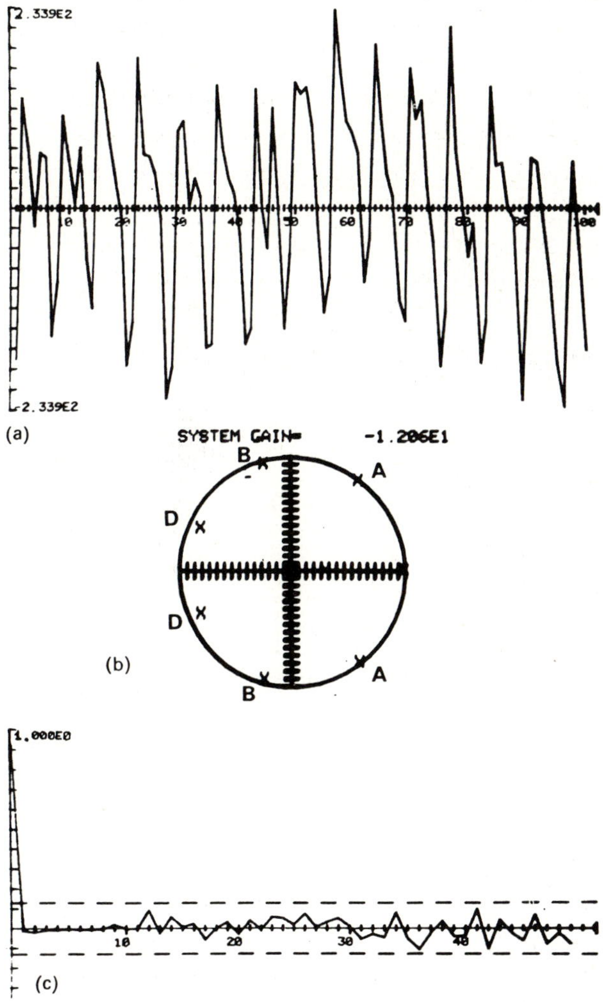

Fig. 2. AR spectral analysis on canine duodenal data: a) 100 points of data; b) z-plane poles; c) residuals.

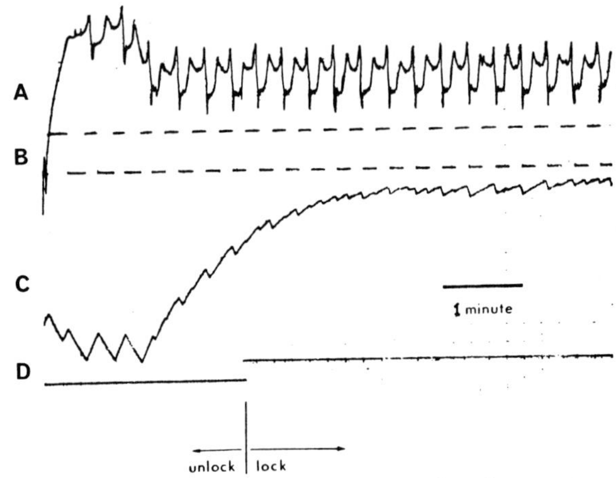

Fig. 3. Phase-lock loop response: a) human gastric signal; b) output from voltage-controlled oscillator; c) filtered output from lock detector; d) lock/unlock indicator.

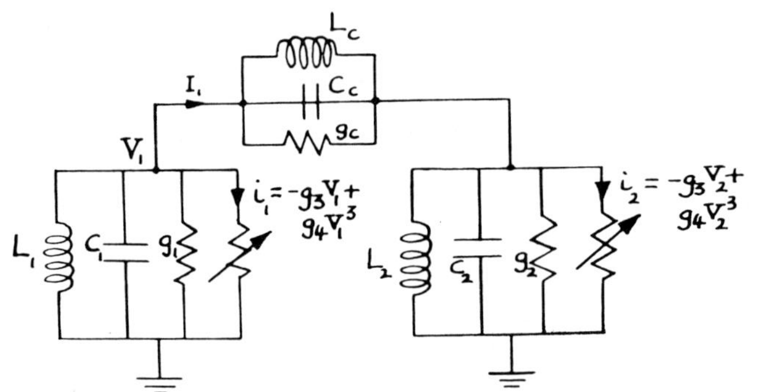

Fig. 4. Equivalent circuit of two coupled van der Pol oscillators.

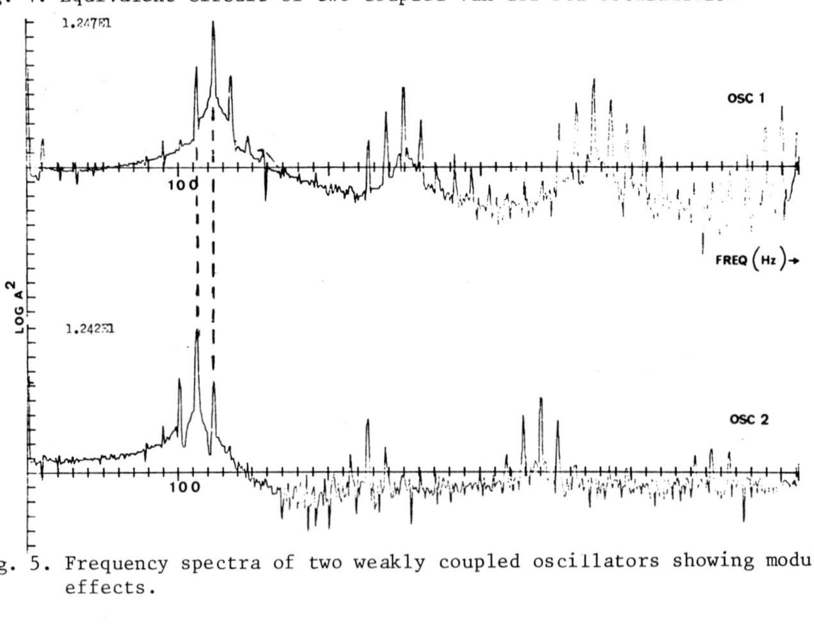

Fig. 5. Frequency spectra of two weakly coupled oscillators showing modulation effects.

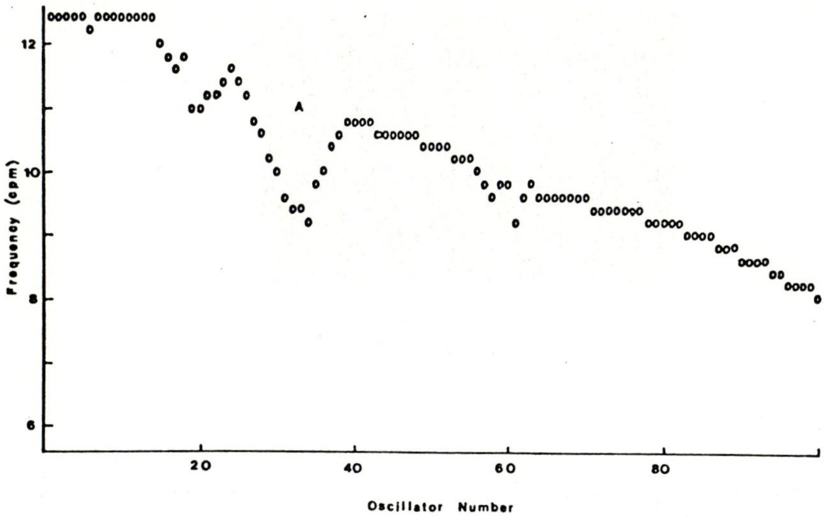

Fig. 6. 100-oscillator small-intestinal model, showing frequency plateau and transient frequency dip at 'A'.

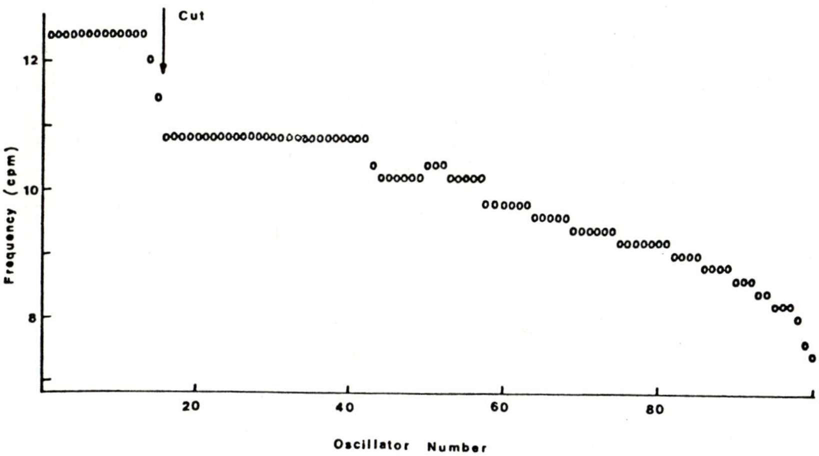

Fig. 7. Small-intestinal model showing a secondary plateau formed below a simulated 'incision'.

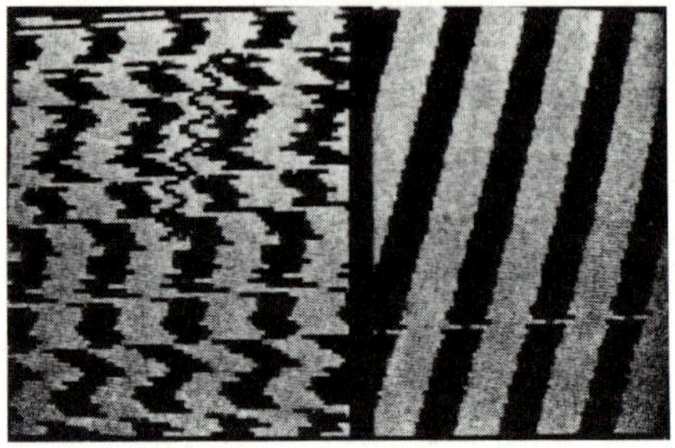

Fig. 8. Calibrated raster scan of Heart Rate Variability (HRV) data.

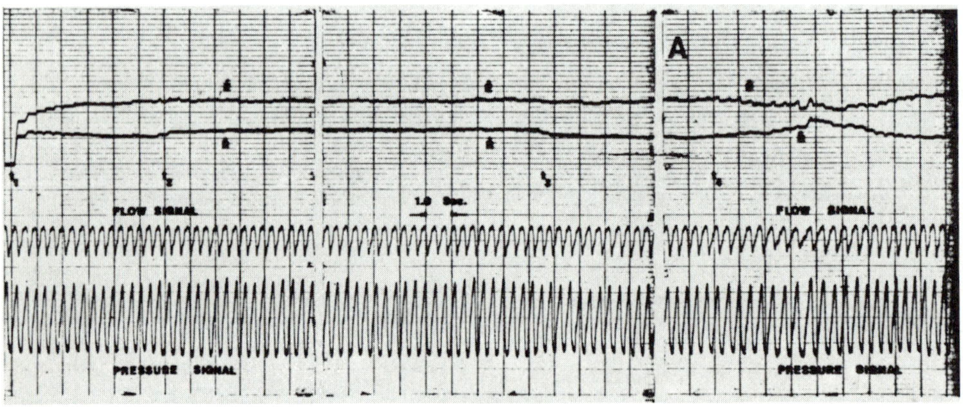

Fig. 9. Transient response of lung parameter tracker on guinea pig data with a bronchoconstrictor drug administered at 'A'.

STOCHASTIC METHODS AND ENGINEERING SURFACES

Dr. D.G. Chetwynd

SUMMARY

Engineering surfaces have in their manufacture a large proportion of random events. It is not therefore surprising that the study of surfaces, either for understanding of tribology or as a means of manufacturing control, is rich in applications for random process theory and spectral estimation. This paper will illustrate a range of such applications.

1. INTRODUCTION

Tribology and surface metrology make much use of random process techniques since many manufacturing processes are essentially a sequence of independent events, for example grinding, grit blasting and spark-erosion. Their study is particularly interesting because the results required of stochastic theory are rather different to those normally presented. This is because historically the theory has developed from the needs of telecommunications engineers where interest is, for example, in zero crossings, while tribologists tend to be more concerned with the behaviour of peaks and valleys. This paper will look at several examples where these differences occur. It is not a formal review but introduces sufficient information to allow the literature to be further pursued. It will not concern itself with the very real and important problems of manufacturing control as a control problem. This is a highly complex subject in its own right, further complicated in that many of its requirements are not functionally driven but come from legal and standardisation requirements.

Surface geometry has functional significance in friction, wear, electrical and thermal contact to name but a few. One way in which an understanding of these phenomena may be attempted is by theoretical modelling in order to predict surface behaviour. Another route to understanding is the attempt to discover the real nature of surfaces through measurement. Clearly these are not independent but in practice they do have separate identities. We will briefly look at both. Alternatively we may be driven to use random process theory by the need to measure and to describe features on manufactured parts, for reasons of control and reproducibility. We may for example need to discover why with two machines which apparently produce similar parts only one produces those with a good lifetime in normal operation. Here we will look briefly at various methods used for such measurements, and at some of the implications and techniques of trend removal from measured data. Both empirical parameterisation and a more systematic approach based on spectral analysis will be examined.

2. MODELLING SURFACES

As there exist good reviews of stochastic modelling of surfaces[1,2] it is fruitful here to concentrate only on the historical perspective of the introduction of increasingly sophisticated modelling techniques. Amonton's laws, roughly, that friction depends on the normal force between surfaces but not on their area, have been known for several centuries. It was not until the 1950's that the idea of real contact areas being small compared to plan-form and friction being caused by local adhesion at contact points became accepted[3]. An early attempt to explain the contact used a concept of a large asperity or bump on which a regular array of similar smaller asperities were sited[4]. This idea of a hierarchy of self similar structures is reminiscent fractal behaviour[5] and indeed there is current debate over whether randomly machined surfaces are fractal in nature. By using hemispherical asperities standard results for stress fields, areas of contact and onset of plasticity due originally to Hertz can be used to estimate the behaviour of contacting surfaces.

The next major development was the transfer from deterministic to stochastic models[6], in which hemispherical asperities are placed with a Gaussian distribution of heights about the mean height of the peaks. It is a very approximate model since all asperities have the same radius and yet it has achieved considerable success in predicting contact phenomena. Plasticity indices were introduced to account for the change in functional constants between situations where contact is predominantly by elastic deformation of asperities and those where it is predominantly plastic.

This simple model was soon replaced by those having a random distribution of curvatures reflecting the observation that there is increased likelihood of high peaks having small curvatures[7]. The correlation between points needed for this model is introduced through techniques much closer to the classical work in time series analysis than through the build up of asperities. It was assumed that a matrix of defined points in the surface have Gaussian amplitude probability density function and an exponential form of autocorrelation function equivalent to a power spectral density which represents white noise subjected to a six dB/octave roll-off. This approach is important for it emphasises that all descriptions of surfaces which can be verified experimentally will depend upon the spacing of the samples taken. It is only plausible to use discrete sampling. This implies that what you see depends upon what you look for. The surface is regarded as a Markov process rather than a fractal one and the peaks are found mathematically rather than defined. Using the simplest three point analysis we find that the probability density of an ordinate of height between y and y + dy being a peak is:

$$f_p(y) = f(y_{-1} < y,\ y < y_o < y + dy,\ y_{+1} < y)$$

$$= \int_{-\infty}^{y} \int_{y}^{y+dy} \int_{-\infty}^{y} f(y_{-1}, y_o, y_{+1})\, dy_{-1}\, dy_o\, dy_1 \qquad (1)$$

For exponential autocorrelation functions, if the correlation of adjacent point is ρ, that between double spaced points is ρ^2 and we find that $P(y_{+1}|y_o, y_{-1}) = P(y_{+1}|y_o)$. This is a condition for a first order Markov process. Then the joint probability density function for three points is

$$f(y_{-1}, y_o, y_{+1}) = \frac{1}{\sqrt{2\pi}} \exp\left(\frac{-y_o^2}{2}\right)\frac{1}{2\pi(1-\rho^2)} \exp\left(\frac{-(y_{-1}-\rho y_o)^2}{2(1-\rho^2)}\right) \cdot \exp \frac{-(y_{+1} - \rho y_o)^2}{2(1-\rho^2)} \qquad (2)$$

Substituting into equation 1 and integrating gives:

$$f_p(y,\rho) = \frac{1}{4\sqrt{2\pi}} \left(1 + \mathrm{erf}\left(\frac{y}{\sqrt{2}}\sqrt{\frac{1-\rho}{1+\rho}}\right)\right)^2 \cdot \exp\left(\frac{-y^2}{2}\right) \qquad (3)$$

where height ordinates are normalised to $\sigma^2 = 1$.

This demonstrates the possibilities and the general strategy for deriving analytical expressions for distributions of peak height, curvatures etc. on which theoretical studies can be built. The model is more realistic than the earlier ones. It requires only two parameters, σ and either ρ (for given sampling) or the independence distance β^* of the autocorrelation function:

$$\rho_y(\tau) = \exp\left(-|\tau|/\beta^*\right) \qquad (4)$$

The earlier models required three: the curvature of peaks, σ and a parameter representing the spatial density of the peaks.

The model is not accurate (merely useful) as is easily demonstrated. The even spectral moments correspond to values of the differentials of the autocorrelation function at $\tau = 0$:

$$m_k = \int_{-\infty}^{\infty} \omega^k S_y(\omega)d\omega = \frac{d^k R(0)}{d\tau^k} \qquad (5)$$

and each differentiation of a profile generates a double differentiation of the correlation function. The variance of profile slope is given by $R''(0)$ which by equation 4 is undefined! (Note in passing this does not mean that we are unable to define expected slopes, curvatures etc. as we may use the strategy outlined in equations 1 to 3). Various modelling attempts have been made to get around this theoretical difficulty. The simplest is the replacement in equation 4 of $|\tau|$ by τ^2. More sophisticated methods use exponential cosine autocorrelation functions[8].

Studies in oceanography, where the random surfaces are also varying in time, had used a different approach to random process modelling[9]. Modifications of this approach were proposed[10] as an alternative model to the Markov approach although the two methods have coalesced in recent years. For a Gaussian process the derivatives are independent Gaussian processes so it is appropriate to set up a model as a multi-variant Gaussian process in height, slope and curvature. For the case of a profile this amounts to specifying the model parameters directly as the spectral moments m_o, m_2 and m_4. This removed the problem of the undefined slopes. In terms of modelling real surfaces the spectral moments can be evaluated directly from measured data or they can be estimated from densities of zero crossings and extrema(9):

$$\sigma^2 = m_o \; ; \; D_{zero} = \frac{1}{\pi}\left(\frac{m_2}{m_o}\right)^{\frac{1}{2}} \; ; \; D_{extr} = \frac{1}{\pi}\left(\frac{m_4}{m_2}\right)^{\frac{1}{2}} \tag{6}$$

From here on, whichever model we adopt, we can calculate, with varying degrees of difficulty, the average behaviour of tribologically significant features such as peak curvatures. This approach can be extended, with difficulty, to model anisotropic surfaces. The effects of a finite sized stylus measuring surfaces has also been included in the models. Very complex interactions between measured parameter values, the algorithm used and the sampling interval relative to the stylus size have been demonstrated. We will not pursue this here but treat it as a timely warning as we turn to examine the processing of real measured data.

3. PROFILE STATISTICS

All the theoretical models make the assumption of Gaussian amplitude statistics. In practice most surface amplitude probability density functions are unimodal, roughly bell-shaped, but often show visible skew. Consider, for example, grinding where the basic random process creates surface, the high peaks of which interact most strongly with the wheel: a negative skew is only to be expected. Simple assumptions about the surfaces are only likely to be valid when they are newly formed by a single process. If a surface has interacted with another, its statistics have probably changed though not necessarily all the way through its structure. This leads to the classic identification of "transition surfaces"(11) as illustrated in Figure 1. Plotting the cumulative probability density function on probability paper for measured data can show closely Gaussian behaviour as in A. Subjecting the surface to mild wear by rubbing against another flat surface under light load and subsequently remeasuring may show the form typified by B. The majority of the surface away from the peaks is not discernably affected but the top few percent change greatly. Quite commonly the section B after its break from A will show a truncated Gaussian distribution in its own right. Clearly there is a need for care in interpreting both data and the predictions of models.

Statistical stationarity is another assumption implicit in both theoretical models and most practical measurements, since they involve some form of ensemble averaging. Strictly stationary statistics can never apply to real data, so all stochastic work is slightly compromised: the question is whether surface statistics may reasonably be considered as a slice out of a stationary process. The mechanism

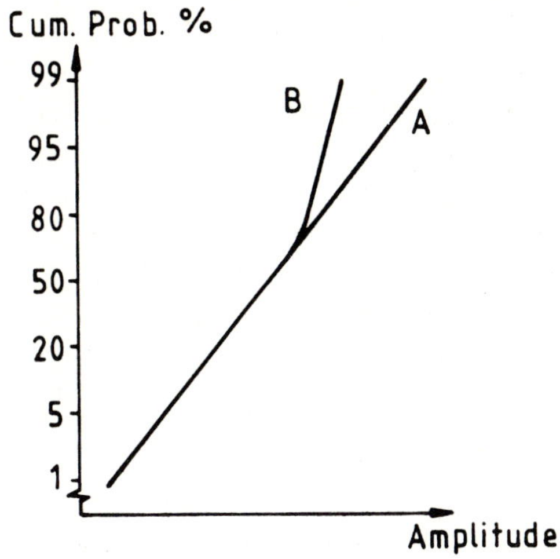

Figure 1 C.A.P.D.F. for a transitional surface.

of creation of machined surfaces does cast some doubts on this. Consider grinding once more. The interaction of grits over dimensions a fraction of a millimetre with the surface will be plausibly random. The general structure of the wheel and dressing process and the stiffness of the machine may impose other random components onto parts over dimensions of a few millimetres. The ways on a machine themselves will be imperfect and their profile may be superposed onto the workpiece over still larger dimensions. It is generally accepted that the larger the machine slides, the lower their precision and this has specific confirmation in work suggesting that roughness depends on the length of the workpiece[12] according to:

$$\sigma^2 = \sigma_t^2 + (kL)^{\frac{1}{2}} \tag{7}$$

where L is the length measured and k a constant for the process. In fact metrologists have long reacted to this situation by using (but studiously avoiding rigorous definitions of) the concepts roughness, waviness and error of form to describe the three mechanisms listed above. Most, but not all, tribological phenomena occur over dimensions where σ_t predominates. For instance, features much larger than the Hertzian contact zone of a wheel are unlikely to affect the friction and wear of it although they may well affect the comfort of vehicle passengers. For this reason there is often little need to worry about stationarity providing proper functionally relevant filtering of the signal is performed. Studies[13] of a large number of different random surfaces on different scales from the surface of the moon to precision bearings show very similar general structures of power spectral density. They may be expressed:

$$S_y(\omega) = \frac{k}{(2\pi\omega)^2} \tag{8}$$

The constant k is termed "topothesy". It seems to be all that need be known for a given process in order reasonably to model its spectral behaviour. The geometrical form of surfaces does appear to consist of similar statistical structures on different scales of size. We return to a notion of fractal-like behaviour. It is interesting to compare the form of equation 8 with the requirements in equation 5 for the spectral moments to be finite. For continuous behaviour it must be assumed that at short wavelengths small topothesies take over so that the integrals converge. Experimental evidence is not clear cut, but generally finer surfaces have smaller topothesy values.

We will return to the modelling of surface structure in a context of understanding machining processes at a later stage. First it is relevant to look at the earlier attempts at parameterisation of surface roughness.

4. ROUGHNESS PARAMETERS

When it first became possible to examine machined surfaces by suitable microscopes or stylus tracing instruments, the desire to place a number to the question "how rough?" was satisfied by estimating the total divergence from a line "parallel to the general trend of the surface". On a microscope the length of surface which could be viewed was limited by the field of view, typically about 1mm or a little less. On profilometers no such restriction occurred and so for consistency the profile was arbitrarily divided into contiguous lengths similar to the field of view of a microscope. Each of these lengths was independently assessed. They could then be averaged to improve the poor repeatibility of peak readings. This process also acted as a crude long wavelength filter, removing the dominant effects of waviness and error of form. The present situation based on visually simple measurements and simple filtrations was soon established. One early attempt to introduce functional significance came with the definition bearing area[14]. The ratio of flat area revealed as a plateau to the total area of measurement when a surface is truncated (an ideal lapping operation) at some depth below the highest point is measured. This was expected to give some indication of the performance of the surface as a bearing. It seems that for many years the metrology community failed to realise this is exactly a cumulative amplitude probability density function. If it had, perhaps the subject would have moved in the direction of modern thinking before rather than after the establishment of standards and customary practices.

The introduction of electronic instrumentation meant that more stable averaged parameters could be used. Variance was the obvious choice but the arithmetic average, $E(|y|)$, has dominated because early instruments did in fact measure the latter by using moving coil meters. For hand assessment, peak based methods, for example the difference between the average height of the five highest peaks and that of the five lowest valleys continued to be used. Digital instruments have caused a large number of new parameters to be introduced in an arbitrary fashion. A few have been concerned with averaged spacing behaviour but most just give amplitude sensitive information. All are ways of measuring the width of the amplitude probability density function and only one is necessary. This illustrates a danger of cheap computers. Easy calculation means that lots of things get calculated and used uncritically. An additional problem for automatic systems has been a requirement that the peak parameters previously used for hand measurement should be included. Definitions of these, not surprisingly, tend to be at least partly intuitive and much effort and ingenuity has gone into the production of algorithms which do what a human would do rather than follow the exact wording of the definition as written. For example if a human is asked to select the five highest peaks on a profile it is almost certain that the five highest peaks would not be selected but that the five judged most significant would be. If functional significance is desired of readings, the human is probably right and rules of threshold and significance must be built into our peak searching algorithms[15].

5. PROFILE FILTERING

In both of the previous two sections there has occurred the need to select the bandwidth of interest from a wide spectrum of information available on the surface. This leads to contemplation of the more general question crucial to surface metrology but common elsewhere: the detemination of a reference figure from which to assess the data. The present context has a reference figure as a simple predetermined geometric shape of uncertain size and position when shape is being considered and as a wavy line of defined minimum wavelength when roughness is being assessed. In either case we need to estimate the optimum placement of the reference line for the given data prior to measuring that data from the reference.

Take first the case of roughness. Polynomial fits are inappropriate because they require too much a priori knowledge of the surface and so a low-pass filter will be used to define the wavy line reference. In practice it is usual to go straight to a roughness measurement referenced to the line by using the complementary high-pass filter. Relevant standards define a simple buffered 2-stage RC filter characteristic to be used with one of a set of predefined cut-off wavelengths. For inspection work with averaging parameters this has proved perfectly satisfactory, subject to the operator understanding how to select a suitable cut-off. However when considering a tribological function, the shape of the profile, particularly near peaks, may be more important than the power (variance) contained within it. The standard filter does not maintain shape. Obviously long wavelengths are attenuated in amplitude relative to short ones for this is the purpose of the filter and is quite acceptable. The problem lies with the phase characteristics of the filter which mean that different frequency components of the profile are shifted in space by different amounts. Ideally we need zero phase-shift filter characteristics, although a linear phase characteristic is acceptable[16]. Such a filter is not realisable as a circuit. Multi-stage filters can approximate to linear phase characteristics but only at the cost of a very sharp roll-off, which is functionally undesirable here. Low cost digital computation has opened new possibilities.

Any filter realisable by analogue circuits can be modelled by a digital computer.

The classical technique for doing this is to convolute the input sequence with the desired impulse response. A more useful real-time technique uses recursion, modelling the Laplace transform of the continuous transfer function by the z-transform of a discrete one. A wide literature exists on digital filter design and so it will not be explored here. Suffice it to mention that some care should be taken over the modelling process as distortions in mapping between continuous and digital regimes can occur and group delays may be introduced. A third technique available if a time delay is acceptable is to compute the Fourier transform of the input sequence, multiply that by the model transfer function and then inverse transform. This leads to a very interesting observation. While we can model any realisable filter, if we are prepared to forego real-time working, we can also implement filters which could never be built in analogue components. This shows most clearly in the third technique for we can multiply in the frequency domain by any amplitude and phase characteristics we wish (remember that these are complex arithmetic operations), although there are limits to what may be done while retaining real data after the inverse transform. Linear and zero-phase characteristics with various roll-off rates can be produced this way but it may not be the most appropriate method to use in practice. Convolution in the spatial domain with a symmetrical weighting function will produce linear phase characteristics. Providing the impulse response decays quickly enough that the weighting function can be truncated to a reasonable length, this allows nearly real-time operation. The output will be delayed but not distorted. The reason we can do this digitally but not by analogue means, is that the symmetrical impulse responses implies knowledge of what the input signal is going to do in the future. We cannot really know this but, within the confines of the delay we are prepared to tolerate, the digital method gives the next best thing. A third method of obtaining zero-phase characteristics is to use time reversal. This can be computationally attractive when the amplitude characteristics is easily set up as a recursive filter. The method is illustrated schematically, as are the others, in Figure 2. The input sequence passes through a recursive filter and into store. It is read out from the store in the reverse order and passed again through the same filter. The overall effect is that the original amplitude characteristic has been squared but the phase shifts introduced in pass 1 are cancelled by those in pass 2 since the data is sequenced in the time reversed sense. In a field such as surface metrology where the size of the data records is usually well defined and where a few seconds delay after the end of the input before obtaining the results is acceptable, time reversal can give very desirable filter characteristics with very low programming overheads.

The fitting of reference figures for shape measurement generally involves estimating a best-fit line, plane, circle or other figure of simple geometry. The most commonly applied criteria for best-fit are least squares deviation or minimax (minimum value of maximum deviation). The least squares line (or plane) is a simple case for linear regression except for one subtlety which is often overlooked. The regression model (for line of slope m, intercept c) minimises the sum of squares of:

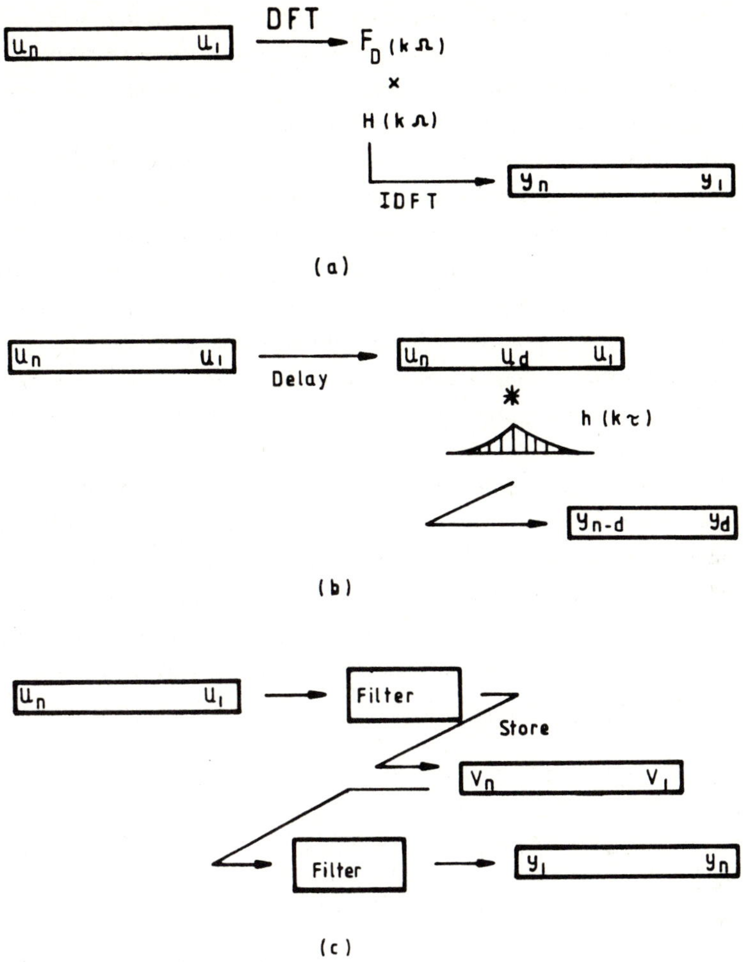

Figure 2. Schemes for linear-phase filters. a) frequency domain multiplication, b) time domain convolution, c) time-reversal.

$$\varepsilon_i = y_i - (mx_i + c) \qquad (9)$$

implying that the residuals are measured in the y-direction. From the metrological point of view, which is the minimisation of deviations from the nominal straight line form, residuals should be measured perpendicular to the reference line (surface). This gives a non-linear formulation:

$$\varepsilon_i = [y_i - c - mx_i]\,(1 + m^2)^{-\frac{1}{2}} \qquad (10)$$

which does not, except for some special cases, have the same minimisation parameters. This could be a serious problem although not for the traditional surface metrological instruments, because their design limits m to less than 0.01 in useful terms. Moreover some instruments measure only y-displacement and the measurement uncertainties should be treated as residuals in the y-direction. The apparent finesse of the more complicated model can be an error: a nice example of the dangers of treating data without full knowledge of how it was obtained.

The linear trend in the data comes mainly from slight mis-setting of the instrument and so is well defined in characteristic. This has aided the practical success of approximations to the least squares line. The reference line is defined as that passing through the centroid of the first n/2 points and the centroid of the last n/2 points. It is intuitively clear that this will give a trend line in the correct direction: with real data it usually gives one very close to the true least squares line. The approximation is not arbitrary. For uniformly spaced data the calculation of the two centroids is equivalent to finding the magnitude of the first order Walsh function (or the zero order Rademacher function, which is identical). The linear trend is being identified with a square wave function above the mean level for the first half and below the mean for the second half or vice versa.

Least squares circle fitting is readily defined, but there is strong motivation for using a linear parameter model in order to avoid iterative techniques. With a circular reference this must clearly be a local linearisation. For a true centre close to the established origin, this model is a limacon, giving:

$$\varepsilon_i = r_i - (a \cos\theta_i + b \sin\theta_i + R) \qquad (11)$$

$\hat{R}$ will be the best fit radius and $(\hat{a}, \hat{b})$ the estimate of the circle centre. The direction of assessment of the residuals is radial from the origin, not normal to the reference surface (radial to the true centre). This, again, proves to be correct for the transducer configuration commonly used on specialised roundness measuring machines.

Rules for minimax fitting for specific cases are well known. It is less widely realised that all such fits have a common basis in optimisation theory[17]. The straight line fit to a data set is readily appreciated to require at least three extreme points equidistant from that line lying alternatively either side of it, for, then, however the line is moved, it will be further from at least one of those points. The minimax criterion can be expressed:

$$\begin{aligned} &\text{minimise} && h \\ &\text{subject to} && mx_i + c + h_i \geqslant y_i \\ & && mx_i + x - h_i \leqslant y_i \end{aligned} \qquad (12)$$

for all i. This is a classical mathematical programme, specifically, here, a linear programme in the parameters (m, c, h). Efficient algorithms can be derived from the well established theory of linear programming, leading to order of magnitude increases in the speed of algorithms for, for example, best fit circles over traditional search methods[18,19]. Without delving deeply into the theory, it is possible to illustrate why this occurs. For any programme:

$$\text{maximise} \quad z = C^T x \qquad \text{subject to} \quad Ax \leqslant b \tag{13}$$

there is a dual programme

$$\text{minimise} \quad z = b^T w \qquad \text{subject to} \quad A^T w \geqslant C \tag{14}$$

which has the same optimum solution. Thus we may solve either primal or dual according to our convenience. But the work required to perform simplex iterations on a linear programme relates to that of inverting an m x m matrix where m is the number of contraints. Thus if A has a few columns and many rows, the usual case for surface metrology, working in the dual will be much more efficient. The primal dual relationship has an interesting geometrically interpretation illustrated for finding the minimum circumscribing circle in Figure 3. The traditional search approach (primal)

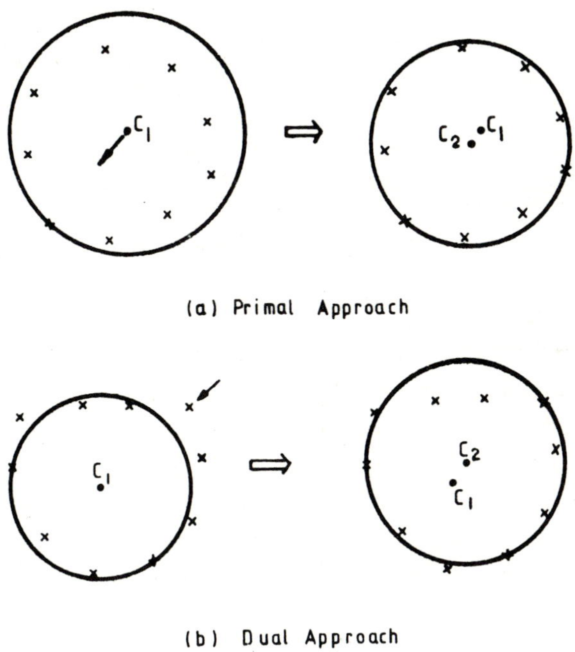

Figure 3. Iterations in finding the minimum circumscribing circle. In (a) must search along direction indicated, but in (b) may jump directly to next stage.

takes an overly large circle around the data and seeks, by shifting the centre and reducing radius, to establish the minimum. Many small steps must be taken to avoid violations of the circumscribing condition during this process. The dual approach takes any three data pointd and constructs a (unique) circle through them. By definition this is the minimum for the points which it does contain, but there will in general be points lying outside it. Choose the largest of these and two of the original points to construct the circle of the next iteration until no more points lie outside. Rather than shrinking a large circle, this method expands an undersize circle as little as possible. Since it is always at a bounding condition on a subset of the data points, it is possible to take large steps during iteration.

6. SURFACE CLASSIFICATION

Most quality control in manufacturing industry is based upon the rejection of workpieces when the average height of the roughness exceeds some specified value. This has some validity in that it is based on practical experience of what works and what does not work in normal use. The control is only valid though while the machining conditions remain as they were when that experience was gained. Switching, say, from a grinding to a turning process could give the same r.m.s. amplitude of roughness but it would give a very different profile shape and different functional behaviour. Clearly much better classification is needed if we are to get control of important surface properties while maintaining necessary flexibility in the manner of their production, in order to exploit new technology or, simply, spare factory capacity.

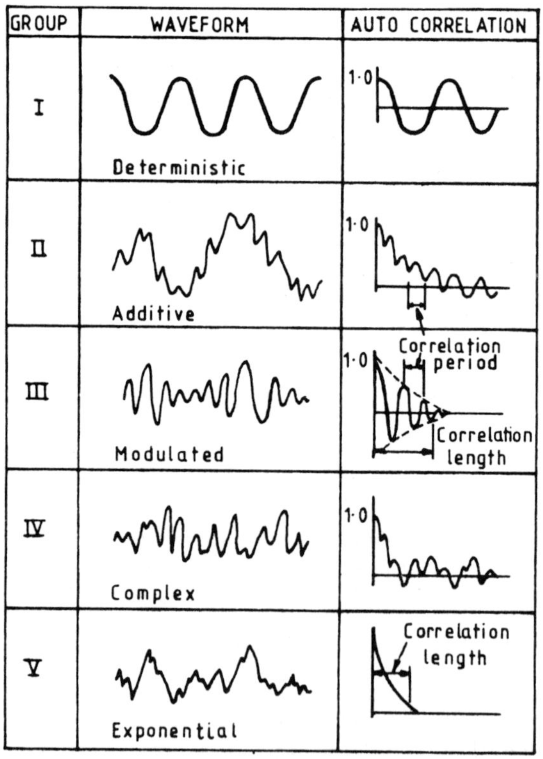

Figure 4. Surface typology due to Peklenik (20).

The earliest attempt at shape classification (surface typology) was the bearing area curve previously discussed. It recognises that the shape of the amplitude probability density function together with its variance allows some judgement of how the surface might behave. The next stage is to bring in wavelength and criteria for the degree of randomness. One way is to classify the autocorrelation function shape and its dimensions as shown in Figure 4[20]. Systems of this nature, while useful for specific problems, have never been widely adopted because they are largely pictorial. Industrial control requires the smallest possible group of numbers upon which go - nogo decisions may be taken. Given that both amplitude and wavelength matter, the simplest could have just two numbers and so the proposal to use average height and average wavelength[21] was introduced. This is more significant than might appear for in defining the parameters it was one of the first practical fruits of random process modelling of surfaces. While definitions readily implementable in instruments were actually used, the theoretical justification was the description of a surface by the first two even moments of the power spectral density, m_o and m_2.

Alone these two parameters do not adequately classify surfaces. They are not readily interpreted functionally, though this may be because our understanding of what is functionally relevant is poor. No models have done other than predict particular functional conditions. Typology has, for the moment, stepped back from its eventual aim and concentrated on the identification of manufacturing processes from the profile generated: a problem important in its own right as a stage of manufacturing functionally specified surfaces. Here also average height and wavelength alone do not discriminate adequately. It would be expected that a surface with some deep, narrow scratches, which hold oil, and relatively large plateau areas would act as a good bearing surface. Honing produces such surfaces and is widely used, for example, in internal combustion engines. Taking numerical descriptors of the shape of the amplitude probability density function is therefore a natural extension of the earlier methods. The obvious and most widely adopted way of doing this is to measure skewness and kurtosis: the normalised third and fourth central moments of the distribution:

$$\text{Skew} = \frac{E[(y-\bar{y})^3]}{\sigma^3}; \quad \text{Kur} = \frac{E[(y-\bar{y})^4]}{\sigma^4} \tag{15}$$

In visual terms, negative skew indicates the presence of more deep scratches than of high peaks. Most processes produce a slight to medium negative skew, though turning because of its cusp-like waveform can give positive values. Kurtosis increases with the "spikiness" of the surface, so honing scratches would give high values. These statistics do have some power in discriminating between processes. Plotting skewness against kurtosis does group the processes although rather weakly, Figure 5. An alternative approach along the same general lines is to model the amplitude probability density function by a suitable two parameter function and to use these parameters as the map variables. A good candidate is the Beta function (a close relative of the better known Gamma function):

$$\beta(a,b) = \int_0^1 y^{a-1} (1-y)^{b-1} dy \tag{16}$$

which according to the values of a and b can show "U-shaped" as well as unimodal distributions of varying degrees of asymmetry[22]. For this distribution treated as a probability density,

$$\bar{y} = \frac{a}{a+b} \quad \text{and} \quad \sigma^2 = \frac{ab}{(a+b+1)(a+b)^2} \tag{17}$$

Only the mean and variance of the profile heights need be known in order to find the β parameters. The system is thus more compact than using the higher central moments. It also better discriminates between processes, Figure 6.

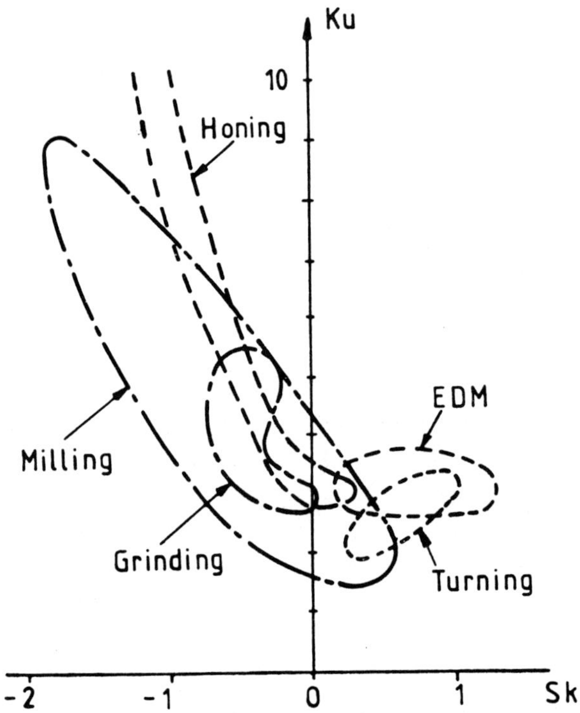

Figure 5. Process identification by central moments. Clustering of profiles of different materials and conditions.

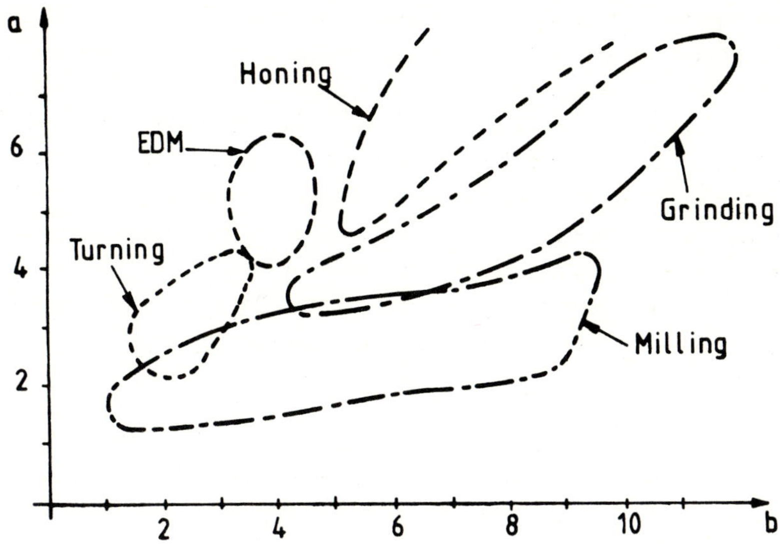

Figure 6. Process identification by beta parameters. Clustering of same data sets as in Figure 5.

7. SPECTRAL ANALYSIS ON SURFACES

The use of techniques based either on the power spectral density or the autocorrelation function has underpinned much of what has been discussed here. However, they are valuable investigative and diagnostic tools in their own right and some examples of this use will be given here. It will be seen elsewhere that estimating confidence limits for these functions is by no means trivial and that surprisingly large uncertainties can arise under what appear, superficially, to be quite good sampling conditions. In surface measurement we commonly have only one or, at best, a few data sequences, each relatively short and having a large bandwidth-duration product. It is necessary to take considerable care in interpeting results of analysis under these conditions. So, unless conditions can be very tightly controlled in specific cases, their use in routine quality control of manufactured parts is problematical. Their value lies in, for example, finding out why things do not behave as they were expected to rather than as go-nogo checks.

An alternative way of presenting the information of an autocorrelation function is to use the structure function:

$$S(\tau) = E(y(x) - y(x+\tau))^2 \tag{18}$$

or

$$S(\tau) = 2\sigma^2(1-\rho_y(\tau)) \tag{19}$$

Clearly, from equation 19, the structure function contains the same information as the autocovariance function, but it can have computational advantages under some conditions and it is insensitive to superimposed d.c. levels. It had been shown to be particularly useful in two areas[23]. It demonstrates easily the presence or absence of long wavelength non-stationarity such as that of equation 7. In studying anisotropy, by expressing statistics along profiles taken in different directions, structure functions are more easily compared than autocorrelation functions. Both these advantages occur basically because its value is 0 at $\tau = 0$. It has also been shown that some of the tribological parameters derived from stochastic models are more neatly expressed in terms of structure function coefficients[8].

Running-in

Correct running-in (a mild wear process during early life which should self-limit) is clearly important to the behaviour of components in normal operation. For instance the wear mechanism involved may influence the amount and type of debris produced. The autocorrelation function changes only slightly in form if the wear is mild. It is only when the surface structure is grossly affected that large changes will occur. Two mechanisms are identified during running in: the cutting of peaks and the redistribution of material by plastic flow of the peaks. The cutting action appears as a truncation of the profile and causes a reduction in correlation length for random surfaces. Plastic deformation tends to smooth the profile, reducing its bandwidth, and so giving increased correlation lengths. Thus not only will the autocorrelation function show whether the wear was mild but it will also show which mechanism is at work[24].

Turning

Spectral analysis can show diagnostic information about many machining operations by following deductive paths similar to that illustrated here for turning[25]. One of the costliest items in routine manufacture is tool replacement. It should not therefore be done too often but a worn-out tool should obviously not be used. It is not possible to take out the tool and examine it on a regular basis, but it is possible to look at the statistics of surfaces which it produces. Figure 7 illustrates the technique. A new, sharp tool should produce a clean cutting profile of relatively low harmonic content. If wear causes small irregularities on the tool these will be imposed on each wavelength and will increase the harmonic richness of the profile. With suitable rules, cutting speeds and tool change can be linked with a certain degree of harmonic degradation before any scrap parts are produced.

Surface condition consists of more than just surface finish. To function properly metallographic conditions must also be correct. Machining at an inappropriate speed or with a built-up edge of metal on the tool can radically alter these properties, for example, by introducing random tearing and microfracture which reduces surface fatigue life. In cases which are not overly severe this may not be

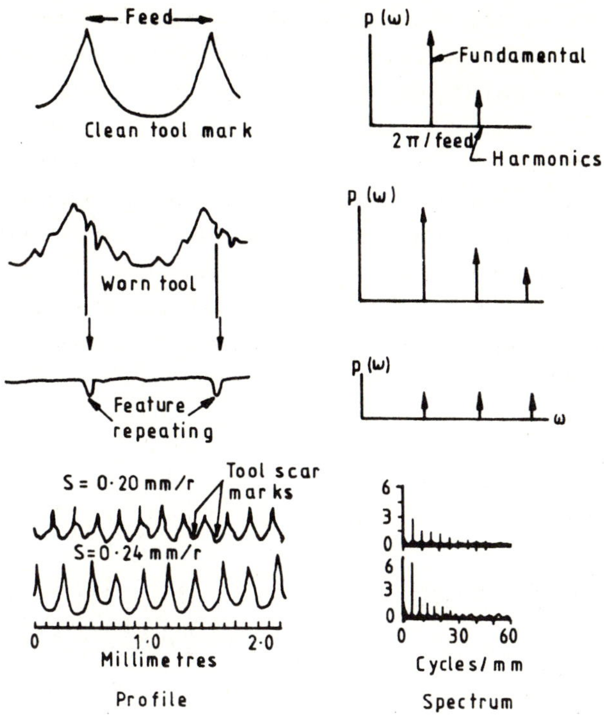

Figure 7. Detection of tool damage.

seen by direct examination of the profile, but increased randomness will show in a general raising of the level of the power spectral density across the frequency range and as a rapid initial decay in the autocorrelation function as shown in Figure 8.

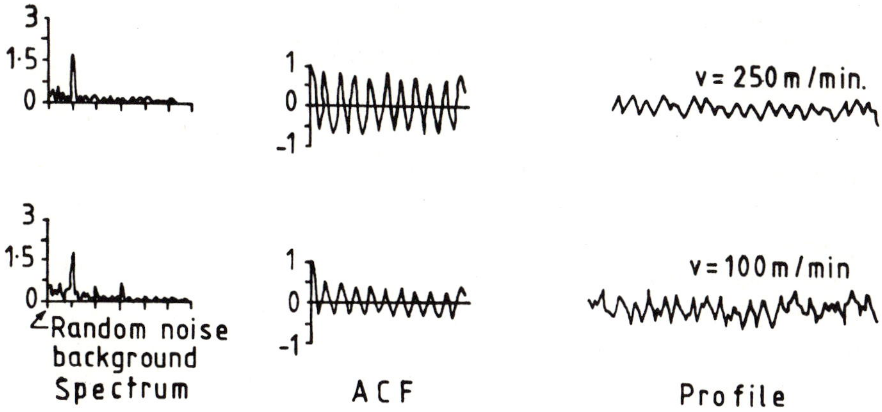

Figure 8. Detection of mild surface damage.

Sometimes it is the longer wavelengths which contain the most useful information as, for example, when the machine itself is malfunctioning. Even when not discernable by eye or ear it may be picked up by means of the autocorrelation function of the surface produced. It is feasible to predict incipient failure of expensive machine tools and to carry out preventive maintenance on them. One such example is shown in Figure 9. Although the peaks in the surface profile have variable heights it is difficult to identify a regular pattern. However, examination of the autocorrelation function reveals one immediately. This pattern resulted from a failure of one roller in the headstock bearing of an automatic lathe. Following the production of this profile, the machine continued to operate for a period of months before its performance deteriorated noticeably. By that time it was too late to save the bearing!

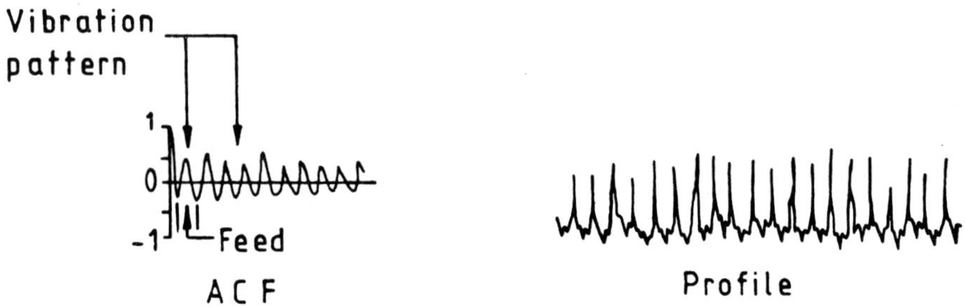

Figure 9. Detection of incipient machine failure.

8. POSTSCRIPT

We have sampled a wide range of cases where random process theory and estimation techniques impinge on the metrology and tribology of engineering surfaces. Sometimes they form the basis of models for fundamental understanding. Sometimes they are used as diagnostic tools. On other occasions its methods can be exploited in the development of efficient algorithms to solve problems already posed. On yet other occasions deeper understanding of the theory leads to the questioning of some current practices. Even so there is much more left unrecorded here than is given, for the whole field of precision engineering and manufacture is rich in examples of control and stochastic processes. With computer power per unit cost still reducing, it is clear that the growth in the use of these techniques will continue apace.

9. REFERENCES

1. T.R. Thomas, ed: "Rough Surfaces", Longman, London, 1982.
2. D.J. Whitehouse and M.J. Phillips: Phil. Trans. R. Soc. Lond. A305, 441-468, 1982.
3. F.P. Bowden and D. Tabor: "Friction and Lubrication of Solids", OUP, 1954.
4. J.F. Archard: Proc. Roy. Soc. Lond. A243, 190, 1957.
5. B.B. Mandelbrot: "Fractals", W.H. Freeman & Co., San Francisco, 1977.
6. J.A. Greenwood and J.B.P. Williamson; Proc. Roy. Soc. Lond. A295, 300, 1966.
7. D.J. Whitehouse and J.F. Archard: Proc. Roy. Soc. Lond, A316, 97-121, 1970.

8. D.J. Whitehouse and M.J. Phillips: Phil. Trans. R. Soc. Lond. A290, 267-298, 1978.
9. M.S. Longuet-Higgins: Proc. Roy. Soc. Lond, A249, 321, 1957.
10. P.R. Nayak: Trans. ASME, J. Lub. Tech., July 1971.
11. J.B.P. Williamson, J. Pullen and R.T. Hunt in FF Ling (ed.) "Surface Mechanics", ASME, New York, 1969.
12. R.S. Sayles and T.R. Thomas: Trans. ASME. J. Lub. Tech. 101F, 409-418, 1979.
13. R.S. Sayles and T.R. Thomas: Nature, 271, 431-4, 1978.
14. F.J. Abbott and F.A. Firestone: Mech. Engng. 55, 569, 1933.
15. D.G. Chetwynd: NELEX, East Kilbride, 1976.
16. D.J. Whitehouse: Proc. Inst. Mech. Engrs. 182 pt. 3k. 306-318, 1968.
17. M.R. Osborne and G.A. Watson: Comput. J., 10, 172-177, 1968.
18. D.G. Chetwynd and P.H. Phillipson: J. Phys. E: Sci. Instrum. 13, 530-538, 1980.
19. D.G. Chetwynd: Proc. Inst. Mech. Engrs. 199 No.82, 93-100, 1985.
20. J. Peklenik: Annals C.I.R.P., 12, 173, 1963.
21. R.C. Spragg and D.J. Whitehouse: Proc. Inst. Mech. Engrs. 185, 47, 1971.
22. D.J. Whitehouse: Annals C.I.R.P., 1978.
23. R.S. Sayles and T.R. Thomas: Wear, 42, 263-76, 1977.
24. D.J. Whitehouse: Int. Conf: Fundamentals of Tribology, M.I.T., 1978.
25. D.J. Whitehouse: Proc. Inst. Mech. Engrs. 192, 1978.

PRACTICAL PROBLEMS IN IDENTIFICATION

Dr. K.R. Godfrey

INTRODUCTION

System identification can be split into four main parts (Box and Jenkins, 1976; Mehra, 1976), which are closely interwoven in any particular application. In the first part, the structure of the proposed model of the system is determined using physical laws and prior knowledge of the system, while in the second, experiments are designed to enable the parameters of the model to be estimated. In the third part, input-output records are processed to estimate the model parameters and finally, the model is validated both by calculating from it any quantities for which values are known independently and by studying its performance on other sets of records. This Case Study lecture will be concerned with practical problems, mainly in the third part of the above procedure, in examples in industrial and biomedical applications with which the author has been associated. The examples emphasise a wide range of different practical problems which need to be taken into account by an experimenter.

In some applications, static (deterministic) models are adequate. An example is the plotting of performance maps for diesel and internal combustion engines which give static relationships, expressed in the form of joint probability density functions, between variables such as mean output torque, mean piston speed and brake specific fuel consumption. From these, the mean fuel consumption rate can be calculated, in turn allowing gearbox design and control strategies to be determined for minimum fuel consumption.

Often, though, the dynamics of a system must be taken into account and that is the situation we will be considering here. In industry, design of effective analogue three-term control requires a reasonable idea of the dynamics of the system to be controlled, a remark which also applies to any form (three-term equivalent or otherwise) of digital feedback control algorithm. Feedforward control is widely used in the chemical and petroleum industries (certainly much more widely than many lecture courses in control engineering would have us believe!) and reasonable dynamic models must be available of both the input-output relationships in question for a feedforward algorithm to get near to cancelling one effect by controlling the other. Dynamics are also particularly important in start-up and shut-down, when sequencing controls may well have to take account of time delays in long pipelines and thermal inertias in heat processes.

Identification and parameter estimation is a large area of study. IFAC (the International Federation of Automatic Control) runs a major international conference on the topic every three years. At the seventh in the series, in York in 1985, there were fifty sessions, with some three hundred papers presented. At the conference in 1973, an Industry-University confrontation took place (reported by Godfrey and Goodwin, 1974) at which the gap between academic research and industry practice was highlighted. Since then, this gap has narrowed somewhat, with new-generation control engineers being well versed in modern control techniques and anxious to apply their knowledge, with low-cost microprocessor technology enabling data to be collected more easily and with the advent of effective time-sharing systems allowing much quicker interaction between engineer and computer. The result is that the use of statistical techniques, such as correlation analysis, spectral analysis and maximum-likelihood methods, is now more routine. As we will see in this lecture, though, effective industrial control can sometimes be designed on the basis of simple step responses, while in many experiments on human beings, it is feasible to perform only impulse response or step response tests.

An important part of the system identification "jargon" is the distinction between non-parametric and parametric models. Examples of non-parametric models are impulse responses, step responses and frequency responses, the latter including several forms such as gain and phase (separately) versus frequencies and Imaginary part versus Real part of the gain as frequency is varied. Such models are of high order so making them difficult to use directly for control system design. Parametric models are of smaller order and, while involving more computation, are simpler to use. Examples are the fitting of curves (often in the form of sums of exponentials) to impulse or step responses and of transfer functions to the frequency response data. The fitting of z-transfer functions to discrete input-output records has been (and remains) an active area of research and the reader is referred to Eykhoff (1974) for an early review and to Norton (1986) for a more recent viewpoint.

EXAMPLE 1 : CONTROL SYSTEM DESIGN USING STEP RESPONSES

When measured responses are relatively noise-free, a step response may be all that is necessary for identification. An example illustrating this concerned a steam reforming plant for producing town gas from naphtha and steam (Godfrey and Shackcloth, 1970). A steam reformer is a chemical reactor in which the reaction is endothermic, requiring heat to be supplied to sustain the reaction. Although town gas is no longer produced in the U.K. by this method, steam reforming is still widely used in other industries, for example in the production of ammonia synthesis gas.

The main requirement in the application being considered here was to implement feedforward control to minimise changes in product quality in the face of throughput changes necessitated by variations in demand. As in many chemical processes, reformer (reactor) outlet temperature was taken as a good guide to product quality. A throughput increase resulted in a decrease of outlet temperature, and to compensate for this, it was necessary to increase the fuel flow to the reformer burners via a feedforward algorithm. As noted in the introduction, effective feedforward control requires a reasonable model of the relevant dynamic responses.

The open-loop responses of the outlet temperature to step changes of burner fuel flow and throughput are shown in Figures 1A and 1B respectively. The step responses of models with single exponential plus time delay are shown by filled-in circles on these Figures and it may be seen that such models provide very good agreement with the measured responses. The optimum values of gain and time constant in the parametri model were estimated by a least-squares procedure as time delay (dead time) was incremented by intervals of 30 seconds. (It is recommended that dead time is taken ou of parameter estimation procedures and that the estimation algorithm is run for several different values of dead time selected by the user on the basis of visual inspection of the responses).

The models were used in the design of the feedforward algorithm (with burner fuel flow being adjusted to compensate for required throughput changes) and the performance as throughput was changed in a ramp from 75% to 43% of full throughput is shown in Figure 2. The rate of decrease was set at the maximum recommended by the plant manufacturer (Haldor Topsøe Ltd.) and it may be seen that during the change, the reactant outlet temperature was kept to within ± 3 deg. C of its set point.

Feedforward is essentially an open-loop procedure, which makes its use without accompanying feedback control inadvisable in most industrial situations. The feedback performs the dual role of compensating for factors other than those being manipulated in the feedforward strategy and of compensating for differences between the assumed and actual plant dynamics. In this example, the ramp is applied to the burner fuel flow, because this input results in the slower dynamic response of the outlet temperature; steam flow is then adjusted by the feedforward algorithm. The burner fuel flow is also adjusted via a feedback control algorithm from outlet temperature, which is why the decrease in fuel flow shown in Figure 2 is not a smooth one. During typical manual throughput changes, with the feedback control still in operation but without the automatic feedforward strategy, the reactant outlet temperature varied by up to ± 20 deg. C from the set point.

In this example, the reforming plant was designed to operate at between 35% and 100% of full throughput, depending on demand. The continued use of feedback from outlet temperature to burner fuel flow during the throughput changes meant that it was possible to achieve highly effective control of outlet temperature over a wide range of throughputs using a simple fixed parameter feedforward algorithm designed on

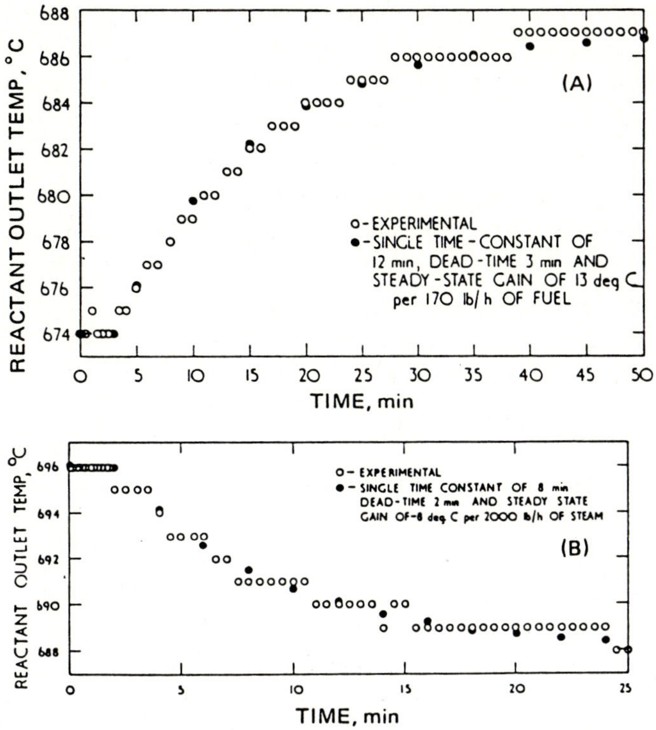

Fig. 1. Open-loop responses of reformer outlet temperature. (A) – to a step increase in burner fuel flow. (B) – to a step increase in steam flow (naphtha flow in ratio)

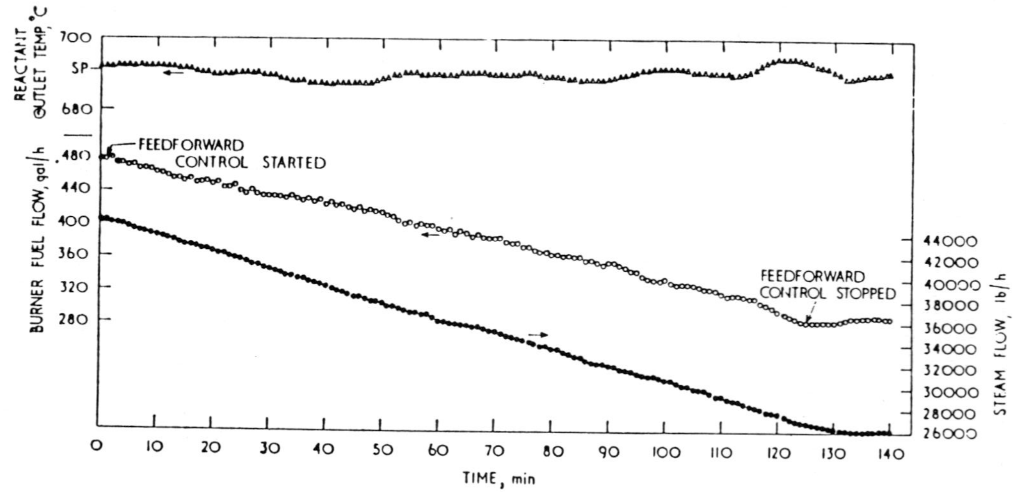

Fig. 2. Demonstration of the feedforward/feedback controller with the throughput decreased from 75% to 43%.

the basis of step responses obtained at the lower end of the operating range.

It was particularly satisfactory to achieve good control using such simple procedures in this example. What is remarkable about Figures 1A and 1B compared with many industrial processes is the noise-free nature of the responses. In the construction of the plant, instrument cables were well shielded and were kept as far apart as possible from power cables. It is clear that the relatively small extra cost which this involved in the plant construction stage paid handsome dividends in the simplicity and effectiveness of the control system design once the plant was operational.

EXAMPLE 2 : A NOISY RESPONSE RECORD - THE USE OF STATISTICAL METHODS TO OBTAIN PROCESS MODELS

The first example has illustrated that a good idea of dynamics can sometimes be obtained using simple step response measurements, but in other instances, process noise or measuring device noise may be so great that little idea of the dynamics can be gained from a single step response. Figures 3 and 4 illustrate a situation where the special features of pseudo-random binary signals have been used to advantage.

The application in question was to one of the distillation columns (the distillate splitter) in the distillation chain of an oil refinery (the British Petroleum Refinery in Belfast). The experiments, described in detail by Godfrey (1969), were conducted not with control system design in mind (indeed, most of the control loops were left closed while the experiments were being carried out) but rather to gain experience of the use of pseudo-random binary signals in an industrial situation.

In the experiments which will be considered in this lecture, the reboiler temperature (T 104) was varied and measurements were made of three column temperatures - at the drawoff to a sidestream benzine stripper (T 1019), at the return from the stripper (T 1052) and the column top temperature (T 1021). The maximum amplitude of perturbation permitted by the plant manager was 5.6 deg. C around a steady level of 204.4 deg. C. Responses to upward and downward steps of 5.6 deg. C are shown in Figure 3, from which it may be seen that the responses of T 1019 and T 1052 are relatively quiet, so that reasonable estimates of the dynamics could be obtained from step responses alone, but the responses of T 1021 are very noisy so that it would be extremely difficult to obtain a meaningful estimate of process dynamics.

A 127-digit pseudo-random binary signal, with clock pulse interval 18.7s and amplitude ± 2.8 deg. C about the steady level was also used to perturb reboiler temperature and recordings of responses during several periods were obtained. While the experiment was in progress, a number of substantial control changes were made, but it was possible to obtain one signal period (39.6 min.) of data relatively unaffected by control changes. The crosscorrelation functions between reboiler temperature and the three column temperatures for this period of data are shown in

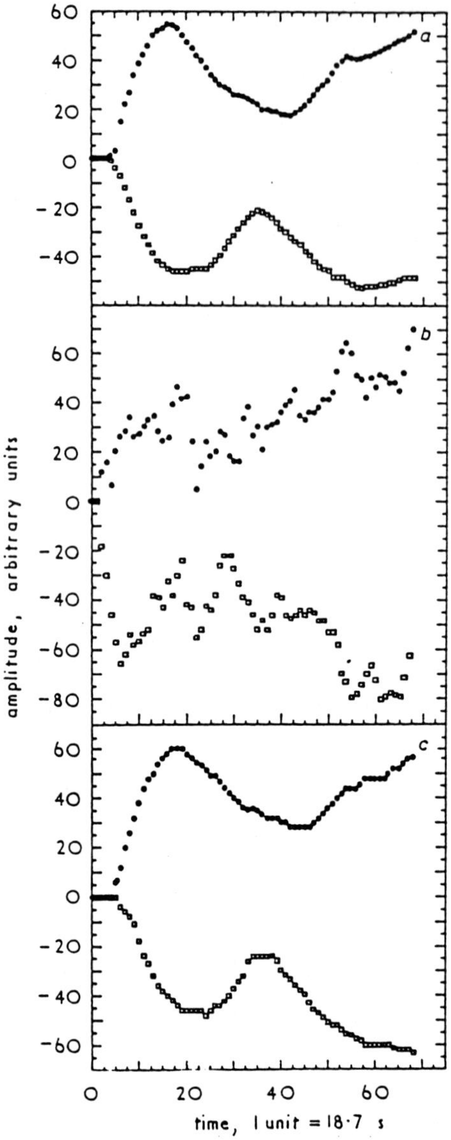

Fig. 3. Responses of column temperatures to upward and downward step changes of 10 deg F (5.6 deg C) in reboiler temperature set point T104.
(a) T1019; (b) T1021; (c) T1052

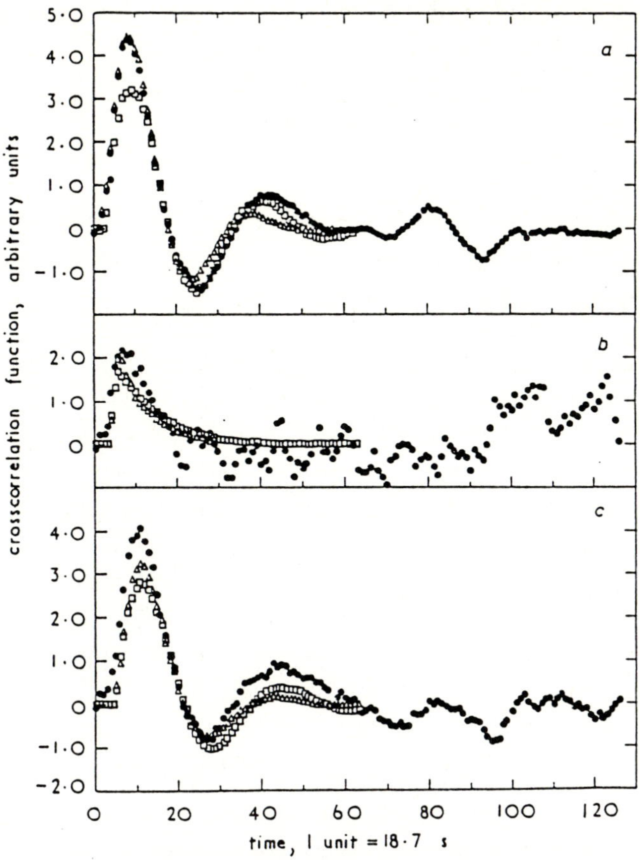

Fig. 4. Crosscorrelation functions measured when perturbing T104 with a 127-digit prbs with basic interval 18.7s.
(a) Between T104 and T1019;
(b) Between T104 and T1021;
(c) Between T104 and T1052.
● Crosscorrelation function
O, Δ Pulse responses of z-transfer functions of orders 2 and 4, respectively.

Figure 4. The pulse responses of input-output z-transfer functions of models with orders n = 2 and n = 4, of the form

$$H(z^{-1}) = \frac{z^{-k}(b_1z^{-1} + b_2z^{-2} + \ldots + b_nz^{-n})}{1 + a_1z^{-1} + a_2z^{-2} + \ldots + a_nz^{-n}} \tag{1}$$

are also shown. For any specified n, the best fit of k (corresponding to the initial time delay), and the parameters $b_1, b_2, \ldots, b_n$, $a_1, a_2, \ldots, a_n$ were computed using a Generalised Least Squares procedure (Clarke, 1966).

The main point to note from Figure 4 is that a much better idea is obtained of the dynamics between T 104 and T 1021 from crosscorrelation than from the step responses. It should also be noted that for all three responses, z-transfer functions of order 2 provide fairly good agreement (in terms of pulse response) to the data. Models of order 4 result in better agreement but this does not necessarily mean that the second-order models would not be adequate for control purposes.

The improvement between step response and crosscorrelation function is in fact more marked than direct comparison between Figures 3 and 4 might indicate, because the crosscorrelation function is proportional to the impulse response which is the derivative of the step response. It is clear that any differencing procedure on the T 1021 step response in Figure 3b would give a very noisy result.

EXAMPLE 3 : THE EFFECT OF AN UNEXPECTED DEPARTURE FROM LINEARITY

Problems arise in interpreting the crosscorrelation between an input perturbation (of impulse-like autocorrelation) and the system response as being proportional to the impulse response if the system departs from the assumed linearity. The usual method of minimising these problems is to keep the amplitudes of the perturbation signals small. Even then, problems can arise if the dynamics in the two directions are different (i.e. the responses differ according to whether the input variable is being increased or decreased). This is a frequently-occurring departure from linearity in industrial situations. The system is said to be bilinear and in a state-space model, there is a multiplication of a state and an input. For a system represented by a single state, for example, the model would be of the form

$$\frac{dx(t)}{dt} = [a + b_1u(t)].x(t) + b_2u(t). \tag{2}$$

If a pseudo-random binary signal based on a maximum-length sequence (m-sequence), as is the case with all commercially available prbs generators, is used to perturb the system, the crosscorrelation function is distorted somewhat for small values of delay and also contains the original dynamics appearing again with diminished amplitude at various values of delay depending on the particular m-sequence used (Godfrey and Briggs, 1972). Crosscorrelation functions for two simulated systems are shown in

Figure 5.

The additional peaks can easily be attributed at first sight to reflections or "echoes" of the dynamics around the original process, particularly when that process is a complex one. This spurious effect has been observed in using prbs to determine the dynamics of gas turbine engines (Moore, 1971; Godfrey and Moore, 1974). It was found that when perturbing fuel flow input, little sense could be made from step responses of the maximum permitted amplitude and even to obtain reasonably noise-free crosscorrelation functions, it was necessary to process data from several periods of a prbs.

The prbs used was based on a 511-digit m-sequence with feedback shift register connections from stages 5 and 9. The crosscorrelation functions between fuel flow input and compressor speed response and temperature response of a Pegasus engine are shown in Figures 6A and 6B respectively. An additional peak is clearly visible on both of these correlation functions and with a process as complicated as a gas turbine engine, it would be all too easy to assume that this is part of the process dynamics, with the main peak reappearing, inverted and attenuated, after a time delay. Further investigation using the theory of Godfrey and Briggs (1972) showed that the extra peak was in exactly the position one would expect with a bilinearity of the type discussed above. Another crosscorrelation function, over a more nonlinear part of the operating range is shown in Figure 7. All the additional peaks are in positions expected on the assumption of direction-dependent dynamics.

The main source of the departure from linearity was thought to be the fuel flow actuator rather than the main engine dynamics, which it was felt should be approximately linear given the small amplitude of the perturbation signals applied. Although it was not possible to conduct tests on the isolated fuel flow actuator of the Pegasus engine, it was possible to test a similar actuator from an Olympus 593 engine (the type used on Concorde). The result (given in Godfrey and Moore (1974)) confirms the view that the actuator was primarily responsible for the bilinear behaviour.

As noted above, the main peak of the crosscorrelation function is also affected to some extent by the bilinearity and the same is true of the estimated parameters of a discrete-time transfer function of the form of equation (1). For control system design for a process of this type, it would be sensible to design for the slower dynamics, accepting a downgraded performance in the opposite direction, but this solution assumes that it is possible to make sense of the dynamics estimated from step responses in the two directions, which was not possible for the system in this example.

A better estimate of the linear dynamics (the first-order kernel in a Volterra series expansion - see Section 3.1 of Lecture L10) can be obtained by using a signal based on an inverse repeat sequence generated by inverting every other digit of a maximum-length sequence (Briggs and Godfrey, 1966). The second half of these signals

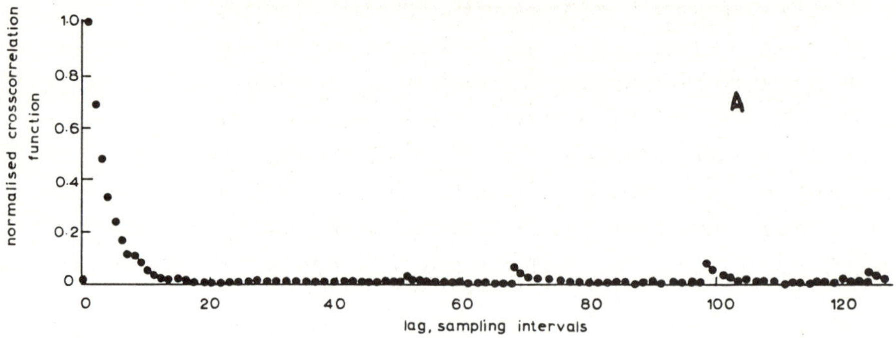

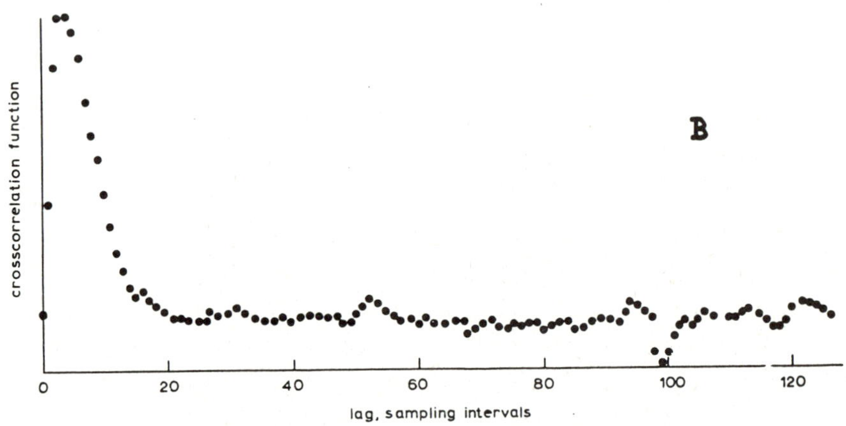

Fig. 5. Input-output crosscorrelation functions for simulated systems with direction-dependent dynamic responses, both with a perturbation signal based on a 127-digit m-sequence with feedback from stages 4 and 7.

(A) System with first-order dynamics, with time constant 2λ in the -1 to +1 direction and 4λ in the +1 to -1 direction (λ = basic interval of the prbs);

(B) System with second-order dynamics, with two time constants each of 4λ in the -1 to +1 direction and two time constants each of 2λ in the +1 to -1 direction.

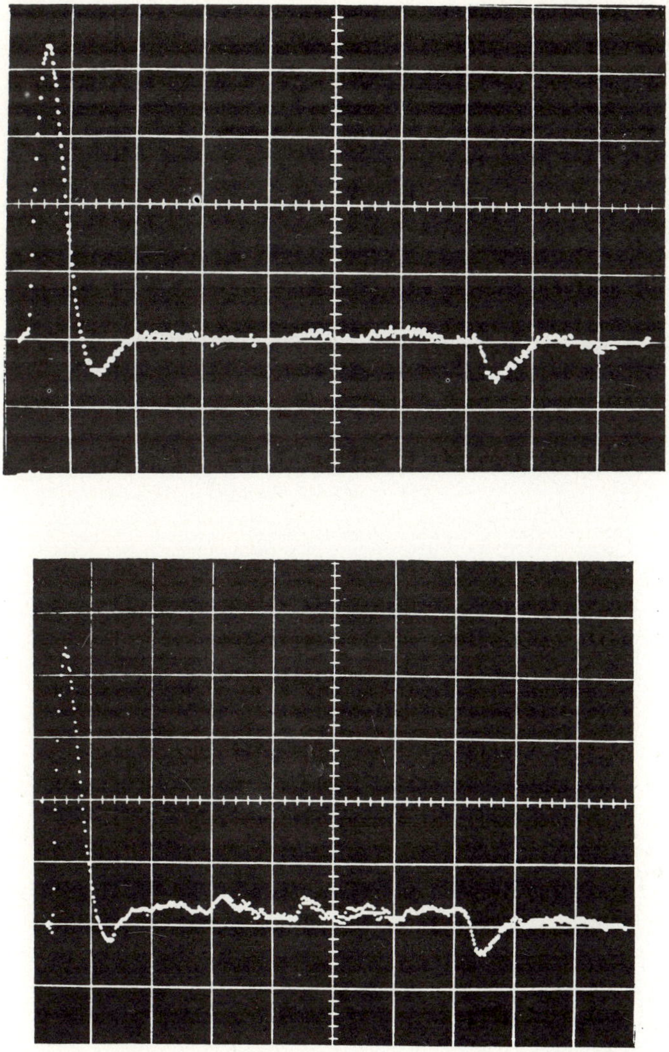

A

B

Fig. 6. Crosscorrelation functions on the Pegasus Vertical Take Off Engine. Perturbation signal based on a 511-digit m-sequence with feedback connections from stages 5 and 9.

A - Fuel flow to compressor speed.

B - Fuel flow to temperature.

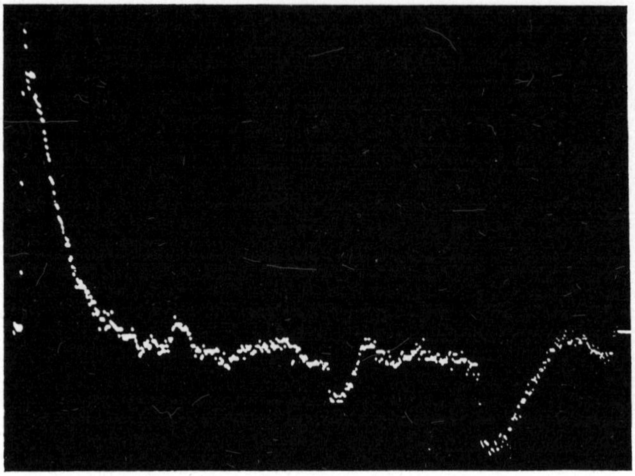

Fig. 7. Crosscorrelation function as for Figure 6A, but at a different, and more nonlinear, point in the operating range.

is the negative of the first half, so that the contribution of even-order kernels to the input-output crosscorrelation function becomes zero.

EXAMPLE 4 : USE OF NORMAL OPERATING RECORDS - THE PROBLEM OF "HIDDEN" FEEDBACK

Data logging of normal operating records can give useful information about the static and dynamic behaviour of a system especially when step disturbances occur. When trying to estimate dynamics, it is very tempting to base models on the analysis of such records, but problems can arise if there are feedback paths, which may mask the cause-effect relationships, or if the normal operating inputs are not persistently exciting, that is if they are zero (or constant) for lengthy periods of time or if their frequency components do not adequately span the passband of the system.

The first of these problems is illustrated here by some results from a blast furnance at the Port Talbot Steelworks of the British Steel Corporation. Normal operating records for the particular furnace concerned were available over several weeks of furnace operation, and from this, a section consisting of 49 casts (just under a week of operation) was selected for analysis on the grounds that it was free from breaks in the records and that casting was done at regular intervals of 3 hours during that period.

Both of the inputs to be examined - ore/coke ratio and blast temperature - appeared from the records to be suitable for identification in that they both varied at frequent intervals while computation revealed that their autocorrelation functions were reasonable approximations to impulse functions and that the crosscorrelation between them was approximately zero. Thus, the crosscorrelation functions between each input in turn and the output variable of interest, percentage silicon in the cast iron, should give values proportional to impulse responses. Percentage silicon was selected as the output variable because it is a good guide to the quality of finished steel from the Works.

It was doubly tempting to use the normal operating records in this instance because of the inherent slowness of response of the furnace, which would have made experiments using externally-applied perturbations very lengthy.

The crosscorrelation functions (or to be strictly accurate, crosscovariance functions, because means of all records were removed) between ore/coke ratio and percentage silicon and between blast temperature and percentage silicon are shown in Figures 8 and 9 respectively. The confidence limits shown were calculated using the methods detailed in Lecture L4 at this Vacation School. Values outside these limits are assumed to denote a statistically significant input-output relationship.

The result for the ore/coke ratio input is very satisfacotry, and a reasonable idea of the dynamic relationship between the ratio and percentage silicon could be obtained from Figure 8. The blast temperature, however, was used as a control variable, with manual adjustment of the temperature according to the value of silicon

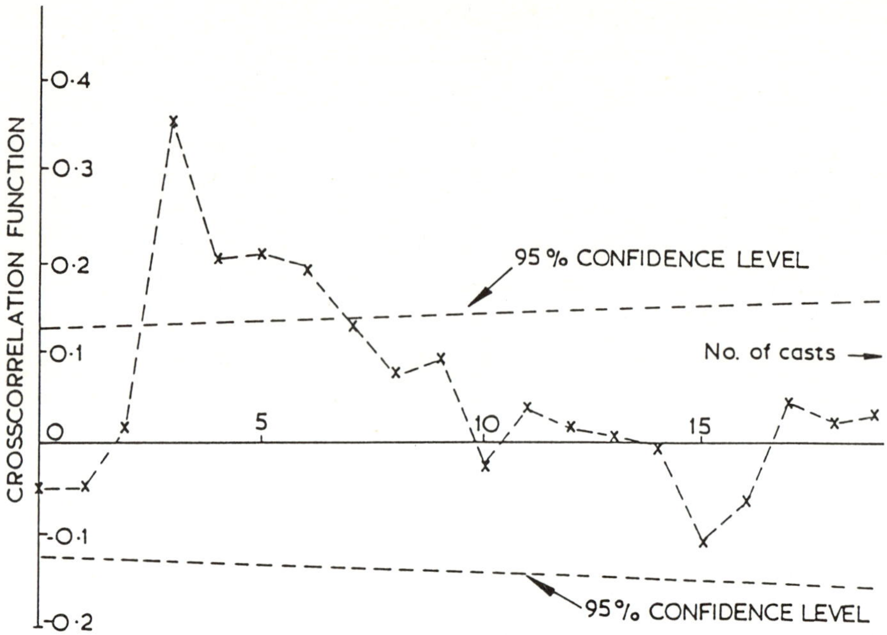

Fig. 8. Crosscorrelation function between ore/coke ratio and percentage silicon in the cast iron for a steelworks blast furnace.

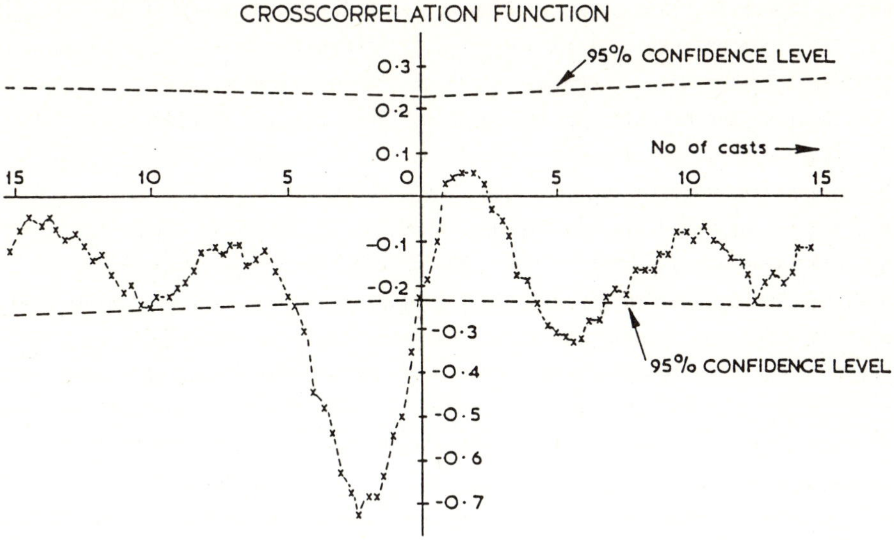

Fig. 9. Crosscorrelation function between blast temperature and percentage silicon in the cast iron for a steelworks blast furnace.

content of the cast. This shows in Figure 9 as a large crosscorrelation for negative time shift (remember that $R_{xy}(-\tau) = R_{yx}(\tau)$). The manual control reduced the blast temperature when the silicon content was high and increased it when the silicon content was low, so the correlation is negative. This swamped the correlation for positive values of τ, which would normally be the estimator of impulse response from blast temperature to percentage silicon. No similar departure from zero crosscorrelation for negative time shifts was found for the ore/coke ratio input, which is not surprising because that is not normally used as a control variable.

Although it is possible to identify the open-loop dynamics of a system from closed-loop experiments under certain conditions, the procedure would have been difficult to apply in this instance because of the variability of the manual feedback. Even for systems with feedback applied automatically, it is most inadvisable to try to estimate the dynamics of a system from closed-loop experiments using normal operating records (Wellstead, 1977). For example, if a non-parametric representation of the dynamics is required in the form of a frequency response function, it is quite possible to estimate the inverse of the feedback transfer function rather than the transfer function of the forward path dynamics. Problems of this nature can be overcome by use of a suitable externally-applied reference signal.

EXAMPLE 5 : PROBLEMS WHICH CAN ARISE IN PARAMETER ESTIMATION DUE TO POOR SPACING OF SAMPLES

In the final two examples in this lecture, we depart from industrial applications and consider instead experiments in pharmacokinetics, i.e. the time course of drugs in the human body. Experiments on human beings are usually extremely limited in both input and output ports and in most cases in pharmacokinetics, administration of a drug is via either an oral or intravenous route, with response measurements restricted to concentration of drug in samples of blood and quantity of drug eliminated via the urine. Input perturbations most often consist of rapidly-administered doses (which may be considered as impulsive inputs) but in the case of intravenous inputs, constant continuous infusions (which may be considered as step functions) are also possible.

The pharmacokinetics generally have to be estimated from sampled records of concentration of drug in systemic blood. The number of samples is usually quite limited, partly because extraction of a large number may make the system depart from assumed stationarily and partly to avoid too much discomfort to the subject participating in the drug trial.

In the first of the examples, we will consider a typical experiment in which a drug is administered orally and blood samples are taken at various intervals after the administration. Absorption of drug into the blood is assumed to be at a first-order rate, with an initial time delay τ and it is assumed that the body behaves, as far as the drug is concerned, as a single well-stirred tank with elimination also at a first-order rate. The quantity, $x(t)$, of drug in the blood is then described by the equations

$$x(t) = 0, \qquad t < \tau \tag{3A}$$

$$\frac{dx(t)}{dt} = -k_e\, x(t) + F.D.\, k_a \exp\,[-k_a(t-\tau)], \quad t \geq \tau \tag{3B}$$

where k_e is the rate constant describing elimination of drug from the blood;

k_a is the rate constant describing absorption of drug from the intestine into the systemic blood;

D is the quantity of drug which is (rapidly) administered (so that the dose rate may be considered to be of form $D.\delta(t)$)

and F, the bioavailability fraction, is the fraction of drug administered that actually reaches the systemic blood. (A fraction 1-F is not absorbed at all, being either eliminated directly via the faeces or absorbed but then transformed into another substance in the liver before reaching the systemic blood).

Observation $y(t)$ is of concentration of drug in the blood and is given by $y(t) = x(t)/V$, where V is the apparent volume of distribution of drug in the blood. The Laplace transform of the observation for $t \geq \tau$ is thus of the form

$$Y(s) = \frac{F.D.k_a.e^{-s\tau}/V}{(s+k_e)(s+k_a)} \quad . \tag{4}$$

For the above (theoretical) model, the corresponding experimental response is of the form

$$y(t) = 0, \qquad t < \tau \tag{5A}$$

$$= A\{\exp[-\alpha(t-\tau)] - \exp[-\beta(t-\tau)]\}, \quad t \geq \tau \tag{5B}$$

where A, α and β are constants which can be estimated from the response data. This has a Laplace transform for $t \geq \tau$ given by

$$Y(s) = \frac{A(\beta-\alpha)e^{-s\tau}}{(s+\alpha)(s+\beta)} \tag{6}$$

which is then compared with equation (4) to give k_e and k_a, quantities often required by Government organisations which authorise the marketing of drugs.

In fact, a structural identifiability problem occurs here, because the fraction F is not usually known so that k_a cannot be obtained by comparing numerators of equations (4) and (6). Comparison of the denominators gives two solutions:

EITHER $k_e = \alpha$, $k_a = \beta$

OR $k_e = \beta$, $k_a = \alpha$.

Structural identifiability problems are nothing to do with *quality* of data, and such identifiability analyses address the question of whether the parameters of any model other than one purely describing data could be estimated uniquely, locally (as in this example, which has two solutions) or not at all, given *perfect* data. It is a many-faceted problem and has been the subject of a good deal of recent research, but it is not the problem being considered in this lecture. For a tutorial on identifiability, the reader is referred to Godfrey and DiStefano (1985).

The local identifiability problem for the particular drug in this example was resolved by prior knowledge that absorption was at a faster rate than elimination. A typical response curve, with the maximum response point normalised to 1, is shown in Figure 10 and this data set was used to obtain estimates of k_e, k_a and the amplitude constant A using a nonlinear least-squares procedure from the NAG (Numerical Algorithms Group) Library. As in Example 1, the initial dead time τ was not estimated by the program, but instead, the program was run for several different values of τ (over a range selected by eye from the data) incremented by 0.1 min. A global minimum for the sum of squared deviations (SSQ) was then sought from the different runs.

The results for four successive values of τ near the global minimum are shown in Table 1, from which it can be seen that, although SSQ scarcely changes at all,

TABLE 1. Optimum curve fits for equations (5) to the data of Figure 10, for four successive values of τ

τ	A	$\alpha(=k_e)$	$\beta(=k_a)$	Sum of squared error, SSQ
1.6	1.19	0.13	3.82	0.0381
1.7	1.17	0.13	5.02	0.0379
1.8	1.15	0.13	7.51	0.0379
1.9	1.14	0.13	14.89	0.0379

there is an enormous variation in k_a as τ is incremented; A varies only a little and k_e not at all (to 2 decimal places). Inspection of Figure 10 shows that the reason for this is that there are too few points on the rise part of the response curve to describe adequately the early part of the curve (often referred to in pharmacokinetics as the "absorption phase").

Optimal sampling strategies have been devised in the context of pharmacokinetics by D'Argenio (1981), for biomedical systems by Mori and DiStefano (1979) and in a

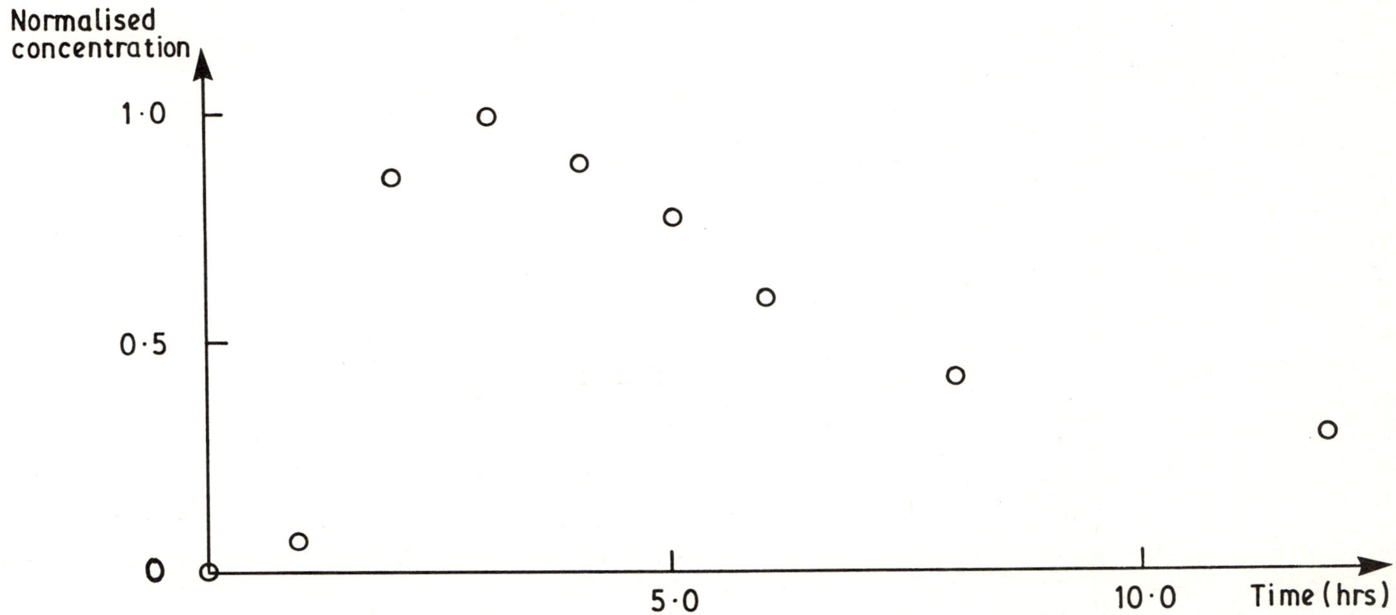

Fig. 10. Normalised concentration of drug in blood in a typical pharmacokinetics trial following an oral dose administered at t=0.

wider context by Goodwin and Payne (1976). Devising these strategies is interesting and challenging theoretically, but their practical use in pharmacokinetics is somewhat limited by the wide inter-subject variability (and indeed quite wide intra-subject variability) experienced in many pharmacokinetics trials. Even so, it is clear that more samples are needed in the early part of response curves (at the expense, if necessary, of a few points later in the curves) if one is to place any credence on reported parameters describing absorption.

EXAMPLE 6 : IDENTIFICATION OF A NONLINEAR CHARACTERISTIC : THE NEED TO CHECK THAT DATA SPANS THE NONLINEARITY

Drug doses are administered in therapeutic quantities, rather than tracer quantities, so that nonlinear effects are sometimes important. For certain drugs (and indeed, alcohol) elimination from the body is capacity-limited and it is vital to recognise this because doubling a dose more than doubles the subsequent concentration of drug in the blood.

Conventionally in pharmacokinetics, such a nonlinearity is modelled by a Michaelis-Menten saturation curve, of the form

$$\text{Elimination rate} = \frac{V_m}{K_m + x(t)} . x(t) \tag{7}$$

where V_m and K_m are constants. This is shown as the full line in Figure 11. Assuming, as in Example 5, that the body acts, as far as the drug is concerned, as a single well-stirred tank, the equation for quantity, $x(t)$, of drug in the systemic blood is given by

$$\frac{dx(t)}{dt} = - \frac{V_m}{K_m + x(t)} x(t) + bu(t) \tag{8}$$

where $u(t)$ is the input and b is the fraction of the administered dose which reaches the systemic blood.

Many authors have proposed experiments in which V_m and K_m are estimated from a single dose-response curve (see Godfrey and Fitch (1984) for a representative list of references). In the experiments considered, either the complete response to a rapidly-administered intravenous dose is measured or that part of the response to an orally-administered dose when the "absorption phase" is assumed to be finished is examined. For either experiment (but with a suitable shift in time scale for the latter), equation (8) can be written as

$$\frac{dx(t)}{dt} = - \frac{V_m}{K_m + x(t)} x(t), \quad t > 0 \tag{9A}$$

$$x(0) = x_0 \tag{9B}$$

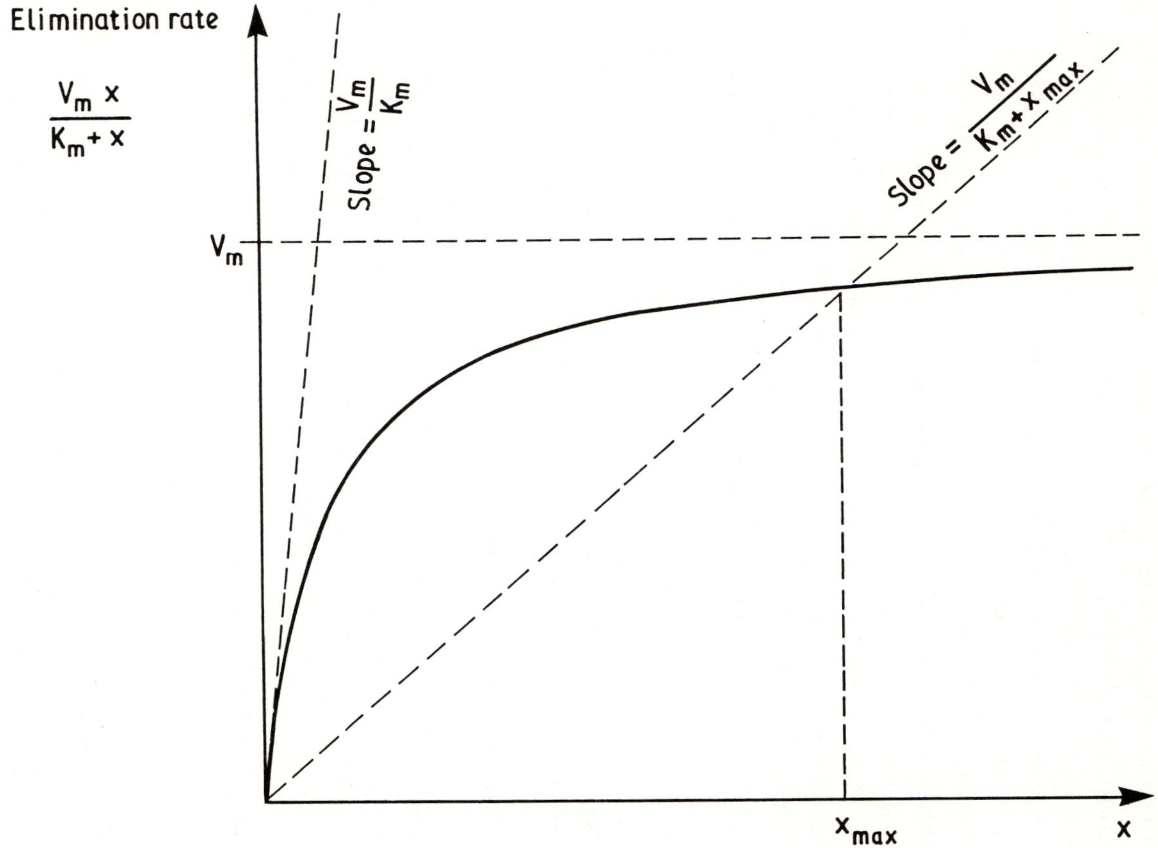

Fig. 11. Michaelis-Menten saturation curve showing bounds corresponding to x=0 and $x = x_{max}$.

where x_0 is known or can be measured. The structural identifiability result for this experiment cannot be obtained by Laplace transformation (since it is not possible to Laplace transform the nonlinear term) and instead has to be obtained by Taylor series expansion of the observation around $t = 0^+$. It is then readily shown that V_m and K_m are uniquely identifiable (Godfrey, 1983, pp. 204-205) from this type of experiment.

The importance of this type of nonlinearity in pharmacokinetics is illustrated in Figure 12, which shows the build up of quantity x(t), in a model described by equation (8) with the particular values $V_m = 15$, $K_m = 12$, with doses (of sizes shown on the curves) administered at regular time intervals of 1 time unit. Note how the quantity at any particular value of t increases considerably more than the increase of dose. (The quantities shown are those just prior to the administration of a dose). It is clear that it would be all too easy to overdose using a dosage regimen designed on the assumption of linear elimination.

While the structural identifiability analysis shows that, given perfect data, V_m and K_m are uniquely identifiable from a single dose response curve, the problems of estimating them from real data become apparent when it is noted that the elimination rate is bounded between two linear elimination rates (Tong and Metzler, 1980):

$$\frac{V_m}{K_m + x_{max}}\, x(t) \le \frac{V_m}{K_m + x(t)}\, x(t) \le \frac{V_m}{K_m + x_{min}}\, x(t) \qquad (10)$$

where x_{max} and x_{min} are the maximum and minimum values of x(t) during the course of the experiment. This is illustrated in Figure 11 for the most usual case when observation is continued for a sufficient length of time that $x_{min} \simeq 0$.

The solution of equations (9A) and (9B) is

$$t = \frac{1}{V_m}\left[x_0 - x(t) + K_m \log_e \frac{x_0}{x(t)}\right]. \qquad (11)$$

The bounds on the elimination rate mean that x(t) is also bounded between two linear responses:

$$\text{Lower linear bound} = x_0 \exp[-V_m t/K_m] \qquad (12A)$$

$$\text{Upper linear bound} = x_0 \exp[-V_m t/(K_m + x_0)]. \qquad (12B)$$

The nonlinear response and upper and lower linear bounds for the particular values $V_m = x_0 = 1$ (so that $x_{max} = 1$) and K_m successively 10, 1, 0.1 and 0.01 are shown in Figure 13A to 13D). From these Figures, it may be seen that as K_m becomes large compared with x_0, the linear bounds get very close together so that it would only be justified to estimate the ratio V_m/K_m rather than V_m and K_m separately.

This is not the complete picture, though, because as K_m becomes small compared with x_0, the elimination rate becomes more nearly constant (at V_m) and the system response more nearly approaches that of a system with response

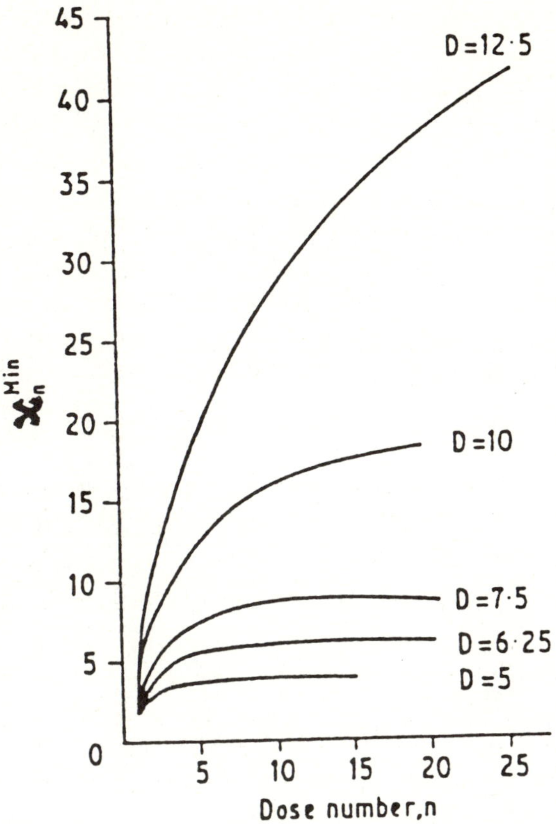

Fig. 12. Build-up of x with time in a model described by equation (8), with V_m = 15, K_m = 12 and b = 1, with doses (shown on curves) administered at intervals of 1 time unit.
(Reproduced, with permission, from J.G. Wagner (1978)).

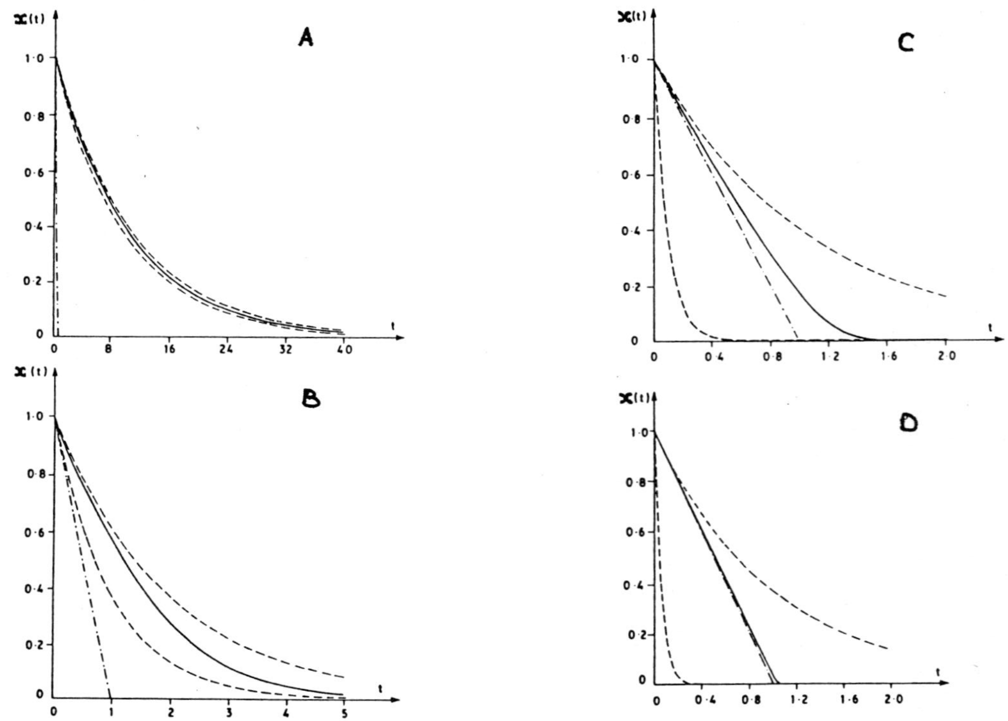

Fig. 13. Unit impulse responses of a model described by equation (8), with $V_m = 1$: (A) $K_m = 10$, (B) $K_m = 1$; (c) $K_m = 0.1$; (D) $K_m = 0.01$.

——— Responses of nonlinear model ----- Linear bounds given by equations (12A) and (12B) —·—·— zero-order response, equation (13).

$$x(t) = x_o - V_m t. \tag{13}$$

These curves are also shown on Figures 13A to 13D from which it may be seen that, as K_m becomes small, very good quality data would be needed for small values of x(t) for K_m to be estimated with any accuracy; estimation of V_m should be relatively unaffected. There is a range, roughly $0.1x_o \leq K_m < x_o$, where fitting the full nonlinear model to single dose response data is likely to be a valid procedure.

CONCLUSIONS

From the examples presented, the following conclusions can be reached and recommendations made. Some are obvious, others less so.

(i) In industrial situations, it is sometimes possible to estimate dynamics reasonably accurately from simple step responses and these should be tried before more sophisticated techniques are used. In experimentation on human beings, it is often not possible to apply anything other than impulsive or step perturbations anyway.

(ii) If the response from an industrial plant is contaminated by noise, useful results can often still be obtained by using an input with impulse-like autocorrelation. Then, provided the system is linear, the input-output crosscorrelation function is proportional to the system impulse response. Parameters of discrete-time transfer functions can be estimated by procedures such as maximum likelihood estimation or generalised least squares.

(iii) System linearity is best assured by the application of small amplitude perturbations, but even then problems can arise when the dynamics are different according to whether the input is being increased or decreased, which is quite a common situation in industrial processes. This departure from linearity gives rise to extra peaks in the crosscorrelation function, which, on a complex piece of plant, can easily be mistaken for parts of the system impulse response.

(iv) Use of normal operating records is tempting when the input is persistently exciting, but problems can arise when feedback is present (and it is not always obvious from visual inspection of records that this is so). For systems with feedback, it is highly advisable to use an externally applied perturbation signal. Without such a signal, the estimated dynamics may bear little resemblance to the dynamics of the forward path.

(v) If data records are sampled, the spacing of the samples may mean that some, but not necessarily all, parameters of a model may be estimated with low accuracy. It is very advisable to do a sensitivity check on the model, varying each parameter in turn and examining its effect on the output. Such a procedure is readily undertaken with the digital simulation packages now available.

(vi) When trying to estimate the parameters of an assumed nonlinearity, it is essential to check that the data available adequately span the nonlinearity. Otherwise, some of the parameters may be estimated with very low accuracy. In some cases, it may only be possible to estimate a reduced form of the nonlinearity, often corresponding to a linear model.

REFERENCES

G.E.P. Box and G.M. Jenkins (1976). Time Series Analysis, Forecasting and Control. (2nd. ed., Holden-Day).

P.A.N. Briggs and K.R. Godfrey (1966). Pseudorandom signals for the dynamic analysis of multivariable systems. (Proc. IEE, 113, 1259-1267).

D.W. Clarke (1966). Generalised least squares estimation of the parameters of a dynamic model. (National Physical Laboratory, Report AUTO 26).

D.Z. D'Argenio (1981). Optimal sampling times for pharmacokinetic experiments. (Jnl. Pharmacokin. Biopharm., 9, 739-756).

P. Eykhoff (1974). System Identification - Parameter and State Estimation. (John Wiley and Sons).

K.R. Godfrey (1969). Dynamic analysis of an oil refinery unit under normal operating conditions. (Proc. IEE, 116, 879-888).

K.R. Godfrey and B. Shackcloth (1970). Dynamic modelling of a steam reformer and the implementation of feedforward/feedback control. (Meas. and Cont., 3, T65-T72).

K.R. Godfrey and P.A.N. Briggs (1972). The identification of processes with direction-dependent dynamic responses. (Proc. IEE, 119, 1733-1739).

K.R. Godfrey and G.C. Goodwin (1974). Industry-University confrontation on process identification. (Automatica, 10, 223-225).

K.R. Godfrey and D.J. Moore (1974). Identification of processes having direction-dependent responses, with gas-turbine engine applications. (Automatica, 10, 469-481).

K.R. Godfrey (1983). Compartmental Models and their Application. (Academic Press).

K.R. Godfrey and W.R. Fitch (1984). Scientific Commentary : On the identification of Michaelis-Menten elimination parameters from a single dose-response curve. (Jnl. Pharmacokin. Biopharm., 12, 193-221).

K.R. Godfrey and J.J. DiStefano, III (1985). Identifiability of model parameters. (7th IFAC/IFORS Symposium on Identification and System Parameter Estimation, York 3rd to 7th. July 1985, pp. 89-114 (Pergamon Press)).

G.C. Goodwin and R.L. Payne (1976). Choice of sampling intervals. (System Identification: Advances and Case Studies (R.K. Mehra and D.G. Lainiotis (eds.)), pp. 251 287.(Academic Press).

R.K. Mehra (1976). Preface to System Identification : Advances and Case Studies (R.K. Mehra and D.G. Lainiotis (eds.)), (Academic Press).

D.J. Moore (1970). Error correction applied to dynamic analysis. (Rolls-Royce (1971) Ltd., Bristol Engine Division, Report EER/5033/70).

F. Mori and J.J. DiStefano III (1979). Optimal nonuniform sampling interval and test-input design for identification of physiological systems from very limited data. (IEEE Trans. Autom. Contr., AC-24, 893-900).

J.P. Norton (1986). An Introduction to Identification. (Academic Press).

D.D.M. Tong and C.M. Metzler (1980). Mathematical properties of compartment models with Michaelis-Menten type elimination. (Math. Biosci., 48, 293-306).

J.G. Wagner (1978). Time to reach steady state and prediction of steady-state concentrations for drugs obeying Michaelis-Menten elimination kinetics. (Jnl. Pharmacokin. Biopharm., 6, 209-225).

P.E. Wellstead (1977). Reference signals for closed-loop identification. (Int. J. Control, 26, 945-962).

SUGGESTIONS FOR FURTHER READING

(a) Books

P. Eykhoff (1974). System Identification-Parameter and State Estimation. (John Wiley and Sons). (Authoritative early review of the subject).

R.K. Mehra and D.G. Lainiotis (eds.) (1976). System Identification : Advances and Case Studies. (Academic Press). (Many interesting papers and case studies. The idea for the volume arose from a special issue of the IEEE Transactions on Automatic Control in December 1974, dealing with system identification and time series analysis).

J.P. Norton (1986). An Introduction to Identification. (Academic Press). (Excellent and readable treatment, with good coverage of practical problems as well as theory).

(b) Papers

An account of early applications of statistical techniques to industrial processes and nuclear power plant:

K.R. Godfrey (1969). The theory of the correlation method of dynamic analysis and its application to industrial processes and nuclear power plant. (Meas. and Cont., 2, T65-T72).

A review of applications of identification up to 1975:

I. Gustavsson (1975). Survey of applications of identification in chemical and physical processes. (Automatica, 11, 3-24).

For a broader account of practical aspects of identification than has been possible in this Case Study Lecture, including discussion of experiment design, the use of prior knowledge, parameter estimation and model validation:

K.R. Godfrey and R.F. Brown (1979). Practical aspects of the identification of process dynamics. (Trans. Inst. Meas. Contr., 1, 85-95).

LQG DESIGN OF SHIP STEERING CONTROL SYSTEMS

Prof. M.J. Grimble and Dr. M.R. Katebi

1. Introduction

A helmsman can steer a ship along a desired course by keeping to a given compass heading. However, the ship can deviate from its intended path if ocean currents and sea waves are present. Thus, frequent position fixes and course corrections are needed to reduce the distance to be travelled.

An autopilot is used to maintain a given set course automatically. The heading angle, as measured by a gyrocompass, is normally fed back through a PID (Proportional, Integral, Derivative) controller to the steering control system. The resulting rudder movement will generate a force normal to the centre of the rudder plane. This force results in a transverse motion of the centre of gravity (CG) of the vessel which changes the heading of the ship. The input to the steering machine represents the control input (δ_c) for the course keeping problem.

The growth of the size of large oil tankers, the need for energy minimization and the poor performance of the conventional PID autopilots has motivated the design of a new generation of course-keeping controllers. This is aided by the advent of cheap digital computing power and new developments in navigation. The energy minimization problem is the main subject of the following discussions.

Energy losses are due to the excess distance sailed, caused by zig-zag course deviations and the added resistance per unit distance which depends upon the sway and yawing motions of the ship (Reid, [3]). The former loss is often small and is neglected here. The added resistance loss is due either to the limit cycle oscillation caused by nonlinearities in the steering system, or due to the effects of wave disturbances. The development of an analogue rudder servo (rather than a bang-bang servo) has significantly reduced the losses due to self-oscillation of the vessel (Blanke, [2]). Hence, the main effort here is concentrated on minimizing the energy loss due to the wave motion.

The frequency of the sea wave disturbance (0.08-2 rad/sec) is usually high compared to the frequency of ship motion (0.005-0.15 rad/sec). The wave frequency encountered by a ship depends on the angle at which the waves hit the ship. In particular, in a following sea, the wave and the ship motions overlap and this results in a more difficult control problem.

For energy saving, the 'added resistance' which can be derived from the equations of motion in the surge direction, is to be minimized. The optimal controller requires access to the system states but the observations are contaminated with white and coloured noise. Hence a state estimator is needed to retrieve the information for feedback control purposes.

The cost criterion, as formulated by (Reid, [3]), is indefinite. The solution to the indefinite LQR problem was studied by (Willems, [10]), (Desai and Weinert, [11]) and (Sivan and Gill, [12]). Some of these results are employed in the LQG problem described in this paper.

The ship sailing in an open sea, is under frequent course deviations due to the effects of wind, second order wave and sea currents. These low frequency disturbances can be offset by introducing integral action into the controller. The filter must also be designed for the effects of low frequency input disturbances which are often represented by integrators driven by white noise.

The paper is organised as follows. The modelling of the ship, wind, wave and steering gear are discussed in §2 and §3. The combined model of ship and the disturbance is given in §4. The cost function is formulated in §5. The existence of the solution to the LQG problem is investigated in §6. The Kalman filter and the controller are designed in §7 and §8. The simulation results are presented in §9. Finally, the conclusions are drawn in §10.

2. MODELLING

The ship, wave, wind and the 'added resistance' models are discussed in the following sections.

2.1 Ship dynamics

The motions of a ship in the horizontal plane are considered for course keeping control problems. The three degrees of freedom, to be controlled, are illustrated in Figure 1. The nomenclature for the following variables are given in Appendix 1. The ship is treated as a rigid body. Newton's laws of motion (Kallstrom [1]) give:

$$m(\dot{u} - vr - x_g r^2) = X \tag{1}$$

$$m(\dot{v} + ur + x_g \dot{r}) = Y \tag{2}$$

$$I_z \dot{r} + m x_g (\dot{v} + ur) = N \tag{3}$$

The external forces (X,Y) and moment is represented as (Blanke 1981 [2]):

$$X = X_{\dot{u}} \dot{u} - Y_{\dot{v}} vr - Y_{\dot{r}} r^2 + X_{waves} + X_s + X_\delta \tag{4}$$

$$Y = Y_{\dot{v}} \dot{v} + X_{\dot{u}} ur + Y_{\dot{r}} r^{\cdot} + Y_{waves} + Y_{vis} + Y_\delta \tag{5}$$

$$N = N_{\dot{r}}\dot{r} + (Y_{\dot{v}} - X_{\dot{u}})uv + Y_{\dot{r}}(\dot{v} + ur) + N_{waves} + N_{vis} + N_{\delta} \quad (6)$$

where $x_{\dot{u}}$, $Y_{\dot{v}}$ and $N_{\dot{r}}$ denote mass and inertia hydrodynamic derivative terms, respectively.

The contribution of shear stresses to the surge resistance can be modelled by a square law relation $X_s = X_{uu}u^2$.

Reaction forces, proportional to $v|v|$ and $r|r|$, add to the hydrodynamic sway force and yaw moments to represent viscous effects (Blanke, 1981 [2]):

$$Y_{vis} = Y_{vv}|v|v + Y_{vr}|v|r + Y_{rv}|r|v + Y_{rr}|r|r \quad (7)$$

$$N_{vis} = N_{vv}|v|v + N_{vr}|v|r + N_{rr}|r|r + N_{rv}|r|v \quad (8)$$

These sway and yaw terms contribute substantially during the manoeuvring a large vessel.

2.2 Rudder forces

The surge x_δ and sway Y_δ forces, and yaw moment N_δ, can be calculated given rudder dimensions and other specifications. The term Y_δ denotes the rudder control force and X_δ represents the drag force due to rudder activity which increases with δ^2. The rudder forces increase with the square of ship speed and $X_\delta = X_{r\delta}V^2\delta^2$, $Y_\delta = Y_{r\delta}V^2\delta$ and $N_\delta = N_{r\delta}V^2\delta$.

Typical upper bounds on executable rudder angle $|\delta|$ and rudder rate $\dot{\delta}$ are 40 degrees and 7 degrees/second, respectively. The rudder rate limit introduces a nonlinearity which often adversely affects the course keeping performance. The control design must therefore ensure the rate saturation condition is avoided.

2.3 Thrust deduction factor

When the flow velocity behind a vessel is increased the pressure balance between the bow and the stern is disturbed and an increased resistance results. The thrust actually available for ship propulsion is not therefore the developed thrust T but (1-t)T, where the thrust deduction factor $t \cong 0.15$, typically. This factor depends upon the fullness of the stern and does not depend upon the ship speed. The rate of propeller revolutions (n) may be assumed to be approximately constant and independent of load, since most marine engines have constant speed regulators. The developed thrust $T(V_a, n)$ is a function of both V_a (inflow velocity at the propeller) and n. The wake fraction $0 < w < 1$ determines the difference between V_a and ships speed V, since

$$V_a = (1 - w)V$$

Typically, $0.2 < w < 0.3$. This fraction, considering only limited excursions in sway velocity at the stern (v_p), can be computed (Blanke [2]) as:

$$w = w_o + w_v(v_f/V)$$

where the coefficients w_o and w_v can be obtained from an experimentally determined wake fraction and relative sway velocity characteristic.

2.4 Nonlinear manoeuvring equations

Keeping the dominant terms, the equations of motion may now be summarised as:

Surge:

$$m(\dot{u} - vr - x_g r^2) = X_{uu}u^2 + X_{vr}vr + X_{rr}r^2 + X_{r\delta}V^2\delta^2 + (1 - t)T + X_w \tag{9}$$

Sway:

$$m(\dot{v} + ur + x_g\dot{r}) = Y_{\dot{v}}\dot{v} + Y_{\dot{r}}\dot{r} + Y_{ur}ur + Y_{uv}uv + Y_{vv}|v|v + Y_{vr}|v|r + Y_{r\delta}V^2\delta + Y_w \tag{10}$$

Yaw:

$$I_z\dot{r} + mx_g(\dot{v} + ur) = N_{\dot{r}}\dot{r} + N_{\dot{v}}\dot{v} + N_{ur}ur + N_{uv}uv + N_{vv}|v|v + N_{vr}|v|r + N_{r\delta}V^2\delta + N_w \tag{11}$$

where X_w, Y_w, N_w denote the external current, wave and wind forces and moment. The left-hand sides of these equations contains the mass forces including centrifugal terms. The right hand sides include fluid forces upon the hull due to acceleration, resistance, viscous effects, rudder angle, propellor thrust and wind and wave forces. Note that $X_{uu} < 0$, $X_{r\delta} < 0$ and $X_{vr} > 0$.

2.5 Linearisation of the non-linear equations

A linearised ship model is adequate to predict ship motion for course keeping. The non-linear equations (10), (11) can be linearised about the steady-state operating point $v = r = \delta = 0$ and $u = u_o = V$. The linearised sway and yaw velocity model becomes:

$$\begin{bmatrix} m - Y_{\dot{v}} & mx_g - Y_{\dot{r}} \\ mx_g - N_{\dot{v}} & I_z - N_{\dot{r}} \end{bmatrix} \begin{bmatrix} \dot{v} \\ \dot{r} \end{bmatrix} = \begin{bmatrix} Y_{uv}V & (Y_{ur} - m)V \\ N_{uv}V & (N_{ur} - mx_g)V \end{bmatrix} \begin{bmatrix} v \\ r \end{bmatrix} + \begin{bmatrix} Y_{r\delta}V^2 \\ N_{r\delta}V^2 \end{bmatrix} \delta + \begin{bmatrix} Y_w \\ N_w \end{bmatrix} \tag{12}$$

The 2x2 square matrices on the left and right-hand side of (12) are called the **mass** and **damping** matrices, respectively. If the non-linear ship equations are normalised using the "prime" system (length unit = ship length L, Time unit = L/V and mass unit = $\rho L^3/2$ where ρ = density of water) then the linearised equations can be expressed in the form:

$$\begin{bmatrix} \dot{v} \\ \dot{r} \end{bmatrix} = \begin{bmatrix} a'_{11}\frac{V}{L} & a'_{12}V \\ a'_{21}\frac{V}{L^2} & a'_{22}\frac{V}{L} \end{bmatrix} \begin{bmatrix} v \\ r \end{bmatrix} + \begin{bmatrix} b'_1\frac{V^2}{L} \\ b'_2\frac{V^2}{L^2} \end{bmatrix} \delta + \begin{bmatrix} d_s \\ d_y \end{bmatrix} \qquad (13)$$

Kallstrom (1979 [1]) has shown that typical values of these coefficients for different vessels and loading conditions are as follows:

coefficient	range of values
a'_{11}	-0.3 , -0.9
a'_{12}	-0.28 , -0.45
a'_{21}	-1.9 , -3.7
a'_{22}	-1.6 , -2.8
b'_1	0.1 , 0.18
b'_2	-0.77 , -1.63

The coefficients in (13) can therefore be determined from scale model or ship test data and vary with both draft and water depth. The disturbance input terms in (13) (d_s and d_y) depend upon the waves, wind and current forces.

2.6 Steering machine

The rudder angle δ is determined by the input to the steering machine δ_c. The bandwidth of the steering gear control system puts a natural limit on the attainable performance of the autopilot. At frequencies in excess of the steering-gear cut-off frequency autopilot loop gain merely amplifies noise and increases steering-gear wear.

A typical steering machine is illustrated in Fig. 2. Position feedback of the actual rudder angle δ is used in the main servo loop. The telemotor system is fast compared with the main servo and it may be approximated by a constant gain. The system also normally contains a rudder "limiter". The maximum rudder speed is determined by this limit and is typically 2.5 to 7 degrees per second. For the present discussion the steering machine will be represented by the first order system $\dot{x}_\delta = -\frac{1}{\tau}x_\delta + \frac{1}{\tau}\delta_c,\ \delta = c_\delta x_\delta.$ (14)

3. External Disturbances

The motions of a vessel are disturbed by wave and wind forces of variable stochastic characteristics. The model for the response of a ship to the waves is rather complex. A ship moving with a certain velocity, experiences the wave excitation at a frequency called the encounter frequency which is a nonlinear function of ship speed and the encounter angle of the waves. Furthermore, it is the wavelength of the waves which determines the force and moment exerted on the hull whereas the excitation frequency is not uniquely associated with the wave length, and with the magnitude of excitation [Blanke [2]]. The modelling of the wind and the wave disturbances are discussed below. Note that the sea currents are offset in the course keeping problem by the use of integral action.

3.1 Wind disturbances

The wind forces on a vessel are usually separated into two components ; mean wind and turbulance. The latter can be modelled as white noise with intensity proportional to the wind speed (Kallstrom, 1979 [1]). The mean wind excitement, although random and non-stationary, gives steady forces over periods much longer than the effective ship time constants (30-250 seconds). This average wind force follows a world wide Rayleigh amplitude distribution with a mean of about 5 m/s. The average wind forces may be modelled using the Davenport wind spectrum.

The amplitude of the force spectral density for the waves is typically three orders of magnitude greater than that for turbulent wind. However, the wind excitment is at low frequencies and cannot be ignored completely.

In state-space form, the wind can be modelled as:

sway force:

$$\dot{x}_{vw} = a_{vw}x_{vw} + w_{vw} \tag{15}$$

yaw moment:

$$\dot{x}_{rw} = a_{rw}x_{rw} + w_{rw} \tag{16}$$

where a_{vw} and a_{rw} are small negative numbers and w_{vw} and w_{rw} are white noises of mean values m_{vw} and m_{rw} and covariances Q_{vw} and Q_{rw}, respectively.

3.2 The sea wave disturbance

The sea wave spectrum is often represented by the Pierson-Moskowitz wave spectrum, i.e.

$$S(\omega) = \frac{A}{\omega^5} \exp\,(-B/\omega^4) \tag{17}$$

where $A = 172.79\ h^2_{1/3}T^4$ and $B = 691.17/T^4$

and $h_{1/3}$ and T denote the significant wave height and average wave period, respectively.

The relationship between the wave frequency (ω) seen at a fixed position and the encounter frequency (ω_e) is (Reid, [3]):

$$\omega_e = \omega + \frac{\omega^2}{g} V \cos(\chi)$$

where V is the ship speed, χ is the encounter angle.

An approximate expression for the net force and moment on hull can be derived (Zuidweg, [4]):

$$X_{waves} = \rho gBLd \cos(\chi)\, k\, \xi_o \sin(\omega_e t) \tag{18}$$
$$Y_{waves} = \rho gBLd \sin(\chi)\, k\, \xi_o \sin(\omega_e t) \tag{19}$$
$$N_{waves} = \frac{1}{24} \rho gBLdk^2 (L^2-B^2)\, \xi_o \sin(2\chi) \cos(\omega_e t) \tag{20}$$

where ρ is mass density of water, L, B and D are Length, Breadth and Depth of the hull, $k = \frac{2\pi}{\lambda}$, is the wave number and ξ_o is the wave amplitude. It is assumed that $kL \ll 1$ and $0.3 < \lambda < 3$. Using the wave model (18-20), the magnitude of force and moment can be plotted as a function of the frequency. This function is called the receptance function. An analytical expression is usually not available. The receptance functions of force and moment for a very large crude carrier (VLCC) vessel, with L = 370 m, B = 65 m and D = 24 m, are shown in Fig. 3 (Blanke, [2]).

The effect of the waves on the vessel can be modelled by fitting a rational transfer-function of the form:

$$H(s) = \frac{b_o + b_1 s + \ldots + b_m s^m}{a_o + a_1 s + \ldots + a_{n-1} s^{n-1} + s^n} \tag{21}$$

tp the receptance functions shown in Fig. 3. A least square technique can be used to find the a and b coefficients. These coefficients, as computed by (Reid, [9]), are shown in Table 1. The model for $\chi = 90^o$, is of order 6, but it is here approximated by a 2nd order transfer-function for simplicity.

The transfer function (21) can be transformed to state space form as follows:

Sway force:

$$\begin{bmatrix} \dot{x}_{1v} \\ \dot{x}_{2v} \end{bmatrix} = \begin{bmatrix} 0 & 1 \\ -a_{ov} & -a_{1v} \end{bmatrix} \begin{bmatrix} x_{1v} \\ x_{2v} \end{bmatrix} + \begin{bmatrix} 0 \\ b'_v \end{bmatrix} \omega_v; \quad y_v = \begin{bmatrix} 1 & 0 \end{bmatrix} \begin{bmatrix} x_{1v} \\ x_{2v} \end{bmatrix} \tag{22}$$

Yaw Moment:

$$\begin{bmatrix} \dot{x}_{1r} \\ \dot{x}_{2r} \end{bmatrix} = \begin{bmatrix} 0 & 1 \\ -a_{or} & -a_{1r} \end{bmatrix} \begin{bmatrix} x_{1r} \\ x_{2r} \end{bmatrix} + \begin{bmatrix} 0 \\ b'_r \end{bmatrix} \omega_r; \quad y_r = \begin{bmatrix} 1 & 0 \end{bmatrix} \begin{bmatrix} \chi_{1r} \\ \chi_{2r} \end{bmatrix} \tag{23}$$

where $b'_v = b_{ov}\sqrt{Q_v}$, $b'_r = b_o\sqrt{Q_r}$, Q_v and Q_r are white noise covariances in the sway and yaw directions, and ω_v and ω_r are white noises of unity covariances.

4. Combined ship and Disturbance Models

The combined ship dynamics (13), steering machine (14), the wind (15,16) and the waves (22, 23) may be represented by the augmented state space equations:

$$\dot{x} = Ax + B\delta_c + D\xi_w \tag{24}$$

$$z = Hx + v \tag{25}$$

where ξ_w and v denote zero-mean, independent, white noise sources of covariances Q_f and R_f, respectively. The system matrices S(A,B,C,D) are obtained by collecting the previous results as:

Ship (13) : $\dot{x}_\ell = A_\ell x_\ell + B_\ell \delta + d, \quad \psi = C_\ell x_\ell$ (26)

where:

$$\begin{bmatrix} \dot{v} \\ \dot{r} \\ \dot{\psi} \end{bmatrix} = \begin{bmatrix} a_{11} & a_{12} & 0 \\ a_{21} & a_{22} & 0 \\ 0 & 1 & 0 \end{bmatrix} \begin{bmatrix} v \\ r \\ \psi \end{bmatrix} + \begin{bmatrix} b_1 \\ b_2 \\ 0 \end{bmatrix} \delta + \begin{bmatrix} d_s \\ d_y \\ 0 \end{bmatrix} \tag{27}$$

$$\psi = \begin{bmatrix} 0 & 0 & 1 \end{bmatrix} x_\ell$$

$$d = d_d + d_h$$

Steering machine (14): $\dot{x}_\delta = A_\delta x_\delta + B_\delta \delta_c = \frac{-1}{\tau} x_\delta + \frac{1}{\tau} \delta_c \qquad \delta = C_\delta x_\delta$

Wind (15, 16): $\dot{x}_d = A_d x_d + D_d \xi_d \quad , \quad d_\ell = C_d x_d$ (29)

$$\begin{bmatrix} \dot{x}_{vw} \\ \dot{x}_{rw} \end{bmatrix} = \begin{bmatrix} a_{vw} & 0 \\ 0 & a_{rw} \end{bmatrix} \begin{bmatrix} x_{vw} \\ x_{rw} \end{bmatrix} + \begin{bmatrix} w_{vw} \\ w_{rw} \end{bmatrix}$$

$$\begin{bmatrix} d_s \\ d_y \end{bmatrix}_{wind} = \begin{bmatrix} 1 & 0 \\ 0 & 1 \end{bmatrix} \begin{bmatrix} x_{vw} \\ x_{rw} \end{bmatrix}$$

Waves (22, 23): $\dot{x}_h = A_h x_h + D_h \xi_h \quad , \quad d_h = C_h x_h$ (30)

$$\begin{bmatrix}\dot{x}_{1v}\\ \dot{x}_{2v}\\ \dot{x}_{1r}\\ \dot{x}_{2r}\end{bmatrix} = \begin{bmatrix}0 & 1 & 0 & 0\\ -a_{ov} & a_{1v} & 0 & 0\\ 0 & 0 & 0 & 1\\ 0 & 0 & -a_{or} & -a_{1r}\end{bmatrix}\begin{bmatrix}x_{1v}\\ x_{2v}\\ x_{1r}\\ x_{2r}\end{bmatrix} + \begin{bmatrix}0 & 0\\ b'_v & 0\\ 0 & 0\\ 0 & b'_r\end{bmatrix}\begin{bmatrix}w_v\\ w_r\end{bmatrix}$$

$$\begin{bmatrix}d_s\\ d_y\end{bmatrix}_{wave} = \begin{bmatrix}1 & 0 & 0 & 0\\ 0 & 0 & 1 & 0\end{bmatrix}\begin{bmatrix}x_{1v}\\ x_{2v}\\ x_{1r}\\ x_{2r}\end{bmatrix}$$

The augmented system model becomes:

$$\begin{bmatrix}\dot{x}_\ell\\ \dot{x}_\delta\\ \dot{x}_h\\ \dot{x}_d\end{bmatrix} = \begin{bmatrix}A_\ell & B_\ell C_\delta & C_h & C_d\\ 0 & A_\delta & 0 & 0\\ 0 & 0 & A_h & 0\\ 0 & 0 & 0 & A_d\end{bmatrix}\begin{bmatrix}x_\ell\\ x_\delta\\ x_h\\ x_d\end{bmatrix} + \begin{bmatrix}0\\ B_\delta\\ 0\\ 0\end{bmatrix}\delta_c + \begin{bmatrix}0\\ 0\\ D_h\xi_h\\ D_d\xi_d\end{bmatrix} \tag{31}$$

$$z = [C_\ell \quad 0 \quad 0 \quad 0]\, x + v$$

the system is illustrated in Fig. 4.

5. Cost-Function or Loss Factor

It is usual to introduce a quadratic cost-function as a criterion for estimating the increase in fuel consumption or speed drop due to course keeping (Clarke, [6]). The thrust T and thrust deduction factor t are assumed to be constant for steering loss calculations. The per-unit added resistance can be obtained from consideration of the surge equation (9):

$$m\dot{u} = (X_{uu}u^2 + (1-t)T) + (X_{vr}+m)vr + X_{r\delta}V^2\delta^2$$
$$+ (X_{rr} + mx_g)r^2 + X_w \tag{32}$$

Recall that the resistance and propeller thrust $X_{uu}u^2 + (1-t)T$ sums to zero in the steady-state when added resistance is zero ($X_{uu} < 0$). The r^2 term can be neglected and noting that the external forces included in X_w are not controllable from the rudder, the drag force to be minimised becomes:

$$X_d = -\,((X_{vr}+m)vr + X_{r\delta}V^2\delta^2) \tag{33}$$

Normalising the drag force expression, using the average surge force due to the propeller $(1-\bar{t})\bar{T}$, gives the average per-unit loss as:

$$J'_o = \lim_{T\to\infty} \frac{1}{2T} E\{\int_{-T}^{T}(-2q_{12}vr + q_{44}\delta^2)dt\} \quad (34)$$

where

$$q_{12} = \frac{1}{2}(X_{vr} + m)/(1-\bar{t})\bar{T} \quad (35)$$

$$q_{44} = -X_{r\delta}V^2/(1-\bar{t})\bar{T} \quad (36)$$

Combining the normalised loss functions for both excess distance travelled and drag force gives the **course keeping energy loss function**:

$$J_o = \lim_{T\to\infty} \frac{1}{2T} E\{\int_{-T}^{T}(-2q_{12}vr + q_{33}(\Delta\psi)^2 + q_{44}\delta^2)\, dt\} \quad (37)$$

5.1 Extension of the cost-function

The total course-keeping loss function must include more than simple energy related terms since the vessel's engineers must be able to tune the controller: (a) to reduce rudder deviations if these are unacceptable, (b) to modify the transient response to course change demands, (c) to affect the robustness of the system when either modelling errors or external disturbances have degraded performance.

Introducing an integral operator $(Lx)(t) \triangleq \int_0^t x(\tau)d\tau$, then if the ficticious state is defined as $x_n(t) \triangleq (L\Delta\psi)(t)$, a term $q_{nn}x_n^2(t)$, $(q_{nn} > 0)$ may be added to the cost-function. This results in an integral action term being introduced into the controller (Grimble [7]) so that steady-state heading errors cannot occur (constant heading errors => $\int_0^\infty x_n^2(\tau)d\tau \to\infty$). A control weighting term $r\delta_c(t)$ can also be introduced to limit rudder excursions. The weighting term can be dynamical to cost high frequency variations but for the present it will be assumed to be a constant, r_c.

Total course-keeping loss function can now be written as:

$$J = \lim_{T\to\infty} \frac{1}{2T} E\{\int_{-T}^{T}(-2q_{12}vr + q_{33}(\Delta\psi)^2 + q_{44}\delta^2 + q_{nn}x_n^2 + r_c\delta_c^2)dt\} \quad (38)$$

The two tuning scalars q_{nn} and r_c give some freedom of action to trade off energy minimisation against other performance and system requirements. The following matrices can now be defined:

$$Q_c = \begin{bmatrix} 0 & -q_{12} & 0 & 0 & 0 \\ -q_{12} & 0 & 0 & 0 & 0 \\ 0 & 0 & q_{33} & 0 & 0 \\ 0 & 0 & 0 & q_{44} & 0 \\ 0 & 0 & 0 & 0 & q_{nn} \end{bmatrix}, \quad R_c = r_c$$

6. The Existence of the Solution to the LQG Problem

The term $-2q_{12}$ vr in the loss-function results in an indefinite Q matrix. Jenssen [8] has shown that the optimal solution exists for LQR problem, ignoring the effects of the waves. The phase relationship between sway velocity and yaw rate caused by the waves is, however, of major importance in consideration of resistance losses. This can be seen by assuming simple sinusoidal swaying, yawing and rudder angle changes as follows (Reid, [9]):

$$\begin{aligned} v &= v_m \sin(\omega t + \phi_v) \\ r &= r_m \sin(\omega t + \phi_r) \\ \delta &= \delta_m \sin(\omega t + \phi_\delta) \end{aligned} \tag{39}$$

From (34)

$$J_o' = -q_{12} v_m r_m \sin(\phi_v - \phi_r) + 1/2\, q_{44} \delta_m^2$$

For $\frac{\pi}{2} < (\phi_o - \phi_r) < \frac{3\pi}{2}$, the resistance would increase and for $0 < (\phi_o - \phi_r) < \pi/2$ and $\frac{3\pi}{2} < (\phi_v - \phi_r) < 2\pi$, it would decrease. The phase shift does, however, depend on the wave direction and frequency. Hence the wave effects should be taken into account in modelling and control design.

An optimal solution to the LQG solution exists if the following two conditions are satisfied:

1) $M(s,\bar{s}) = [R_c + B^T(\bar{s}I - A^T)^{-1} C^T Q_c C (sI-A)^{-1} B] > 0$ (40)

2) $N(s,\bar{s}) = [R_f + H(sI-A)^{-1} D Q_f D^T (\bar{s}I - A^{-1})] > 0$ (41)

The condition $M(s,\bar{s}) > 0$, for all s, is the necessary and sufficient condition for the existence of a solution to the control Riccati equation (Willems, [10]). Note that Q_c is indefinite but M can be positive-semidefinite. Thus condition (1) guarantees an optimal solution to the control problem.

The condition $N(s,\bar{s}) > 0$, for all s, is the necessary and sufficient conditions for the existence of a solution to the filter Riccati equation. This can be shown by using the duality principle applied to the previous result. It is, however, very cumbersome to show analytically that the conditions (40) and (41) are satisfied for the ship steering problem. Instead a numerical approach is taken and M and N are computed for the frequency range of interest. Note that $M(s, \bar{s})$ is a scalar and its real part should be always positive. This is shown in Fig. 5 for different encounter angles of the waves. If it is assumed that the sway velocity and heading angle measurement are available $N(s, \bar{s})$ is a 2x2 matrix and its positive-definiteness can be tested by looking at its eigenvalues where it should be positive and real. The variation of the eigenvalues with frequency are shown in Fig. 6.

The results presented in Fig. 5 and 6 show that an optimal LQG solution exists to the course keeping problem of interest and more over the optimal filter and control gains can be found by solving Riccati equations.

7. Kalman Filter Design

It was shown in the previous section that a solution exists to the filtering problem provided a sway measurement signal is available. The filter is required to estimate the sway velocity and yaw rate for feedback purposes in order to minimize the energy and also remove the large high frequency wave motion signals from the low frequency motion estiamtes. The use of low freqency state estimate model feedback ensures that the rudder does not respond to the wave motion and reduces wear and tear in the rudder. The filter is defined as:

$$\dot{\hat{x}} = A\hat{x} + B\delta_c + K_f(z-H\hat{x}) \tag{42}$$

where

$$K_f = P_f H^T R_f^{-1}$$

and P_f is the solution of the Riccati equation:

$$AP + PA^T - PH^T R_f^{-1} HP + DQ_f D^T = 0 \tag{43}$$

Note that the high frequency model (wave model) changes with encounter angle. Hence, the Riccati equation should ideally be solved for each encounter angle. Clearly there is a need for either a robust or adaptive estimator to cope with such variations.

7.1 Filter integral action

The integral action introduced in the loss function (38) is needed to compensate for the effects of low frequency disturbances. This ensures that the low frequency heading angle deviation is controlled and consequently, reduces the distance to be sailed. Augmenting the integral action state to the system model (31) does of course increase the system order, and the computational load. The integral state is however, completely independent of the high and low frequency subsystems in the Kalman filter model. The filter gain can therefore be computed using the following procedure (Grimble, [7]).

Decompose the system as:

$$A = \begin{bmatrix} A_s & 0 \\ A_I & 0 \end{bmatrix}, \quad B = \begin{bmatrix} B_s \\ 0 \end{bmatrix}, \quad D = \begin{bmatrix} D_s \\ 0 \end{bmatrix}, \quad H = \begin{bmatrix} H_s \\ 0 \end{bmatrix}$$

The Ricatti equation (43) can now be decomposed as:

$$A_s P_{11} + P_{11} A_s^T - P_{11} H_s^T R_f^{-1} H_s P_{11} + D_s Q_f D_s^T = 0 \tag{44}$$

$$A_I P_{11} + P_{21} A_s^T - P_{21} H_s^T R_f^{-1} H_s P_{11} = 0 \tag{45}$$

$$A_I P_{12} + P_{21} A_I^T - P_{21} H_s^T R_f^{-1} H_s P_{12} = 0 \tag{46}$$

where

$$P = \begin{bmatrix} P_{11} & P_{12} \\ P_{21} & P_{22} \end{bmatrix} \quad \text{and} \quad K_f = \begin{bmatrix} K_{f1} \\ K_{f2} \end{bmatrix} = \begin{bmatrix} P_{11} H_s R^{-1} \\ P_{21} H_s R^{-1} \end{bmatrix}$$

Note that P_{22} is not required for computation of the Kalman gain.

Now define:

$$A_f = (A_s - K_{f1} H_s)^T \tag{47}$$

From (45)

$$P_{21} A_f = - A_I P_{11}$$

$$P_{21} = -A_I P_{11} A_f^{-1} \tag{48}$$

Note that the system (A_s, B_s, D_s, H_s) is assumed to be stabilisable and detectable. Hence, equation (43) has a unique positive semidefinite solution. The resulting filter is then asymptotically stable and the eigenvalues of the matrix A_f all have negative real parts. Thus A_f^{-1} exists (Grimble [7]) and P_{21} follows from (48). To compute the Kalman filter gain the reduced Riccati equation (44) must be solved and (48), and hence the addition of integral action introduces only a small increase in the computational load.

8. Control design

Using the separation principle, the control gain can be found by solving the following Riccati equation:

$$PA + A^T P - PBR_c^{-1} B^T P + C^T Q_c C = \dot{P} \tag{49}$$

where C is the partitioned matrix $C = [C_\ell \quad 0]$ where $C_\ell = I$. For indefinite Q_c, the solution exists if and only if the condition (40) is satisfied. The solution can be found by integrating (49) backward in time or by eigenvalue-eigenvector method. The control law is then defined as:

$$u = -K_c \hat{x}$$

where

$$K_c = R^{-1}B^TP$$

and x is the estimate states from the Kalman filter.

9. Simulation Results

The simulation studies carried out here is a feasibility study to investigate the performance of the LQG controller for the ship steering problem. The following model of a 250,000 dwt tanker at full load is chosen [Reid, (9)]:

$$\begin{bmatrix} \dot{v} \\ \dot{r} \\ \dot{\phi} \\ \dot{\delta} \end{bmatrix} = \begin{bmatrix} -0.013 & -8.6 & 0 & 7.03\times10^{-2} \\ -9.3\times10^{-5} & -4.3\times10^{-2} & 0 & -4.59\times10^{-4} \\ 0 & 1.0 & 0 & 0 \\ 0 & 0 & 0 & -3.333 \end{bmatrix} \begin{bmatrix} v \\ r \\ \phi \\ \delta \end{bmatrix} + \begin{bmatrix} 0 \\ 0 \\ 0 \\ 0.333 \end{bmatrix} \delta_c$$

$$+ \begin{bmatrix} 2.72\times10^{-8} & 0 \\ 0 & 3.65\times10^{-13} \\ 0 & 0 \\ 0 & 0 \end{bmatrix} \begin{bmatrix} d_1 \\ d_2 \end{bmatrix} \qquad (50)$$

The wave model is the same as shown in Table 1. The low frequency motion is assumed to be a constant bias added to the input of the system. The control and filter gains (K_c and K_f) are found by solving Riccati equations (43) and (49). The closed loop system was simulated on a VAX 11/750 and the following loss functions were computed for encounter angle 30°, 45°, 60° and 90°.

$$\overline{vr}_{loss} = - \int_o^\infty q_{12} vr \, dt \qquad (51)$$

$$\overline{\phi}^2_{loss} = \int_o^\infty q_{33} \, \phi^2 \, dt \qquad (52)$$

$$\overline{\delta}^2_{loss} = \int_o^\infty q_{44} \, \delta^2 \, dt \qquad (53)$$

$$\delta^2_{c\ loss} = \int_o^\infty R_c \delta_c^2 \, dt \qquad (54)$$

These losses are shown in Table 2. The r.m.s. values and the mean values of the sway velocity, yaw rate, heading angle and the rudder angle are also computed and shown in Tables 3 and 4.

Table 2 shows that as the sea changes from following sea (30°) to head seas (90°) the energy loss increases. The dominant losses are the sway/yaw cross coupling loss ($\overline{vr}$) and the rudder loss ($\overline{\delta}^2$).

The rms and mean values also show that the magnitudes of the variables do not reach the saturation level. The measurements, system states and estimates, using the Kalman filter, are shown in Figs. 7 to 11.

The open loop uncompensated gain and phase are also shown in Fig. 12. The frequency range of interest for a large tanker is between 0.01 and 0.5

(rad/sec). The system is open loop unstable and this can be seen from the phase variation which is greater than 180° for the frequencies between 0.025 and 0.1 rad/sec. The controller and filter gain and phase plots, are shown in Fig. 13. The stability of the system can be seen from the phase plots which has at least 30° phase margin over the frequency range of interest. The gain plot has shifted upward and the closed loop system has a positive gain in the desired frequency range.

10. Conclusions

The feasibility of using an LQG controller to minimise the energy loss in the ship course keeping problem was investigated. It was shown that one of the main problems was to design a Kalman filter which would estimate the ship motions. The disturbance model changes significantly in different sea conditions. Hence, a fixed gain Kalman filter may not give an adequate estimation accuracy. The performance of the closed-loop system was also studied in the frequency domain and it was noted that a reasonable stability was obtained. The study requires further investigation on the implementation aspects of the designed controller.

References

1. Kallstrom, G.G. "Identification and adaptive control applied to ship steering', PhD Thesis, Lund Institute of Technology, May 1979.
2. Blanke M., 'Ship propulsion losses related to automatic steering and prime mover control', PhD Thesis, Technical University of Denmark, 1981.
3. Reid R.E. 'Identification and minimisation of propulsion losses related to steering', Report No. MA-RD-940-82038, University of Illinois, USA.
4. Zuidweg J.K. 'Automatic Guidance of ships as a control problem', Thesis, Tenchnische Hugeschool, Delft, The Netherlands, 1970.
5. Reid R.E. 'A proposal for performance criteria for propulsion losses due to ship steering', Trans ASME, Vol 104, June 1982.
6. Clarke, D., 'Development of a cost function for autopilot optimisation' Proc. Symp. on ship steering Automatic Control, Genoa, Italy, 1980.
7. Grimble, M.J. 'Controllers for LQG self-tuning applications with coloured measurement noise and dynamic costing' to be published in IEE, 1985.
8. Jenssen, J.R., 'A case of a linear quadratic optimal design with a weighting matrix which is not non-negative definite', Int. J. Control, Vol 39, No 6, 1984.
9. Reid, R.E. A.K. Tugcu and B.C. Mears, 'An optimal controller arising from minimisation of a quadratic performance criterion of indefinite form', Prc. 21st Conf. on Decision and Control, Orlando, Dec. 1982.
10. Willems, J.C., 'On the existence of a non-positive solution to the Riccati Equation' IEEE, on Ac, Vol AC-19, 1974.

11. Desai V.B. and H.L. Weinert, 'A vector space approach to the indefinite LQG problem', Int. J. Control, 1984, 39, 507-515.

12. Gill A, and R. Sivan, 'Optimal control of linear systems with quadratic costs which are not necessarily positive definite', IEEE, Trans. Auto. Control, 14, 1983.

Appendix 1

x = Surge distance

y = Sway distance

ϕ = Heading deviation

δ = Rudder angle

m = mass of ship

x_g = centre of gravity

u = surge velocity

v = sway velocity

r = rate of turn

V = ship speed

I_z = moment of inertia

X = surge force

Y = sway force

N = torque

250,000 dwt Tanker, Full Load at 15 Knots

Sway Force

	30°	45°	60°	90°
b_o	1.95×10^{3}	1.65×10^{3}	4.57×10^{2}	9.33×10^{4}
a_o	8.59×10^{-2}	8.91×10^{-2}	1.01×10^{-1}	1.20×10^{-3}
a_1	5.64×10^{-9}	5.35×10^{-9}	7.04×10^{-12}	1.47×10^{-4}
a_2	1.	1.	1.	3.74×10^{-1}

Yaw Moment

	30°	45°	60°	90°
b_o	1.46×10^{6}	7.98×10^{5}	2.18×10^{5}	3.35×10^{7}
a_o	8.25×10^{-2}	9.14×10^{-2}	1.47×10^{-1}	1.49×10^{-3}
a_1	9.30×10^{-10}	4.34×10^{-4}	2.04×10^{-9}	1.57×10^{-4}
a_2	1.	1.	1.	5.92×10^{-2}

Table 1 : The wave model for different encounter angles (Reid: (5))

	vr	δ^2	$\beta\delta_c^2$	x^2
30 degs	0.0025	0.0026	0.56×10^{-4}	0.37×10^{-6}
45 degs	0.0021	0.0026	0.56×10^{-4}	0.35×10^{-6}
60 degs	0.0024	0.0027	0.57×10^{-4}	0.38×10^{-6}
90 degs	0.0997	0.7037	0.0091	0.12×10^{-4}

Table 2 The losses at different encounter angles

	$\sqrt{\bar{v}^2}$	$\sqrt{\bar{r}^2}$	$\sqrt{\bar{x}^2}$	$\sqrt{\bar{\delta}^2}$
30 degs	0.158×10^{-3}	0.366×10^{-6}	0.251×10^{-4}	0.429×10^{-4}
45 degs	0.144×10^{-3}	0.341×10^{-6}	0.221×10^{-4}	0.431×10^{-4}
60 degs	0.155×10^{-3}	0.359×10^{-6}	$.253 \times 10^{-4}$	0.432×10^{-4}
90 degs	0.471×10^{-2}	$.464 \times 10^{-5}$	0.959×10^{-4}	0.699×10^{-3}

Table 3 Root-mean-squared values of sway velocity, yaw rate heading angles and rudder angles

	$\bar{v}$	$\bar{r}$	$\bar{x}$	$\bar{\delta}$
30 degs	$-.672 \times 10^{-3}$	0.174×10^{-5}	$-.241 \times 10^{-3}$	$-.681 \times 10^{-4}$
45 degs	-0.585×10^{-3}	0.152×10^{-5}	$-.139 \times 10^{-3}$	$-.619 \times 10^{-4}$
60 degs	$-.574 \times 10^{-3}$	0.153×10^{-5}	$-.243 \times 10^{-3}$	$-.656 \times 10^{-4}$
90 degs	0.475×10^{-3}	0.656×10^{-6}	0.135×10^{-3}	0.296×10^{-3}

Table 4 Mean values of sway velocity, yaw rate, heading angle and rudder angle at different encounter angles

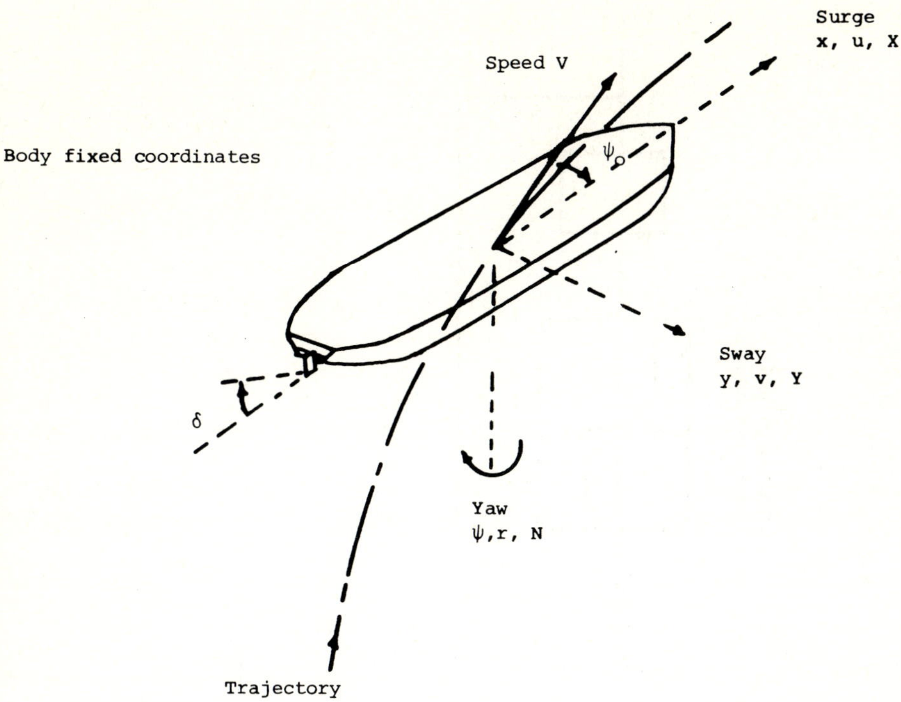

Fig. 1: Ship fixed body axis

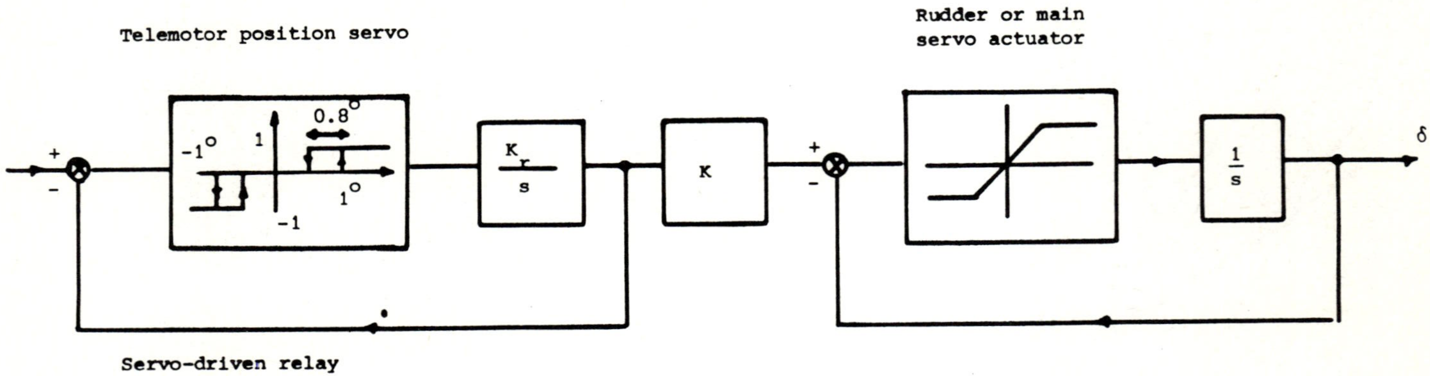

Fig. 2: Nonlinear steering gear model

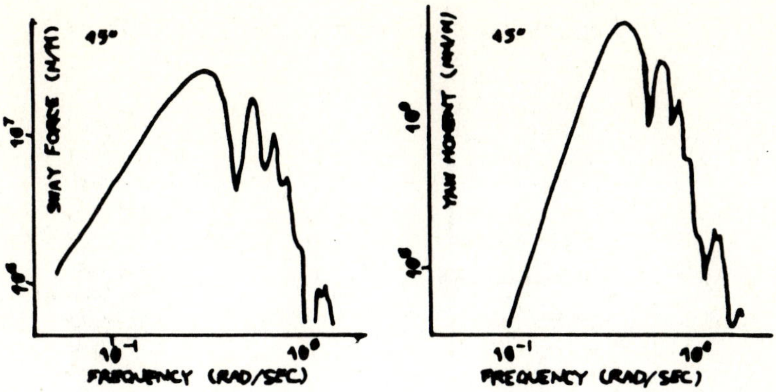

Fig 3 The Receptance Functions for a Tanker
(Blanke, [2])

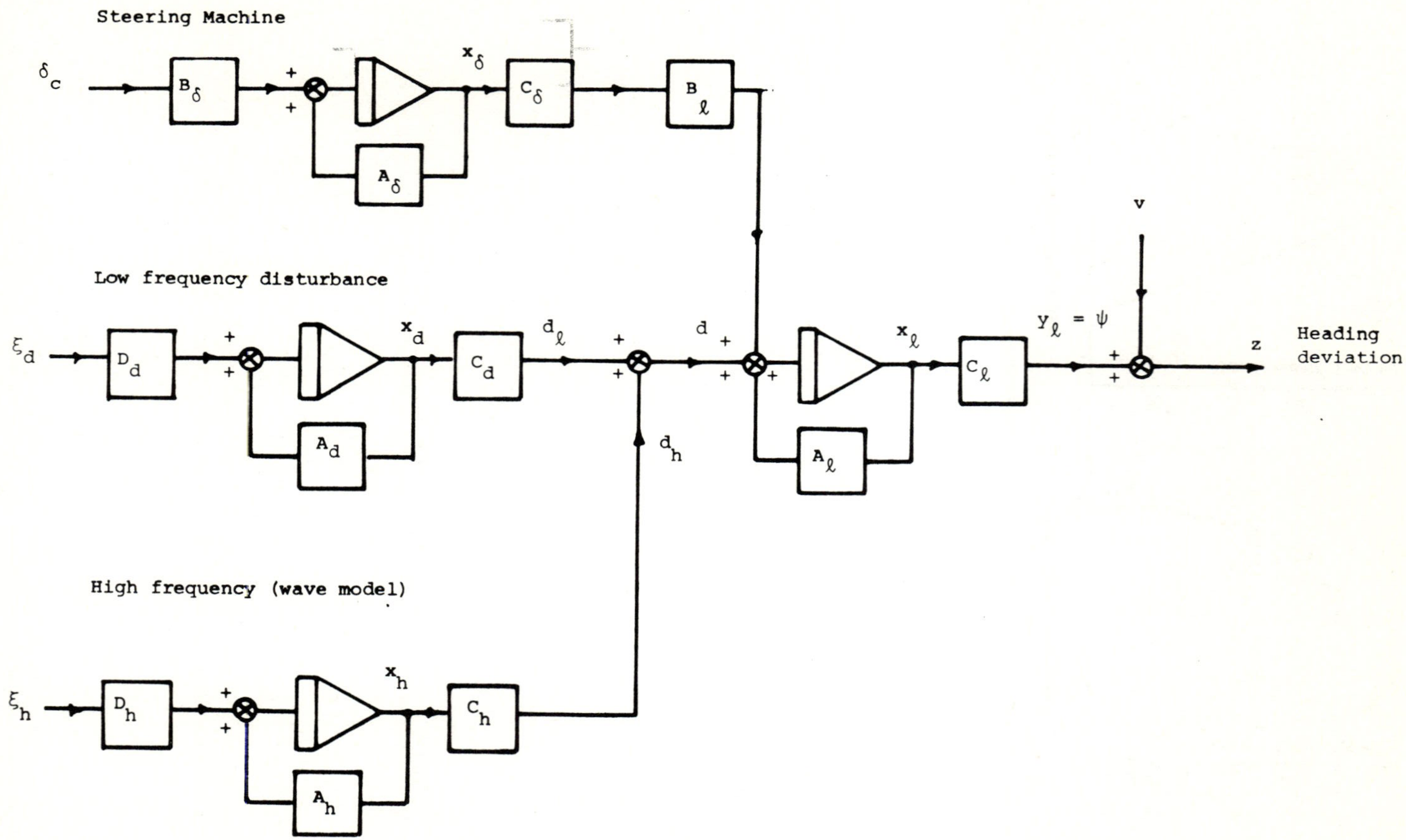

Fig. 4: Ship, Wind and Wave Models

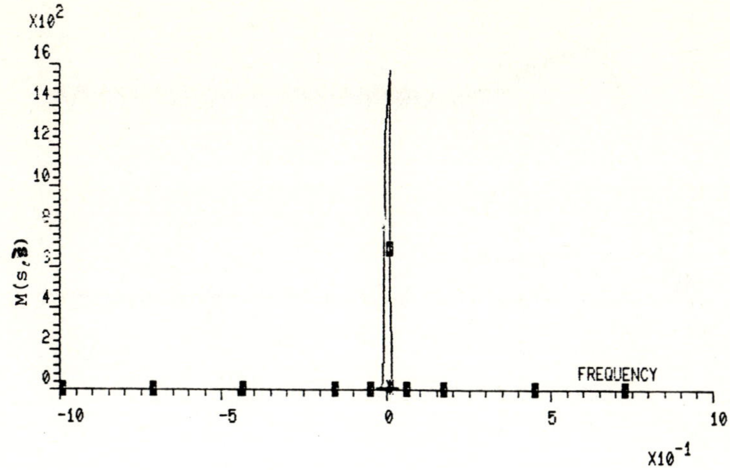

Fig 5 Real $M(s_1\bar{s})$

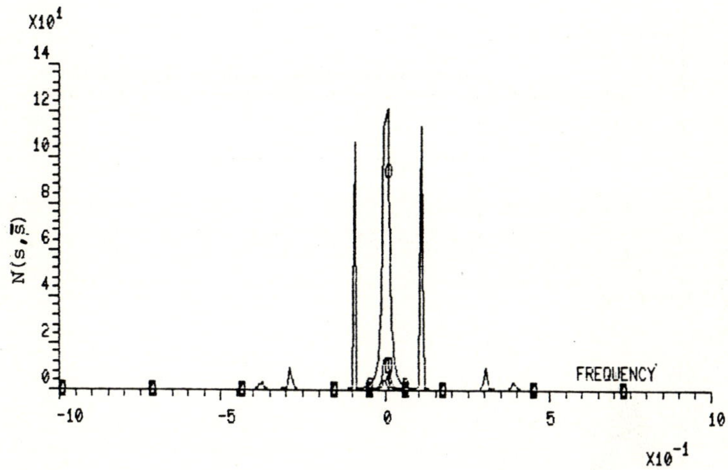

Fig 6 λ_i Real $N(s_1\bar{s})$, $1 = 1,2$

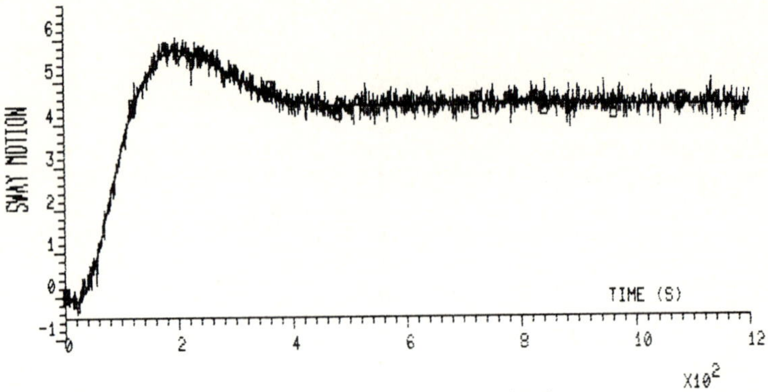

Fig. 7 Sway Velocity and Estimate

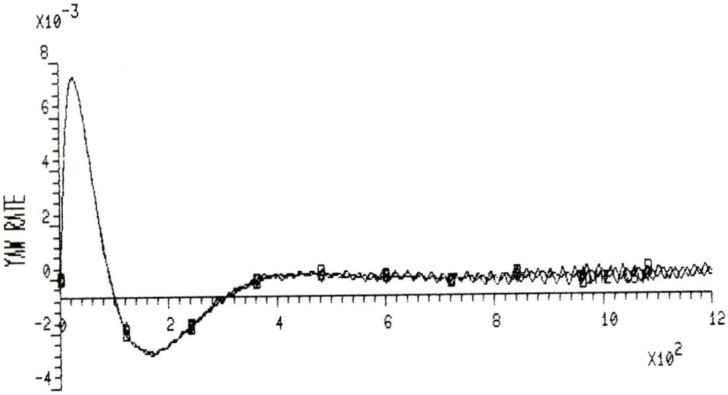

Fig. 8 Yaw Rate and Estimate

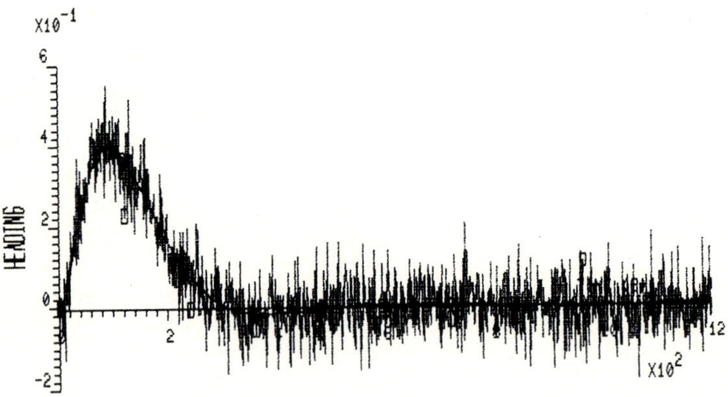

Fig. 9 Heading angle and Estimate

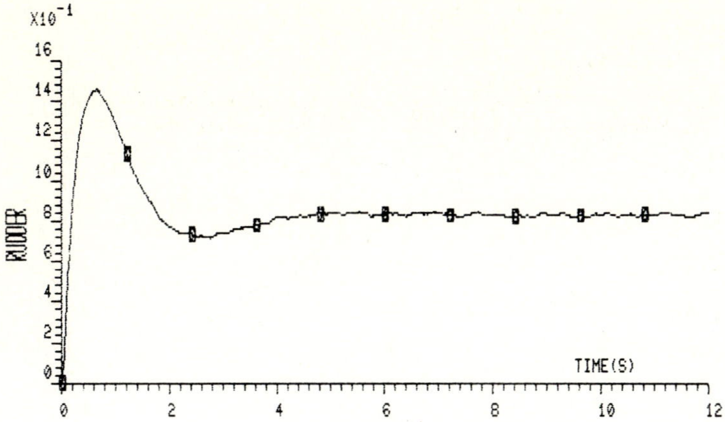

Fig.10 Rudder Angle and Estimate

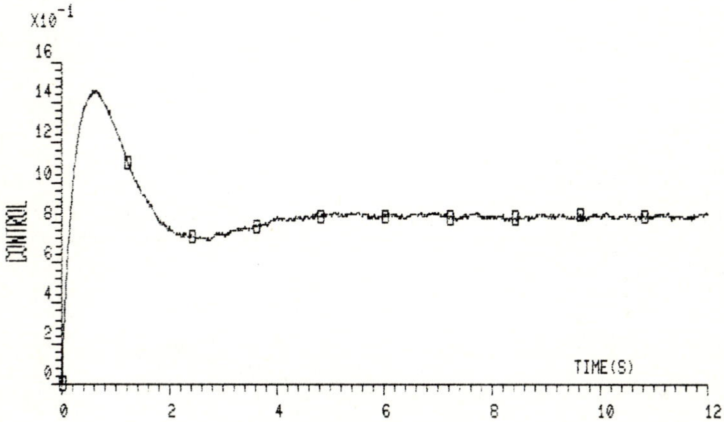

Fig.11 Rudder Command Signal

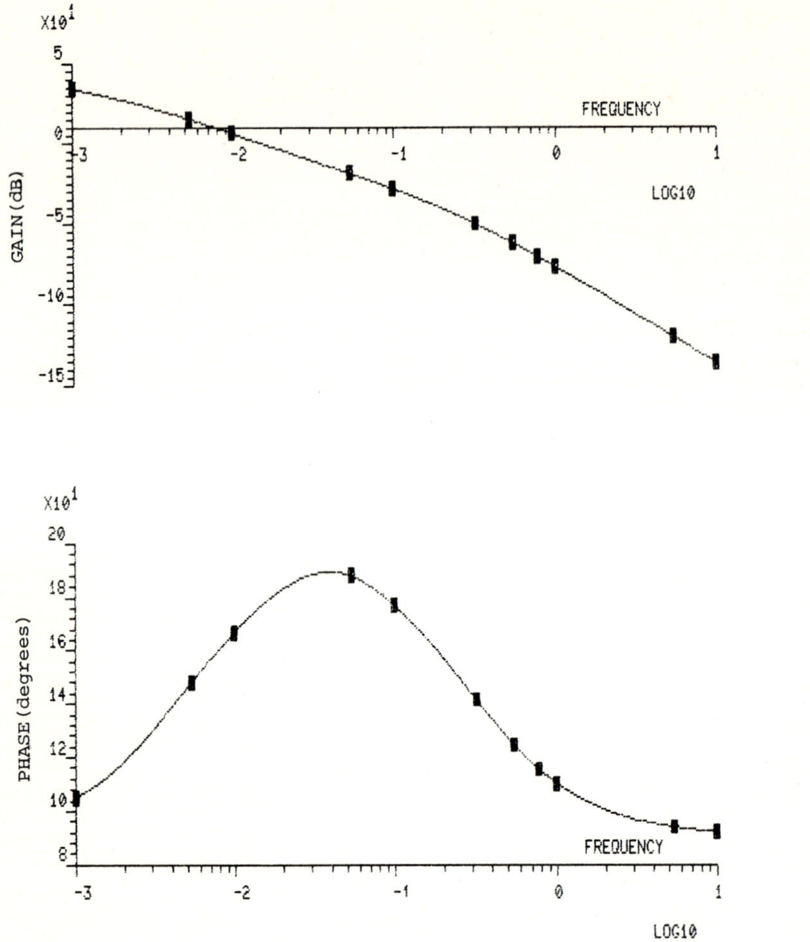

Fig.12 Uncompensated Open Loop Gain and Phase Plots for a Tanker

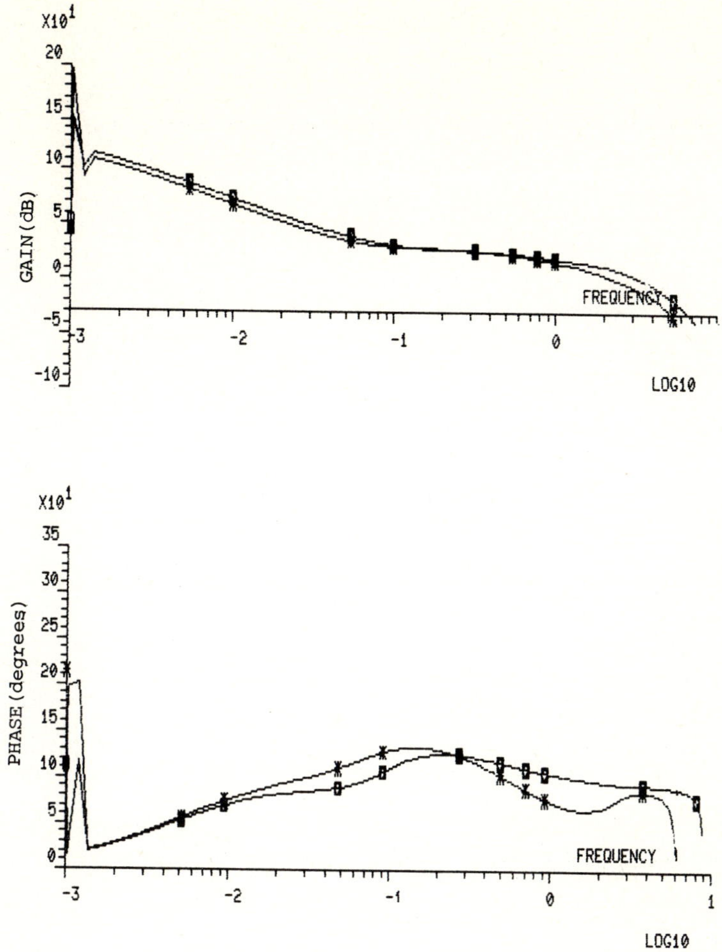

Fig.13 <u>Compensated Open-Loop Gain and Phase Plot for a Tanker</u>

SUBJECT INDEX

Lecture Notes in Control and Information Sciences

Edited by M. Thoma

Vol. 43: Stochastic Differential Systems
Proceedings of the 2nd Bad Honnef Conference
of the SFB 72 of the DFG at the University of Bonn
June 28 – July 2, 1982
Edited by M. Kohlmann and N. Christopeit
XII, 377 pages. 1982.

Vol. 44: Analysis and Optimization of Systems
Proceedings of the Fifth International
Conference on Analysis and Optimization of Systems
Versailles. December 14–17, 1982
Edited by A. Bensoussan and J. L. Lions
XV, 987 pages, 1982

Vol. 45: M. Arató
Linear Stochastic Systems
with Constant Coefficients
A Statistical Approach
IX, 309 pages. 1982

Vol. 46: Time-Scale Modeling of Dynamic Networks
with Applications to Power Systems
Edited by J. H. Chow
X, 218 pages. 1982

Vol. 47: P. A. Ioannou, P. V. Kokotovic
Adaptive Systems with Reduced Models
V, 162 pages. 1983

Vol. 48: Yaakov Yavin
Feedback Strategies for Partially
Observable Stochastic Systems
VI, 233 pages, 1983

Vol. 49: Theory and Application of Random Fields
Proceedings of the IFIP-WG 7/1
Working Conference
held under the joint auspices of the
Indian Statistical Institute
Bangalore, India, January 1982
Edited by G. Kallianpur
VI. 290 pages. 1983

Vol. 50: M. Papageorgiou
Applications of Automatic Control Concepts
to Traffic Flow Modeling and Control
IX, 186 pages. 1983

Vol. 51: Z. Nahorski, H.F. Ravn, R.V.V. Vidal
Optimization of Discrete Time Systems
The Upper Boundary Approach
V, 137 pages 1983

Vol. 52: A. L. Dontchev
Perturbations, Approximations and Sensitivity Analysis
of Optimal Control Systems
IV, 158 pages. 1983

Vol. 53: Liu Chen Hui
General Decoupling Theory of Multivariable
Process Control Systems
XI, 474 pages. 1983

Vol. 54: Control Theory for Distributed
Parameter Systems and Applications
Edited by F. Kappel, K. Kunisch,
W. Schappacher
VII, 245 pages. 1983.

Vol. 55: Ganti Prasada Rao
Piecewise Constant Orthogonal Functions
and Their Application to Systems and Control
VII, 254 pages. 1983.

Vol. 56: Dines Chandra Saha, Ganti Prasada Rao
Identification of Continuous
Dynamical Systems
The Poisson Moment Functional
(PMF) Approach
IX, 158 pages. 1983.

Vol. 57: T. Söderström, P. G. Stoica
Instrumental Variable Methods
for System Identification
VII, 243 pages. 1983.

Vol. 58: Mathematical Theory of
Networks and Systems
Proceedings of the MTNS-83 International
Symposium
Beer Sheva, Israel, June 20–24, 1983
Edited by P. A. Fuhrmann
X, 906 pages. 1984

Vol. 59: System Modelling and Optimization
Proceedings of the 11th IFIP Conference
Copenhagen, Denmark, July 25-29, 1983
Edited by P. Thoft-Christensen
IX, 892 pages. 1984

Vol. 60: Modelling and Performance
Evaluation Methodology
Proceedings of the International Seminar
Paris, France, January 24–26, 1983
Edited by F. Bacelli and G. Fayolle
VII, 655 pages. 1984

Vol. 61: Filtering and Control of Random
Processes
Proceedings of the E.N.S.T.-C.N.E.T. Colloquium
Paris, France, February 23–24, 1983
Edited by H. Korezlioglu, G. Mazziotto, and
J. Szpirglas
V, 325 pages. 1984